T0231126

CRC SERIES IN RADIOTRACERS IN BIOLOGY AND MEDICINE

Editor-in-Chief

Lelio G. Colombetti, Ph.D.
Loyola University
Stritch School of Medicine
Maywood, Illinois

STUDIES OF CELLULAR FUNCTION USING RADIOTRACERS
Mervyn W. Billinghurst, Ph.D.
Radiopharmacy
Health Sciences Center
Winnipeg, Manitoba, Canada

RECEPTOR-BINDING RADIOTRACERS
William C. Eckelman, Ph.D.
Department of Radiology
George Washington University School of Medicine
Washington, D.C.

GENERAL PROCESSES OF RADIOTRACER LOCALIZATION
Leopold J. Anghileri, D.Sc.
Laboratory of Biophysics
University of Nancy
Nancy, France

BIOLOGIC APPLICATIONS OF RADIOTRACERS
Howard J. Glenn, Ph.D.
University of Texas System Cancer Center
M.D. Anderson Hospital and Tumor Institute
Houston, Texas

RADIATION BIOLOGY
Donald Pizzarello, Ph.D.
Department of Radiology
New York University Medical Center
New York, New York

BIOLOGICAL TRANSPORT OF RADIOTRACERS
Lelio G. Colombetti, Ph.D.
Loyola University
Stritch School of Medicine
Maywood, Illinois

RADIOBIOASSAYS
Fuad S. Ashkar, M.D.
University of Miami/Jackson Memorial Hospital Medical Center
University of Miami School of Medicine
Miami, Florida

Additional topics to be covered in the series include Basic Physics, Nuclear Measurements, Radionuclides Production, Data Analysis in Radiotracer Studies, Dosimetry and Radiation Protection, and Mechanisms of Localization of Radiotracers: Compartmental Distribution.

Studies
of
Cellular Function
Using
Radiotracers

Editor

Mervyn W. Billinghurst, Ph.D.

Director
Radiopharmacy
Health Sciences Center
Winnipeg, Manitoba, Canada

Editor-in-Chief
CRC Series in Radiotracers in Biology and Medicine

Lelio G. Colombetti, Ph.D.

Loyola University
Stritch School of Medicine
Maywood, Illinois

CRC Press

Taylor & Francis Group
Boca Raton London New York

CRC Press is an imprint of the
Taylor & Francis Group, an **informa** business

First published 1982 by CRC Press
Taylor & Francis Group
6000 Broken Sound Parkway NW, Suite 300
Boca Raton, FL 33487-2742

Reissued 2018 by CRC Press

© 1982 by Taylor & Francis
CRC Press is an imprint of Taylor & Francis Group, an Informa business

No claim to original U.S. Government works

A Library of Congress record exists under LC control number: 81010214

Publisher's Note
The publisher has gone to great lengths to ensure the quality of this reprint but points out that some imperfections in the original copies may be apparent.

Disclaimer
The publisher has made every effort to trace copyright holders and welcomes correspondence from those they have been unable to contact.

ISBN 13: 978-1-138-50690-9 (hbk)
ISBN 13: 978-1-138-56196-0 (pbk)
ISBN 13: 978-0-203-71021-0 (ebk)

Visit the Taylor & Francis Web site at http://www.taylorandfrancis.com and the CRC Press Web site at http://www.crcpress.com

FOREWORD

This series of books on Radiotracers in Biology and Medicine is on the one hand an unbelievably expansive enterprise and on the other hand, a most noble one as well. Tools to probe biology have developed at an accelerating rate. Hevesy pioneered the application of radioisotopes to the study of chemical processes, and since that time, radioisotopic methodology has probably contributed as much as any other methodology to the analysis of the fine structure of biologic systems. Radioisotopic methodologies represent powerful tools for the determination of virtually any process of biologic interest. It should not be surprising, therefore, that any effort to encompass all aspects of radiotracer methodology is both desirable in the extreme and doomed to at least some degree of inherent failure. The current series is assuredly a success relative to the breadth of topics which range from in depth treatise of fundamental science or abstract concepts to detailed and specific applications, such as those in medicine or even to the extreme of the methodology for sacrifice of animals as part of a radiotracer distribution study. The list of contributors is as impressive as is the task, so that one can be optimistic that the endeavor is likely to be as successful as efforts of this type can be expected to be. The prospects are further enhanced by the unbounded energy of the coordinaing editor. The profligate expansion of application of radioisotopic methods relate to their inherent and exquisite sensitivity, ease of quantitation, specificity, and comparative simplicity, especially with modern instrumentation and reagents, both of which are now readily and universally available. It is now possible to make biological measurements which were otherwise difficult or impossible. These measurements allow us to begin to understand processes in depth in their unaltered state so that radioisotope methodology has proved to be a powerful probe for insight into the function and perturbations of the fine structure of biologic systems. Radioisotopic methodology has provided virtually all of the information now known about the physiology and pathophysiology of several organ systems and has been used abundantly for the development of information on every organ system and kinetic pathway in the plant and animal kingdoms. We all instinctively turn to the thyroid gland and its homeostatic interrelationships as an example, and an early one at that, of the use of radioactive tracers to elaborate normal and abnormal physiology and biochemistry, but this is but one of many suitable examples. Nor is the thyroid unique in the appreciation that a very major and important residua of diagnostic and therapeutic methods of clinical importance result from an even larger number of procedures used earlier for investigative purposes and, in some instances, procedures used earlier for investigative purposes and, in some instances, advocated for clinical use. The very ease and power of radioisotopic methodology tempts one to use these techniques without sufficient knowledge, preparation or care and with the potential for resulting disastrous misinformation. There are notable research and clinical illustrations of this problem, which serve to emphasize the importance of texts such as these to which one can turn for guidance in the proper use of these powerful methods. Radioisotopic methodology has already demonstrated its potential for opening new vistas in science and medicine. This series of texts, extensive though they be, yet must be incomplete in some respects. Multiple authorship always entails the danger of nonuniformity of quality, but the quality of authorship herein assembled makes this likely to be minimal. In any event, this series undoubtedly will serve an important role in the continued application of radioisotopic methodology to the exciting and unending, yet answerable, questions in science and medicine!

Gerald L. DeNardo, M.D.
Professor of Radiology, Medicine,
Pathology and Veterinary Radiology
University of California, Davis-
Sacramento Medical School
Director, Division of Nuclear Medicine

THE EDITOR-IN-CHIEF

Lelio G. Colombetti, Sc.D., is Professor of Pharmacology at Loyola University Stritch School of Medicine in Maywood, Ill. and a member of the Nuclear Medicine Division Staff at Michael Reese Hospital and Medical Center in Chicago, Ill.

Dr. Colombetti graduated from the Litoral University in his native Argentina with a Doctor in Sciences degree (summa cum laude), and obtained two fellowships for postgraduate studies from the Georgetown University in Washington, D.C., and from the M.I.T. in Cambridge, Mass. He has published more than 150 scientific papers and is the author of several book chapters. He has presented over 300 lectures both at meetings held in the U.S. and abroad. He organized the First International Symposium on Radiopharmacology, held in Innsbruck, Austria, in May 1978. He also organized the Second International Symposium on Radiopharmacology which took place in Chicago in September, 1981, with the active participation of more than 500 scientists, representing over 30 countries. He is a founding member of the International Association of Radiopharmacology, a nonprofit organization, which congregates scientists from many disciplines interested in the biological applications of radiotracers. He was its first President (1979/1981).

Dr. Colombetti is a member of various scientific societies, including the Society of Nuclear Medicine (U.S.) and the Gesellschaft für Nuklearmedizin (Europe), and is an honorary member of the Mexican Society of Nuclear Medicine. He is also a member of the Society of Experimental Medicine and Biology, the Coblenz Society, and the Sigma Xi. He is a member of the editorial boards of the journals *Nuklearmedizin* and *Research in Clinic and Laboratory*.

PREFACE

Radiotracers have been widely used during the last few decades to study the function of organs in humans. An important part of these studies is the investigation of the direct function of cellular entities by the localization and metabolization of radiotracers by the cells. The scope of this volume is restricted in several ways since it deals only with cellular function.

Since radiotracers are used in minutes quantities, no pharmacological or therapeutic effects are conceivable; therefore radiotracers will not interfere with the function of the cellular activity being studied. Using radiolabeled compounds the normal or abnormal function of the cells can be followed. In this volume we take an in-depth look at some of these functions, partially in animal experimentation with possible application to humans and partially in man himself.

In some studies lower animals are used as models for specific systems in developing a radiotracer technique. Animal studies employing radiotracers unsuitable for in vivo studies in humans because of their radionuclidic properties have also been included because of their value in elucidating the function of specific organs.

The subject of cellular function has been relatively broadly defined since a rigid interpretation would be very limiting and give a somewhat disjointed view of the topic. For example, in the area of uptake in bone, the formation of new bone and the laying down of hydroxyapatite crystals is a function of the osteoblasts and osteocytes. However, it is evident that both ^{47}Ca and ^{99m}Tc-labeled pyrophosphate, or their analogues, are not incorporated into the bone by the direct action of the osteoblasts and osteocytes, but rather by an ion exchange process on the crystal surface. Hence a strict interpretation would exclude this subject. On the other hand, ^{59}Fe, which has been shown by autoradiography to be associated with the osteocytes, would be included.

This volume is the result of the concerted effort of a number of scientists to summarize in a succinct way the current understanding of the mechanisms of these localizations. The editors of the book gratefully acknowledge this combined effort.

Lelio G. Colombetti
Mervyn W. Billinghurst

THE EDITOR

Mervyn W. Billinghurst, Ph.D., is Director, Radiopharmacy at the Health Science Centre, Winnipeg, Manitoba, Canada.

Dr. Billinghurst obtained his B.Sc. and his M.Sc. with honors in Inorganic Chemistry at Auckland University, Auckland, New Zealand. In 1970 he received his Ph.D. in analytical chemistry from McMaster University, Hamilton, Ontario, Canada. He is a member of The Society of Nuclear Medicine, a member of the New Zealand Institute of Chemistry, and the Canadian Association of Radiopharmaceutical Scientists.

Dr. Billinghurst has presented invited and submitted papers at international meetings. He has published more than 20 research papers, written chapters for several books and co-authored one book. His current research interests are the development, quality control and the improved understanding of the methods of localization of radiopharmaceuticals.

CONTRIBUTORS

James S. Arnold, M.D.
Department of Nuclear Medicine
Veteran's Administration Medical
  Center
Iron Mountain, Michigan

William H. Beierwaltes, M.D.
Professor of Medicine
Director of Nuclear Medicine
University of Michigan Medical Center
Ann Arbor, Michigan

M. Donald Blaufox, M.D., Ph.D.
Professor of Radiology (Nuclear
  Medicine)
Professor of Medicine
Director, Division of Nuclear Medicine
Albert Einstein College of Medicine
Bronx, New York

L. Rao Chervu, Ph.D.
Associate Professor of Radiology
  (Nuclear Medicine)
Chief, Nuclear Medicine Central
  Laboratory
Albert Einstein College of Medicine
Bronx, New York

Mrinal K. Dewanjee, Ph.D.
Director, Radiopharmaceutical
  Laboratory
Associate Professor of Pathology
Mayo Medical School
Department of Laboratory Medicine
The Mayo Clinic
Rochester, Minnesota

Alan R. Fritzberg, Ph.D.
Associate Professor of Radiology
Division of Nuclear Medicine
University of Colorado Health Sciences
  Center
Denver, Colorado

Brian M. Gallagher, Ph.D.
Section Leader, Pharmacological
  Sciences Department
Radiopharmaceutical Research
  Department
New England Nuclear Corporation
North Billerica, Massachusetts

Milton D. Gross, M.D.
Assistant Professor of Internal
  Medicine
Department of Internal Medicine
Division of Nuclear Medicine
University of Michigan Medical Center
Chief
Nuclear Medicine Service
Veteran's Administration Medical
  Center
Ann Arbor, Michigan

D. M. Lyster, Ph.D.
Associate Professor of Nuclear
  Pharmacy
Vancouver General Hospital
Faculty of Pharmaceutical Sciences
University of British Columbia
Vancouver, British Columbia, Canada

Dennis P. Swanson, R.Ph., M.S.
Director, Nuclear Pharmacy
University of Michigan Medical Center
Ann Arbor, Michigan

Donald M. Wieland, Ph.D.
Director of Radiopharmaceutical
  Research
Nuclear Pharmacy
University of Michigan Medical Center
Ann Arbor, Michigan

Yukio Yano
Chemist
Donner Laboratory
Lawrence Berkeley Laboratory
University of California
Berkeley, California

TABLE OF CONTENTS

Chapter 1

THE MEASUREMENT OF BRAIN GLUCOSE METABOLISM USING RADIOTRACERS

Brian M. Gallagher

TABLE OF CONTENTS

I. INTRODUCTION

Under normal circumstances, the mammalian brain uses glucose as its primary metabolic substrate.[1,2] Although the human brain represents roughly 2% of the total body weight (with a weight of $\sim$ 1400 g), it receives about 15% of the cardiac output ($\sim$ 800 mℓ/min or 57 mℓ/100 g/min) and utilizes $\sim$ 20% of the resting total body oxygen consumption ($\sim$ 250 mℓ/min).[3] The brain at rest extracts $\sim$ 10% of the glucose flowing through it ($\sim$5 mg or 28 μmol glucose per 100 g per minute) with a K_m of transport of 7 to 9 mM which is approximately equal to the circulating plasma glucose concentration.[3] The carrier-mediated transport of glucose across the blood-brain barrier has been amply demonstrated by several investigators.[4,5] The hexose carrier displays saturable uptake, stereospecificity, and competitive inhibition with other hexoses. This carrier system would appear to be the major limiting factor in the cerebral utilization of glucose by determining the rate of sugar transport, and, hence, substrate availability.[6] This chapter will discuss the major techniques that employ radiotracers for the measurement of glucose transport and metabolism by the mammalian brain.

II. GLUCOSE TRANSPORT

The kinetics of glucose transport and of certain analogs across the blood-brain barrier have been measured by use of a water reference technique in the anesthetized rat.[4] Briefly, the method employs the coinjection of a [^{14}C]-labeled hexose and [^{3}H]-H$_2$O (as a freely diffusible internal reference) into the common carotid artery of the anesthetized rat followed by decapitation 15 sec post injection. This permits a single passage of the bolus through the brain microcirculation. The cerebral hemisphere ipsilateral to the injection site and an aliquot of the injection mixture are counted for radioactivity and the so-called "brain uptake index" (I_b) is then calculated as follows:

$$I_b = \frac{[^{14}C]\text{-hexose}/[^{3}H]\text{-water (brain)}}{[^{14}C]\text{-hexose}/[^{3}H]\text{-water (injectate)}} \times 100$$

The brain uptake index is equivalent to the fractional extraction of [^{14}C]-hexose (E_{hexose}) relative to the fractional extraction of the [^{3}H]-H$_2$O reference (E_{H_2O}) following algebraic rearrangement:

$$I_b = \frac{[^{14}C]\text{-hexose (brain)}/[^{14}C]\text{-hexose (injectate)}}{[^{3}H]\text{-water (brain)}/[^{3}H]\text{-water (injectate)}} \times 100$$

or

$$I_b = \frac{E_{hexose}}{E_{H_2O}}$$

This method has been utilized to determine the Michaelis-Menton kinetic parameters (K_m and V_{max}) that describe the carrier-mediated transport of sugars between the plasma and the brain.[6] The K_m is defined as the concentration of unlabeled hexose that decreases the brain uptake index (I_b) to one-half the value observed for the same [^{14}C]-hexose present in tracer concentrations. Once the I_b for a tracer concentration of [^{14}C]-hexose is determined, the characteristics of saturability (or self-inhibition) and cross-inhibition (or competition) between different hexoses for the same carrier, can be measured by adding varying concentrations of the appropriate unlabeled hexose to the injectate. The cross-inhibition experiments employ increasing concentrations of a hexose or competing substrate in addition to [^{14}C]-glucose in the injectate to permit the

Table 1
BLOOD BRAIN BARRIER HEXOSE
CARRIER AFFINITY CONSTANTS FOR
RATS

Substrate	K_m(mM)	K_i(mM)
Glucose	9.0	—
2-Deoxy-D-glucose	6.0	—
3-O-Methyl D-glucose	—	10
D-Mannose	22	21
D-Galactose	42	40
Phlorizin	—	0.40
Phloretin	—	0.016

From Pardridge, W. M. and Oldendorf, W. H., *Biochim. Biophys. Acta*, 382, 377, 1975. With permission.

calculation of K_i. This is defined as the concentration of competing substrate which reduces the uptake of [^{14}C]-glucose by one-half. The values obtained in rats utilizing this approach appear in Table 1. The equivalence of the experimentally determined K_m (from self-inhibition studies) and K_i (from cross-inhibition) values for D-mannose and D-galactose would lend support to the concept of a shared common carrier molecule with glucose. A maximum transport velocity (V_{max}) was calculated for these sugars and found to be constant for all the hexoses at 1.56 μmol/g tissue per minute. The finding that the V_{max} was constant for these hexoses suggests that the Michaelis-Menton equilibrium assumption is valid in this model and that the rate-limiting step of transport, which is probably the movement of the carrier across the membrane, is independent of the hexose structure.[6] The inhibition of [^{14}C]-glucose uptake by the glycoside, phlorizin and its aglycone, phloretin, demonstrates the potency with which certain drugs can influence brain glucose uptake. Several reports have demonstrated that 2-deoxyglucose is transported from the plasma to the brain tissues by the same saturable carrier that transports glucose.[4,5,7,8] The significance of this will become apparent later in this chapter.

Since the brain is a highly complex and heterogeneous organ with a variety of structural and functional components, it has been of interest to measure the rate of glucose utilization as an indicator of functional activity. These studies have shown that these different structures can independently regulate the levels of functional activity and energy metabolism and that these processes are closely linked. Much of the current knowledge on cerebral energy metabolism in vivo has been determined by means which measure the average rates of energy metabolism for the whole brain.[9-12] Although these methods have permitted the measurement of the changes in the overall cerebral metabolic rate as the result of pathological conditions and pharmacological interventions, they do not permit the determination of the relationship between functional activity and energy metabolism in discrete foci that occur as the result of normal physiological activity.

III. [^{14}C]-DEOXYGLUCOSE METHOD

Methods have recently been proposed for quantitating both whole brain and localized glucose utilization using [^{14}C]-glucose[13,14] and [^{14}C]-2-deoxy-D-glucose[15] an analogue of glucose with metabolic properties that are particularly suited to metabolic studies. As discussed above, deoxyglucose shares the same saturable carrier for entry into the brain as does glucose. In the tissues, deoxyglucose competes with glucose for

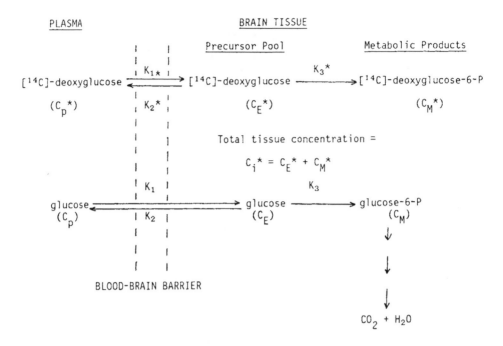

FIGURE 1.   Diagrammatic representation of the deoxyglucose model. (From Sokoloff, L., Reivich, M., Kennedy, C., Des Rosiers, M. H., Patlak, C. S., Pettigrew, K. D., Sakurada, O., and Shinohara, M., *J. Neurochem.*, 28, 897, 1977. With permission.)

hexokinase, which phosphorylates both compounds to their respective hexose-6-phosphates.[16] Unlike glucose-6-phosphate, which is converted to fructose-6-phosphate and then undergoes both anaerobic and aerobic catabolism, 2-deoxyglucose-6-phosphate is a poor substrate for subsequent metabolic steps. This is due to the absence of the hydroxyl group on carbon-2.[16,17] The net result of this special feature is that the 2-deoxyglucose-6-phosphate formed intracellularly is metabolically trapped over a reasonable time period during which measurements of the quantity of a labeled deoxyglucose and deoxyglucose-6-phosphate can be made.

The basis for the [14C]-deoxyglucose technique which has been developed by Sokoloff and colleagues[15] is illustrated in Figure 1. The model is based on the premise that [14C]-deoxyglucose (C_p^*) in the plasma and circulating plasma glucose (C_p) share the same carrier for transport into the brain with the rate constant K_1^* for [14C]-deoxyglucose and K_1 for glucose and the rate constants K_2^* and K_2 for the reverse process for deoxyglucose and glucose, respectively (i.e., back transport of either compound from the brain to the plasma). Both compounds equilibrate in a tissue precursor pool defined by C_E for glucose and C_E^* for deoxyglucose and then compete for the intracellular hexokinase to be phosphorylated to [14C]-deoxyglucose-6-phosphate and glucose-6-phosphate with the rate constants of K_3^* and K_3, respectively. The model is also based on some additional assumptions including a steady state carbohydrate metabolism in the brain, and that the tissue glucose and deoxyglucose are present in a single, common compartment. An important premise is the lack of redistribution of [14C]-activity following phosphorylation of deoxyglucose due to phosphatase activity or further metabolism over the duration of the measurement. This latter assumption would appear to be valid.[15]

The mathematical derivation of the working equations will not be reiterated here as this has been published recently;[15] however, some basic concepts upon which the model is based are necessary to appreciate the final working equation of the method. The total content of [14C] per mass of tissue (C_i^*) is equal to the sum of the concentrations of [^{14}C]-deoxyglucose ([^{14}C]-DG) in the precursor pool C_E^* and its metabolic product [^{14}C]-deoxyglucose-6-phosphate ([^{14}C] DG-6-P) (C_M^*):

$$C_i^* = C_E^* + C_M^*$$

The change in these concentrations with time, t, is expressed as

$$\frac{dC_i^*}{dt} = \frac{dC_E^*}{dt} + \frac{dC_M^*}{dt}$$

and is influenced by the rates of transport into the tissue or loss from the tissue back into the plasma and the hexokinase-catalyzed phosphorylation to [^{14}C]-DG-6-P:

$$\frac{dC_E^*}{dt} = K_1^* * C_p^* - K_2^* * C_E^* - K_3^* * C_E^*$$

where K_1^*, K_2^*, and K_3^* are the rate constants as shown in Figure 1.

The model also employs the Fick principle which states that the rate of uptake or loss of a compound by a tissue is equal to the difference between the rates of delivery to the tissue via the arterial blood and the rate of removal in the venous blood:

$$F (C_A^* - C_V^*) = \frac{dC_i^*}{dt}$$

where F = rate of blood flow per mass of tissue, C_A^* and C_V^* are the [^{14}C]-deoxyglucose concentrations in the arterial and venous blood, respectively.

Since,

$$\frac{dC_i^*}{dt} = \frac{dC_E^*}{dt} + \frac{dC_M^*}{dt}$$

then,

$$F (C_A^* - C_V^*) = \frac{dC_E^*}{dt} + \frac{dC_M^*}{dt}$$

Since the assumption is made that steady state conditions apply, the net uptake of glucose from the blood is equal to the rate of glucose utilization by the tissue. Therefore, for glucose,

$$F (C_A - C_V) = R_i$$

where C_A and C_V are the arterial and venous glucose conentrations, respectively, and R_i is the net rate of glucose utilization by the brain tissue.

Another consideration in the model is the fact that [^{14}C]-deoxyglucose and glucose are competitive substrates for hexokinase with rates of phosphorylation of v^* and v, respectively. The relationship of v^*/v can be expressed in the modification of the Michaelis-Menton equation to take into account the influence of competitive substrates[18] where V_m^* and V_m are the maximum velocities, K_m^* and K_m the apparent Michaelis constants, and C_E^* and C_E the substrate concentrations (intracellular precursor pools) for [^{14}C]-deoxyglucose and glucose, respectively:

$$v^*/v = \cfrac{\cfrac{C_E{}^* V_m{}^*}{K_m{}^* (1 + C_E/K_m) + C_E{}^*}}{\cfrac{C_E V_m}{K_m (1 + C_E/K_m) + C_E}}$$

This equation can be simplified since tracer amounts of [^{14}C]-deoxyglucose are employed, therefore, $C_E{}^*$ is a negligible amount. Lastly, the assumption is made that glucose utilization is a steady state process from the point of [^{14}C]-deoxyglucose administration throughout the time course of the actual process to the point of measurement of the tissue radioactivity. By taking into account the appropriate kinetic constants, arterial blood [^{14}C]-activity, a constant arterial glucose concentration and a proportional rate of [^{14}C]-deoxyglucose utilization to the rate of glucose metabolism, the following operational equation has been derived (see Reference 15 for full derivation and experimental bases for the assumptions employed).

$$R_i = \cfrac{C_i{}^*(T) - K_1{}^* e^{-(K_2{}^* + K_3{}^*)T} \displaystyle\int_0^T C_p{}^* e^{(K_2{}^* + K_3{}^*)t}\, dt}{\left(\cfrac{\lambda\, V_m{}^*\, K_m}{\phi\, V_m K_m{}^*}\right) \left[\displaystyle\int_0^T (C_p{}^*/C_p)\, dt - e^{-(K_2{}^* + K_3{}^*)T} \displaystyle\int_0^T (C_p{}^*/C_p) e^{(K_2{}^* + K_3{}^*)t}\, dt\right]}$$

Although somewhat overwhelming in appearance (since in the course of the derivation of the equation, numerous processes, each governed by its own mathematical representation, were considered), the concept expressed by this equation is relatively simple. It has incorporated the standard relationship that determines the rate of enzyme-catalyzed processes as determined with radiotracers. The numerator represents the quantity of [^{14}C]-deoxyglucose-6-phosphate formed over time T and the quantity of [^{14}C]-DG plus [^{14}C]-DG-6-P is equal to $C_i{}^*$ for the reasons given earlier (no further metabolism or back diffusion). This quantity is measured using a quantitative autoradiography technique employing standards and determining the relative radioactivity by densitometry. Subtracted from $C_i{}^*(T)$ is a term that represents free, unmetabolized [^{14}C]-deoxyglucose still in the tissue. If time T is sufficiently long, this term is essentially zero and can be dropped. The denominator represents the integrated specific radioactivity of the precursor pool times a factor, termed the lumped constant,

$$\frac{\lambda\, V_m{}^* K_m}{\phi\, V_m K_m{}^*}$$

Where λ = ratio of distribution volume of [^{14}C]-DG to that of glucose in the tissue, and

ϕ = fraction of glucose that, once phosphorylated, is glycolytically and oxidatively metabolized. The other terms are described earlier.

which is essentially a correction factor for differences between [^{14}C]-DG and glucose and is determined in separate experiments (described in Reference 15). The term with the exponential factor in the denominator accounts for the time lag in the achievement of equilibrium between the tissue precursor pool and the plasma.

All of the terms in the operational equation that contain rate constants approach zero with increasing time as the plasma [^{14}C]-deoxyglucose concentration approaches zero due to blood clearance. An interval of 30 to 45 min post injection of a bolus of [^{14}C]-DG has been shown to effectively equal the zero concentration. This time has also been shown to be an adequate period of time for most of the [^{14}C]-activity in the tissue to be present as [^{14}C]-deoxyglucose-6-phosphate such that one can measure the optical density of an autoradiograph and relate the value to [¹⁴C]-deoxyglucose-6-phosphate content and calculate directly the relative rates of glucose utilization in various cerebral regions.

In summary of the method, a pulse of [¹⁴C]-deoxyglucose is injected, timed arterial blood samples are taken for the measurement of plasma [¹⁴C]-deoxyglucose and glucose concentrations, the animals are decapitated at time T which is 30 to 45 minutes post injection, sections of brain are prepared for autoradiography and the [¹⁴C] concentrations, C_i*, are measured for time T using the quantitative autoradiographic technique. Local cerebral glucose metabolism for each region of brain studied is then calculated using the operational equation. The application of this technique has permitted the mapping of functional neural pathways in the brains of rats and monkeys[19] by measuring changes in the glucose utilization determined autoradiographically. For example, electrical stimulation of one sciatic nerve in the rat caused pronounced increases in glucose consumption (increased optical density in the autoradiographs) in the ipsilateral dorsal horn of the lumbar spinal cord. The local injection of penicillin (which will produce seizures when applied to the motor cortex) into the hand-face area of the motor cortex of the monkey, produced a selective increase in [¹⁴C]-deoxyglucose utilization adjacent to the area of drug application and in discrete foci including the putamen, globus pallidus, caudate nucleus and thalamus of the same side.[19] These regions are also believed to be selectively activated during seizure activity as measured using electrophysiological methods. Studies by these same investigators using the rat, in which the visual system is 80 to 85% crossed at the optic chiasma, showed that unilateral enucleation showed a marked decrease in [¹⁴C]-deoxyglucose utilization in the contralateral visual centers compared with the ipsilateral visual centers. The olfactory centers in the rat brain have been similarly mapped using this technique.[20]

This method provides a potentially powerful tool by which the neural sites of action of neuropharmacological agents such as γ-butyrolactone, apomorphine, amphetamine, morphine, naloxone, and D-lysergic acid diethylamide have been measured.[20]

IV. [¹⁴C]-GLUCOSE METHOD

Based on a procedure published by Gaitone,[21] Hawkins et al.[13] described a method for estimation of the cerebral metabolic consumption of glucose (CMR_{gl}) in the whole rat brain. In this method, [¹⁴C]-glucose is employed rather than [¹⁴C]-deoxyglucose and the metabolic rate is determined globally rather than a regional utilization. The method employs [¹⁴C]-glucose which is injected intravenously and the CMR_{gl} is estimated from the determination of the specific activity of glucose in the blood and the accumulation of [¹⁴C] in the non-glucose, acid-soluble fraction of the brain. This method has been extended and modified recently[14] and used to measure the glucose consumption of the rat cerebral cortex in conditions of normoxia, hypoxia, and hypercapnia in vivo. The method is based on the principle that the rate of [¹⁴C]-glucose accumulation into the acid-soluble metabolite pool in the brain can be described according to the equation

$$\frac{d[^{14}C]\,activity}{dt} = CMR_{gl} \times G_{SA} - K_{CO_2}[^{14}C]$$

where G_{SA} is the specific activity of glucose in the brain tissue and $K_{CO_2}[^{14}C]$ represents the loss of [¹⁴C] as $^{14}CO_2$. Again, the assumption of a constant glucose metabolism is made and that the loss of [¹⁴C]-activity is negligible since the glucose is labeled in the C-2 position. The above equation can be then integrated to give

Table 2

CEREBRAL METABOLIC RATE FOR GLUCOSE MEASURED IN RATS USING THE [^{14}C]-GLUCOSE

	Pa_{CO_2} (mmHg)	Pa_{O_2} (mmHg)	CMR_{gl} (μmol-g^{-1} · min^{-1})
Normal	37.5 ± 0.4	134 ± 10	0.77 ± 0.06
Hypoxia	32.4 ± 0.8	22.1 ± 0.4	1.52 ± 0.14
Hypercapnia	80.0 ± 1.6	111 ± 3	$0.22 \pm .09$

From Borgstrom, L., Norsberg, K., and Siesjo, B. K., *Acta Physiol. Scand.*, 96, 569, 1976. With permission.

$$CMR_{gl} = \frac{\Delta [^{14}C]_i}{\int_0^T (G_{SA})\, dt}$$

where $\Delta[^{14}C]_i$ is the total amount of [^{14}C] accumulated into the acid-soluble, non-glucose metabolic pool. This is determined in acid homogenates of brain tissue using chromatographic techniques. The G_{SA} in the tissues is estimated to be similar to that actually measured in the plasma at 2 and 4 min post injection and a linear decrease for the labeled glucose over this time period was found. Therefore, instead of using the plasma specific activity for estimating

$$\int_0^T (G_{SA})\, dt \qquad \text{only} \qquad \int_2^4 (G_{SA})\, dt$$

need be calculated for short study intervals. Hypoxia or hypercapnia was induced by ventilating immobilized rats with various gas mixtures 2 min after the bolus injection of [2-^{14}C-]-glucose. The CMR_{gl} in the non-steady state could be calculated as follows:

$$_2^4 CMR_{gl} = \frac{\Delta_0^4 [^{14}C]_i - \Delta_0^2 [^{14}C]_i}{\int_0^4 (G_{SA})\, dt}$$

where $_2^4 CMR_{gl}$ denotes the average glucose consumption during the first 2 min of the experimental period during gas exposure. $\Delta_0^2 [^{14}C]_i$ is measured in control animals and the mean value for the group was used to derive individual values of $\Delta_0^4 [^{14}C]_i$ in the experimental groups. This same approach can be used for the determination of CMR_{gl} in the "steady state" condition.[14]

Using this method, it has been demonstrated that the CMR_{gl} was stimulated approximately twofold during the first two min of hypoxia (Table 2). This is believed to be due to an acceleration of the glycolytic flux due to activation of phosphofructokinase (the enzyme that catalyzes the conversion of fructose-6-phosphate to fructose-1, 6-diphosphate). During the first 2 min of hypercapnia ($Pa_{CO_2} \sim 80$ mmHg) the CMR_{gl} was reduced to less than one-third that of normal, believed again to be due to an inhibition of phosphofructokinase which is a key controlling enzyme of the glycolytic flux.[14]

V. [^{11}C]-GLUCOSE METHOD

Although the methods discussed above have provided much of what is currently known about brain glucose metabolism, these techniques are limited to use in experimental animals, since sacrifice of the animal is necessary to determine the distribution of radioactivity either in the whole brain or in brain sections employing autoradiographic techniques. The extension of these methods, in particular the deoxyglucose method, toward the in vivo measurement of glucose metabolism would allow studies to be carried out in humans. Recent studies employing a positron-emitting substrate, ^{11}C-glucose, for measuring glucose metabolism in vivo have proven successful.[22-24] The method employs positron emission tomography that can measure the absolute radioactivity content of a given volume of tissue noninvasively[25] and is analogous to the quantitative autoradiographic technique discussed earlier. In addition, a suitable mathematical model was developed[22] to permit calculation of the cerebral metabolic rates. The method is relatively simple in that the radiotracer is biochemically identical to circulating glucose which obviates the need for correction factors due to differences in transport properties and enzyme affinities as with 2-deoxy-D-glucose. Since the measurement of ^{11}C-activity in the brain requires a relatively short time (between 4 and 6 min after injection) using the newer "fast" tomographic instruments, it is proposed that only 5 to 6% of ^{11}C-activity is lost from the tissue as [11]CO$_2$ although this has not been rigorously proven. The relatively short time required for the in vivo determinations of activity in the brain permits repeated measurements during the same experiment.

The method involves the injection of ^{11}C-glucose with arterial blood sampling at 15 sec intervals for the first minute and every 30 sec thereafter for 6 min for the determination of ^{11}C-activity and the determination of the radioglucose specific activity in the blood.[24] The efficiency of the detectors used on the emission tomograph are determined using calibrated phantoms. Cerebral blood volume is also measured by the inhalation of [11]CO by the test subject forming the intravascular marker, ^{11}C-carboxyhemoglobin. The tissue blood volume (V_b) is determined from the equilibrium image of ^{11}C-carboxyhemoglobin in the brain (counts·sec^{-1}·mℓ^{-1} of tissue and venous blood samples (counts·sec^{-1}·g^{-1} blood) during the scan. The V_b (mℓ/100 g) was calculated as follows:

$$V_b = \frac{[\text{counts} \cdot \text{sec}^{-1} \cdot \text{m}\ell \text{ tissue}^{-1}] \times 100}{[\text{counts} \cdot \text{sec}^{-1} \cdot \text{g blood}^{-1}] \times \rho \times \delta \times f}$$

where ρ = the density of blood; δ = tissue density; and f = ratio of mean tissue hematocrit to large vessel hematocrit. The measurement of the cerebral blood volume is used to correct the scan data for the ^{11}C-glucose present in the brain vascular compartment (and therefore is not tissue-associated radioactivity). This value is used in conjunction with the glucose concentration in the plasma and its specific activity and subtracted from the total counts in the field of the detector.[24] This method has permitted the determination of the glucose cerebral metabolic rate (Φ) noninvasively (5.21 ± 0.80 mg glucose·100 g^{-1}·min^{-1}) in monkeys and the values obtained compare favorably with those obtained by other means in both monkeys and humans.

VI. [^{18}F]-FLUORODEOXYGLUCOSE METHOD

Since glucose rapidly enters several metabolic pathways, the tendency toward equilibrium would constantly affect net glucose influx and the subsequent redistribution of labeled metabolites might result in a continuous loss of radioactivity from the brain at short measurement times. Although this loss has been estimated to be small,[24] it was of interest to develop a glucose analog labeled with a gamma-emitting nuclide which would be transported and phosphorylated as is glucose but which, similar to 2-deoxy-D-glucose, would not undergo subsequent metabolic steps with a redistribution

Table 3

DISTRIBUTION OF RADIOACTIVITY IN MOUSE TISSUES FOLLOWING THE INTRAVENOUS INJECTION OF ^{18}FDG

% Injected dose per gram tissue[a]

	1 min	15 min	30 min	60 min	90 min
Blood	7.1 ± 0.7	1.5 ± 0.2	0.9 ± 0.1	0.6 ± 0.1	0.4 ± 0.1
Heart	39.8 ± 3.4	27.5 ± 6.4	32.7 ± 8.6	31.7 ± 2.7	32.2 ± 2.7
Lungs	5.4 ± 0.4	2.7 ± 0.1	2.5 ± 0.1	2.5 ± 0.2	2.4 ± 0.1
Liver	9.6 ± 0.4	1.6 ± 0.2	1.1 ± 0.1	0.8 ± 0.1	0.9 ± 0.1
Kidneys	13.3 ± 1.4	2.5 ± 0.1	2.1 ± 0.5	1.1 ± 0.1	0.5 ± 0.1
Brain	4.3 ± 0.5	6.1 ± 0.1	5.3 ± 0.9	4.6 ± 0.8	3.4 ± 0.3

[a]   Mean ± SDM of 4 — 8 mice per determination.

From Gallagher, *J. Nucl. Med.*, 18, 990, 1977. With permission.

of activity. Because the hydroxyl groups at carbons 1, 3, 6, and perhaps 4, of glucose are involved in the binding of glucose to hexokinase, glucose analogs with substituents on only the relatively non-critical 2 position are substrates for hexokinase.[26] The extension of the deoxyglucose method to positron emission tomographic studies required an analog that could be conveniently labeled and the resulting product should behave biochemically similar to deoxyglucose. The use of F-18 ($t_{1/2}$ 110 min) as the label was chosen since (1) the hexokinase reaction is relatively insensitive to structural modification at this position;[26] (2) the C-F bond is strong resulting in a biologically stable label; and (3) the characteristics of F-18 for positron emission tomography and for radiopharmaceutical synthesis are convenient. Furthermore, it was known that 2-deoxy-2-fluoro-D-glucose (FDG) was a good substrate for hexokinase[32] and that 2-deoxy-2-fluoro-D-glucose-6-phosphate (FDG-6-P) is a relatively poor substrate for subsequent metabolic steps.[17] While hexoses are readily transported across cell membranes, the permeability of hexose phosphates is very low. Thus, measurements of the tissue content of FDG should be reflective of the integrated transport and phosphorylation of FDG:

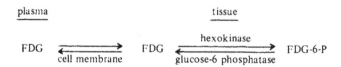

Initial studies[27] have shown that following intravenous injection of FDG into mice, the radioactivity distributes to all of the well-perfused organs, but rapidly clears from all organs except the heart and the brain where it remains relatively constant for up to 2 hr in the heart and decreases slowly in the brain (Table 3). The relative hexokinase activity toward D-glucose in various tissues had been previously measured.[28] These results showed that the hexokinase activity of brain > heart > kidney > lung > liver in terms of the activity per weight of tissue. This same relative pattern was determined for ^{18}FDG.[29] The brain and heart rapidly phosphorylated ^{18}FDG both in vitro and in vivo (Figure 2). Virtually all of the activity present in the brain between 1 and 120 min was in the form of ^{18}FDG-6-P. Organ homogenates of brain showed the highest hexokinase activity on a per weight basis. In addition, there was no observed conversion

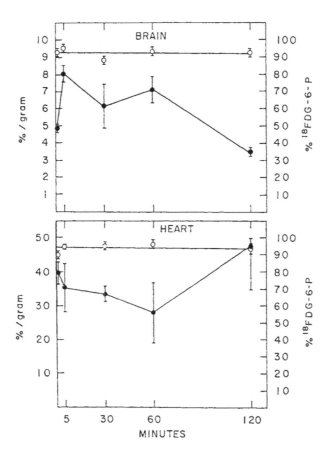

FIGURE 2. The distribution of ^{18}F radioactivity ($\bullet$——$\bullet$) and ^{18}FDG-6-P (0-0) in mouse brain and heart expressed as percent of injected dose per gram and percent of total activity present in tissue as ^{18}FDG-6-P, respectively. Each point represents the mean ± SDM from 4 animals. (From Gallagher, B. M., Fowler, J. S., Gutterson, N. I., MacGregor, R. R., Wan, C. N., and Wolf, A. P., *J. Nucl. Med.*, 19, 1154, 1978. With permission.)

of FDG-6-P to free FDG, a factor that contributes to the retention of radioactivity over the course of the measurement.

The distribution of FDG in both the heart and the brain[28,30] is not uniform, but rather, is regionally localized. The cerebral cortex of the dog showed the highest % injected dose per gram followed by the cerebellum and medulla (Table 4). Within the cerebral cortex, the gray matter had approximately two to three times the activity per gram than the white matter. This is in accord with results obtained using the [^{14}C]-deoxyglucose method in which the deoxyglucose utilization by gray matter in rats was $\sim$ 3 times that of the white, although the individual values vary for different regions.[15]

Studies were carried out to directly compare [^{14}C]-deoxyglucose and [^{14}C]-fluoro-deoxyglucose using autoradiography in the rat to demonstrate the equivalence of these tracers[31] (Figure 3). The pattern of radioactivity within the brain is identical for both compounds and radically different from the pattern displayed by [^{14}C]-3-0-methyglucose which shares the same carrier system for transport with glucose but is not phosphorylated by hexokinase and, hence, is not metabolically trapped. When local cerebral glucose utilization was calculated using the equations discussed earlier, the values obtained for the local rates of glucose metabolism for various structures were distributed

Table 4

DISTRIBUTION OF
RADIOACTIVITY IN DOG BRAIN
FOLLOWING INJECTION OF ^{18}FDG

	60 min	135 min	
	%/g	%/g	%/g
Cerebral cortex	0.029	0.047	0.027
Gray matter	0.028	0.059	0.027
White matter	0.017	0.018	0.015
Cerebellum	0.025	0.032	0.026
Medulla	0.023	0.026	0.021

From Gallagher, B. M., Ansari, A., Atkins, H., Casella, V., Christman, D. R., Fowler, J. S., Ido, T., MacGregor, R. R., Som, P., Wan, C. N., Wolf, A. P., Kuhl, D. E., and Reivich, M., *J. Nucl. Med.*, 18, 990, 1977. With permission.)

similarly (r = 0.989, $P < 0.001$) although FDG gave a consistently somewhat lower value ($\sim$ 18%) than deoxyglucose[31] (Table 5). This is believed to be due to a lower value for the lumped constant for FDG.

These findings in experimental animals and in vitro formed the basis for the extension of the [^{14}C]-deoxyglucose technique to the [^{18}F]-fluorodeoxy-glucose method for the measurement of local cerebral glucose utilization in man.[31-33] Representative scans from one subject are shown in Figure 4. The level at which these slices were made are indicated in relation to the obito-meatal line. The regional distribution of ^{18}F-activity within the brain is quite apparent. The relationship of the heterogeneous activity distribution measured in vivo to functional anatomical regions is shown in Figure 5. The numbers to the left of the scans represent the calculated cerebral metabolic rate for glucose in mg/100 g tissue per minute in the structure labeled to the right. Since there are no local cerebral glucose data available for humans, the results obtained by these preliminary studies cannot be compared to literature values.

The exciting potential that this elegant technique provides may form the basis for future studies of regional brain glucose metabolism in man. The extension of this technique should permit the mapping of functional neural pathways in man, both in the normal subject and in a wide variety of neural disorders. The method clearly has application in understanding the biochemical basis of psychiatric disorders and drug effects on energy utilization in discrete foci. The gap between experimental studies carried out in animals and what is true for the human brain can now be bridged by the combined technologies of rapid syntheses of suitable glucose analogs pioneered at the Brookhaven National Laboratory[34-36] and the application of rapid and quantitative positron emission transaxial tomographic machines. The future contributions to our understanding and perhaps ultimate treatment of the wide array of disease processes that may be due, in part, to aberrant cerebral glucose metabolism can now only begin to be estimated. The future would appear to be most promising in this regard.

VII. SUMMARY

This chapter has briefly discussed some of the major techniques that have been employed to measure glucose metabolism with radiotracers. Although much of our current understanding of brain glucose metabolism derives from the earlier studies employing measurements of glucose extraction and metabolism using [^{14}C]-labeled

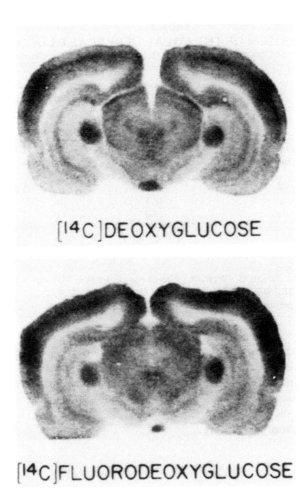

[14C]DEOXYGLUCOSE

[14C]FLUORODEOXYGLUCOSE

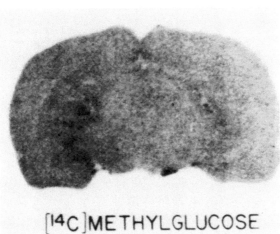

[14C]METHYLGLUCOSE

FIGURE 3. Autoradiographs of brain sections from rats 30 min after an intravenous bolus injection of [14C]-deoxyglucose, [14C]-fluorodeoxyglucose or [14C]-3-O-methylglucose. The deoxyglucose and fluorodeoxyglucose autoradiographs are very similar whereas the 3-O-methylglucose does not demonstrate a regional localization. This indicates that local metabolism rather than merely blood flow and transport determines local tissue [14C]-concentrations at long intervals after pulse injections. (From Reivich, M., Kuhl, D., Wolf, A., Greenberg, J., Phelps, M., Ido, T., Casella, V., Fowler, J., Hoffman, E., Alavi, A., Som, P., and Sokoloff, L., *Circ. Res.*, 44, 127, 1979. With permission.)

Table 5
COMPARISON OF THE LOCAL CEREBRAL GLUCOSE METABOLISM IN THE RAT

	[^{14}C] - deoxyglucose (mg · 100 g^{-1} · min^{-1})	[^{14}C] - fluorodeoxyglucose (mg · 100 g^{-1} min^{-1})
Gray Matter		
Visual cortex	20.3	17.0
Auditory cortex	29.3	21.0
Lateral geniculate	16.9	14.0
Superior olive	26.5	21.5
Lateral lemniscus	20.5	17.9
Inferior colliculus	36.5	27.4
Superior colliculus	18.0	13.2
White Matter		
Internal capsule	5.9	5.5
Cerebellar white matter	6.7	6.1

From Reivich, M., Kuhl, D., Wolf, A., Greenberg, J., Phelps, M., Ido, T., Casella, V., Fowler, J., Hoffman, E., Alavi, A., Som, P., and Sokoloff, L., *Circ. Res.*, 44, 127, 1979. With permission.

substrates, these studies must be limited to use in experimental animals. The recent developments in positron emission tomography and the rapid syntheses of a variety of suitable gamma-emitting substrates allows the extension of the earlier techniques to use in man. The ultimate impact of these new technologies in understanding this important area of physiology, both in the normal and the disease state, is only beginning. The future should be most rewarding.

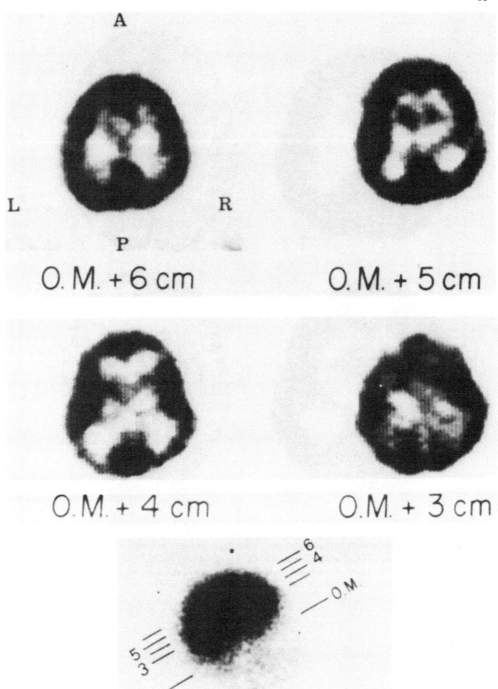

FIGURE 4. Section scans of the head beginning 30 min after the bolus injection of ¹⁸FDG to a normal subject. The level of each scan is indicated in relation to the orbito-meatal (o.m.) line on a rectilinear scan of the subject's head. A = anterior, P = posterior, L = left, and R = right. (From Reivich, M., Kuhl, D., Wolf, A., Greenberg, J., Phelps, M., Ido, T., Casella, V., Fowler, J., Hoffman, E., Alavi, A., Som, P., and Sokoloff, L., *Circ. Res.*, 44, 127, 1979. With permission.)

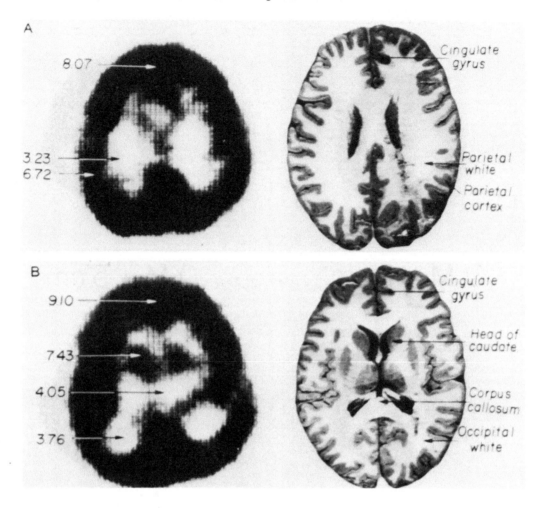

FIGURE 5.   Section scans of the head and photograph of a slice through the human brain at approximately the same anatomical level. The numbers represent the local cerebral metabolic rate for glucose in mg/100 g/min in the structure labeled: (A) scan at O.M. + 6cm level; (B) scan at O.M. + 5cm level; (C) scan at O.M. + 4cm level; and (D) scan at O.M. + 3cm level. (From Reivich, M., Kuhl, D., Wolf, A., Greenberg, J., Phelps, M., Ito, T., Casella, V., Fowler, J., Hoffman, E., Alavi, A., Som, P., and Sokoloff, L., *Circ. Res.*, 44, 127, 1979. With permission.)

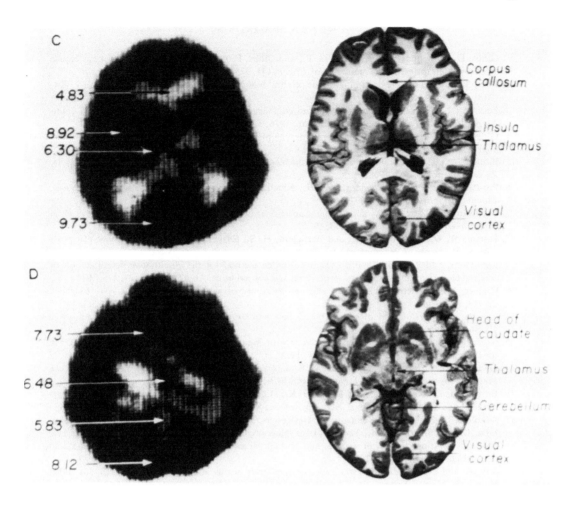

FIGURE 5C and D.

REFERENCES

1. Gibbs, E. L., Lennox, W. G., Nims, G. F., and Gibbs, F. A., Arterial and cerebral venous blood: arterial-venous differences in man, *J. Biol. Chem.*, 144, 325, 1942.
2. Kety, S. S., Quantitative determination of cerebral blood flow in man, *Methods Med. Res.*, 1, 204, 1948.
3. Sokoloff, L., in *Basic Neurochemistry*, 2nd ed., Siegel, G. J., Albers, R. W., Katzman, R., and Agranoff, B. W., Eds., Little, Brown, Boston, 1976.
4. Oldendorf, W. H., Measurement of brain uptake of radiolabeled substances using a tritrated water internal standard, *Brain Res.*, 24, 372, 1970.
5. Bidder, T. G., Hexose translocation across the blood brain barrier interface: configurational aspects, *J. Neurochem.*, 15, 867, 1968.
6. Pardridge, W. M. and Oldendorf, W. H., Kinetics of blood-brain barrier transport of hexoses, *Biochim. Biophys. Acta*, 382, 377, 1975.
7. Bachelard, H. S., Specificity and kinetic properties of monosaccharide uptake in guinea pig cerebral cortex *in vivo*, *J. Neurochem.*, 18, 213, 1971.
8. Horton, R. W., Meldrum, B. S., and Bachelard, H. S., Enzymic and cerebral metabolic effects of 2-deoxy-D-glucose, *J. Neurochem.*, 21, 1973.
9. Kety, S. S. and Schmidt, C. F., The nitrous oxide method for the quantitative determination of cerebral blood flow in man: theory, procedure, and normal values, *J. Clin Invest.*, 27, 476, 1948.
10. Cremer, J. E., Selective inhibition of glucose oxidation by triethyltin in rat brain *in vivo*, *Biochem. J.*, 119, 95, 1970.
11. Nilsson, B., Measurement of overall blood flow and oxygen consumption in the rat brain, *Acta Physiol. Scand.*, 92, 142, 1974.
12. Norberg, K. and Siesjo, B. K., Quantitative measurement of blood flow and oxygen consumption in the rat brain, *Acta Physiol. Scand.*, 91, 154, 1974.
13. Hawkins, R. A., Miller, A. L., Cremer, J. E., and Veech, R. L., Measurement of the rate of glucose utilization by rat brain *in vivo*, *J. Neurochem.*, 23, 917, 1974.
14. Borgstrom, L., Norberg, K., and Siesjo, B. K., Glucose consumption in rat cerebral cortex in normoxia, hypoxia, and hypercapnia, *Acta Physiol. Scand.*, 96, 569, 1976.
15. Sokoloff, L., Reivich, M., Kennedy, C., DesRosiers, M. H., Patlak, C. S., Pettigrew, K. D., Sakurada, O., and Shinohara, M., The [^{14}C]-deoxyglucose method for the measurement of local cerebral glucose utilization: theory, procedure, and normal values in the conscious and anesthetized albino rat, *J. Neurochem.*, 28, 897, 1977.
16. Sols, A. and Crane, R. H., Substrate specificity of brain hexokinase, *J. Biol. Chem.*, 210, 581, 1954.
17. Bessell, E. M. and Thomas, P., The effect of substitution of C-2 of D-glucose-6-phosphate on the rate of dehydrogenation by glucose-6-phosphate dehydrogenase from yeast and rat liver, *Biochem. J.*, 131, 83, 1973.
18. Dixon, M. and Webb, E. C., *Enzymes*, 2nd ed., Academic Press, New York, 1964, 84.
19. Kennedy, C., DesRosiers, M. H., Jehle, J. W., Reivich, M., Sharpe, F., and Sokoloff, L., Mapping of functional neural pathways by autoradiographic survey of local metabolic rate with [^{14}C]-Deoxyglucose, *Science*, 187, 850, 1975.
20. Sokoloff, L., Relationship between physiological function and energy metabolism in the central nervous system, *J. Neurochem.*, 29, 13, 1977.
21. Gaitone, M. K., Rate of utilization of glucose and "compartmentation" of oxoglutarate and glutamate in rat brain, *Biochem. J.*, 95, 803, 1965.
22. Raichle, M. E., Larson, K. B., Phelps, M. E., Grubb, R. L., Welch, M. J., and Ter-Pogossian, M. M., In vivo measurement of brain glucose transport and metabolism employing ^{11}C-Glucose, *Am. J. Physiol.*, 228, 1936, 1975.
23. Raichle, M. E., Larson, K. B., Higgins, C. S., Grubb, R. L., Eichling, J., Welch, M. J., and Ter-Pogossian, M. M., Three-dimensional in vivo mapping of brain metabolism and acid-base status, in *Cerebral Function, Metabolism and Circulation*, Ingvar, D. H. and Lassen, N. A., Eds., Munksgaard, Copenhagen, 1977, 188.
24. Raichle, M. E., Welch, M. J., Grubb, R. L., Higgins, C. S., Ter-Pogossian, M. M., and Larson, K. B., Measurement of regional substrate utilization rates by emission tomography, *Science*, 199, 986, 1978.
25. Phelps, M. E., Emission computed tomography, *Semin. Nucl. Med.*, 7, 337, 1977.
26. Coe, E. L., Inhibition of glycolysis in ascites tumor cells preincubated with 2-deoxy-2-fluoro-D-glucose, *Biochem. Biophys. Acta*, 264, 319, 1972.
27. Gallagher, B. M., Ansari, A., Atkins, H., Casella, V., Christman, D. R., Fowler, J. S., Ido, T., MacGregor, R. R., Som, P., Wan, C. N., Wolf, A. P., Kuhl, D. E., and Reivich, M., ^{18}F-Labeled 2-deoxy-2-fluoro-D-glucose as a radiopharmaceutical for measuring regional myocardial glucose metabolism *in vivo*, tissue distribution and imaging studies in animals, *J. Nucl. Med.*, 18, 990, 1977.

28. Long, C., Studies involving enzymic phosphorylation. I. The hexokinase activity of rat tissues, *Biochem. J.*, 50, 407, 1952.
29. Gallagher, B. M., Fowler, J. S., Gutterson, N. I., MacGregor, R. R., Wan, C. N., and Wolf, A. P., Metabolic trapping as a principle of radiopharmaceutical design: some factors responsible for the biodistribution of [^{18}F]-2-deoxy-2-fluoro-D-glucose, *J. Nucl. Med.*, 19, i154, 1978.
30. Gallagher, B. M., Fowler, J. S., MacGregor, R. R., Lambrecht, R. M., Wolf, A. P., Crawford, E. J., and Friedkin, M. E., *In vivo* measurement of transport and metabolic processes using radiotracers, in *Principles of Radiopharmacology*, Vol. 2, Colombetti, L. G., Ed., CRC Press, Boca Raton, 1979, 135.
31. Reivich, M., Kuhl, D., Wolf, A., Greenberg, J., Phelps, M., Ido, T., Casella, V., Fowler, J., Hoffman, E., Alavi, A., Som, P., and Sokoloff, L., The [^{18}F]-Fluorodeoxyglucose method for the measurement of local cerebral glucose utilization in man, *Circ. Res.*, 44, 127, 1979.
32. Reivich, M., Kuhl, D., Wolf, A., Greenberg, J., Phelps, M., Ido, T., Casella, V., Fowler, J., Gallagher, B., Hoffman, E., Alavi, A., and Sokoloff, L., Measurement of local cerebral glucose metabolism in man with [^{18}F]-2-fluoro-2-deoxy-D-glucose, *Acta Neurol. Scand.*, 56 (Suppl. 64), 190, 1977.
33. Kuhl, D. E., Hoffman, E. J., Phelps, M. E., Ricci, A. R., and Reivich, M., Design and application of the Mark IV scanning system for radionuclide computed tomography of the brain, in *Medical Radionuclide Imaging*, Vol. 1, IAEA Symposium on Medical Radionuclide Imaging, International Atomic Energy Agency, Vienna, 1977, 309.
34. Ido, T., Wan, C. N., Fowler, J. S., and Wolf, A. P., Fluorination with F$_2$, a convenient synthesis of 2-Deoxy-2-Fluoro-D-Glucose, *J. Org. Chem.*, 42, 2341, 1977.
35. Ido, T., Wan, C. N., Casella, V., Fowler, J., Wolf, A., Reivich, M., and Kuhl, D., Labeled 2-deoxy-2-fluoro-D-glucose, 2-deoxy-2-fluoro-D-mannose and [^{14}C]-2-deoxy-2-fluoro-D-glucose, *J. Labeled Compd. Radiopharm.*, 14, 175, 1978.
36. Shiue, C. Y., MacGregor, R. R., Lade, R. E., Wan, C. N., and Wolf, A. P., The synthesis of 1-^{11}C-2-Deoxy-D-Glucose for measuring regional brain glucose metabolism *in vivo*, *J. Nucl. Med.*, 19, 676, 1978.

Chapter 2

MECHANISM OF LOCALIZATION OF RADIOTRACERS IN MYOCARDIUM IN HEALTH AND DISEASE*

Mrinal K. Dewanjee

TABLE OF CONTENTS

I. INTRODUCTION

In 1927, Blumgart and Weiss[1] pioneered the use of radioactive materials in evaluating the cardiovascular system and studied the circulation time with the radioisotope ^{214}Bi obtained from the ^{226}radium- ^{222}radon-^{214} bismuth cascade generator. Recently, the widespread availability of artificial radionuclides that have various chemical and physical properties; and the interdisciplinary efforts of nuclear physicists, radiochemists, and physicians, have led to the present status of cardiovascular nuclear medicine in providing noninvasive methods for the diagnosis of heart disease. The recent success of these cardiovascular techniques was based on the development of the high-resolution scintillation camera, the knowledge of the physiology of circulation,[1-7] the availability of mathematical models, the development of tracers, and the knowledge of the mechanism of localization of these tracers in myocardial cells in health and disease.

Recently, several articles have reviewed the different aspects of cardiovascular nuclear medicine,[8-25] including radiotracers, instrumentation, and data processing, and the experimental and clinical evaluation of these techniques. The scintigraphic methods in cardiovascular nuclear medicine can be divided into seven types of imaging (Table 1): imaging of functional myocardial cells, acute myocardial infarct, regional myocardial perfusion and blood flow, catecholamine pool and turnover, labeled cells in myocardial diseases, platelet-vessel-wall interaction and arterial and venous thrombosis.

II. SCINTIGRAPHIC STUDIES OF MYOCARDIUM IN HEALTH AND DISEASE

The most important parameter we want to measure by tracer and other techniques is how well the individual, as well as total muscle cells of myocardium, perform metabolically. The coronary blood flow could be determined in the animal model by implanting a Doppler ultrasound probe or electromagnetic flow meter, and the diameter of coronary vessel is measured by angiography. We assume that if the muscle mass is well perfused as measured by blood flow, it is healthy.

Large molecular weight serum proteins and cellular components of blood, due to their large size, could not diffuse out of vascular space and are called nondiffusible tracers, e.g., ^{131}I- or ^{99m}Tc-labeled human serum albumin, ^{99m}Tc-labeled red blood cells, ^{111}In-labeled transferrin, and ^{15}O-carboxyhemoglobin. Tracers which diffuse through the interstitial and intracellular space are called diffusible tracers, e.g., ^{133}Xe, ^{85}Kr, ^{15}O-labeled water ^{131}I-antipyrine, and ^{11}C-labeled alcohol. Alkali metal ions, e.g., ^{43}K, ^{38}K, and potassium analogs ^{82}Rb, ^{86}Rb, ^{201}Tl, are diffusible tracers that are extracted by myocardium for metabolic process. Metabolic substrates are diffusible indicators which are extracted and metabolized by myocardium, e.g., fatty acid, ^{11}C-labeled palmitate, ^{123}I-labeled hexadecenoic acid, glucose, ^{11}C-labeled glucose, or ^{18}F-labeled deoxyglucose, amino acid, ^{11}C or ^{13}N-labeled glutamate. Tracer uptake in an organ is limited by flow or diffusion through a cell membrane. Tracers, e.g., ^{85}Kr, ^{133}Xe, which are extracted from blood in proportion to flow are flow-limited tracers, e.g., the organ uptake is limited by the amount delivered. Tracers, e.g., ^{43}K, fatty acid, amino acid, glucose, which are transported across cell membranes by enzyme or carrier are diffusion limited. In myocardium, the vascular capillary membrane and myocyte membrane are the diffusion barriers. These barriers are partially or totally inactive in myocardial diseases.

The use of extractable diffusible tracers, for example the potassium ion analogs, is based on Sapirstein's principle[2] that the tracer distributed in accordance with regional blood flow will maintain the distribution (reflecting the flow) during an early period

Table 1

SCINTIGRAPHIC STUDIES IN EVALUATION OF DISEASE OF THE HEART

Group	Type of Imaging	Radiotracers
I.	Functioning myocardial cells	
	A. Potassium ion analogs	
	1. K$^+$, Rb$^+$, Cs$^+$	^{42}K, ^{43}K, ^{82}Rb, ^{86}Rb, ^{129}Cs, ^{131}Cs, ^{134m}Cs
	2. Tl$^+$	^{199}Tl, ^{201}Tl
	3. NH$_4^+$	^{13}NH$_4^+$
	B. Manganese ion: Mn^{2+}	^{52}Mn, ^{54}Mn
	C. Substrates	
	1. Fatty acids and analogs	^{131}I-oleic acid, ^{11}C-palmitic acid
		^{123}I-hexadecenoic acid, ^{123}Tc-tellurahexadecenoic acid, ^{123}I-iodo-phenyl fatty acid
		^{123}I-arachidonic acid
	2. Glucose and analogs	^{18}F-deoxyglucose, ^{11}C-glucose
	3. Amino acids and analogs	^{13}N-alanine (L) and ^{13}N-asparagine (L)
		^{13}N-glutamate (L)
II.	Acute myocardial infarct	
	A. Divalent and trivalent cations and anions	^{113}Sn^{2+}, ^{169}Yb^{3+}, ^{18}F$^-$
	B. Organometallic compounds	^{203}Hg- and ^{197}Hg-chlormerodrin
		^{203}Hg- and ^{197}Hg-hydroxymercurifluorescein
	C. Metal chelates	^{99m}Tc-tetracyclines, ^{99m}Tc-imidophosphates
		^{99m}Tc-pyrophosphate, ^{99m}Tc-diphosphonates
		^{99m}Tc-glucoheptonate, ^{67}Ga-citrate
		^{68}Ga-EDTMP^a
		^{113m}In-EDTMP
	D. Labeled antibodies	^{131}I-labeled myosin antibody, Fab$_2$ fragment
		^{131}I-labeled myoglobin antibody
III.	Regional myocardial perfusion and blood flow	
	A. Labeled microspheres or macroaggregates	^{99m}Tc-microspheres or ^{99m}Tc- or ^{111}I-macroaggregates (HSA)
		^{113m}In- or ^{68}Ga-labeled microspheres and macroaggregates
	B. Diffusible inert gases	^{133}Xe, ^{81m}Kr (daughter of ^{81}Rb), ^{85}Kr
IV.	Catecholamine pool and turnover	^{11}C-norepinephrine
		^{123}I-benzyl guanidine
V.	Labeled cells and serum proteins	^{99m}Tc-red blood cells, ^{111}In-transferrin, ^{131}I-serum albumin
VI.	Labeled cells in myocardial infarction, cardiomyopathy, and cardiac rejection	^{111}In-labeled white blood cells and lymphocytes ^{67}Ga-citrate
VII.	Platelet de-endothelialized vessel wall interaction and arterial thrombosis	^{111}In-labeled platelets

^{a} EDTMP = ethylenediaminetetramethylenephosphonate.

after injection, and that the relationship between the myocardial blood flow (MBF) and myocardial uptake (MU) at critical time (t_c) is given by the following equation:

$$MBF = MU_{t_c} \times \frac{CO}{ID} \qquad (1)$$

in which CO is the cardiac output and ID is the total injected dose. Thus, myocardial blood flow is proportional to myocardial uptake when the integrated venous efflux equals the integrated recirculated arterial influx, and the myocardial uptake equals the product of myocardial blood flow, extraction, and tracer concentration. Equation 1 could be applied to two types of tracers: (1) potassium ion analogs, and (2) inert gases and iodoantipyrine. The inert gases reflect regional blood flow only for a brief period before washout from the myocardium. Because potassium ion and its analogs are

Table 2

DECAY CHARACTACTERISTICS AND WHOLE-BODY RADIATION DOSE OF DIFFUSIBLE EXTRACTABLE TRACERS

Ionic tracer	Mode of decay	Photon energy (keV) and abundance[a]	Physical half-life	Whole-body radiation dose (mrad/mCi)
$^{38}K^+$	β^+	511 (200%) 2170 (100%)	7.7 min	31
$^{42}K^+$	β^-	1525 (18%)	12.4 hr	—
$^{43}K^+$	β^-	373 (85%) 619 (81%)	22.2 hr	622
$^{81}Rb^+$	β^+, Electron capture	190 (57% from ^{81m}Kr) 446 (24%) 511 (67%)	4.6 hr	222
$^{82}Rb^+$	β^+, Electron capture	511 (192%) 776 (13%)	1.3 min	2
$^{84}Rb^+$	β^+, Electron capture	511 (38%) 822 (73%)	33.0 day	16,000
$^{86}Rb^+$	β^-	1079 (9%)	18.7 day	—
$^{127}Cs^+$	β^+, Electron capture	125 (18%) 411 (64%)	6.2 hr	—
$^{129}Cs^+$	Electron capture	372 (32%) 412 (25%)	33 hr	240
$^{130}Cs^+$	β^+, Electron capture	511 (92%) 35 (2%)	30.0 min	—
$^{131}Cs^+$	β^-, Electron capture	31—36 (88%)	9.7 day	—
$^{134m}Cs^+$	Isomeric transition	128 (14%)	2.9 hr	—
$^{201}Tl^-$	Electron capture	69—83 (98%) 135 (2%) 167 (8%)	73.0 hr	200
$^{13}NH_4^+$	β^+	511 (200%)	10 min	5

[a]   Since each positron after annihilation produces 2 photons, 100% decay by β^+ emission corresponds to an intensity of 200% of 511 keV photons.

trapped intracellularly, they maintain their initial flow-dependent distribution for longer periods than do inert gases, making imaging possible in different orientation. Table 1 shows the list of all tracers used in cardiovascular nuclear medicine for experimental and clinical studies. The high-resolution portable gamma camera makes it possible to obtain acceptable images with the Hg X-ray (69 to 80 keV) of ^{201}thallous ion. In spite of expense and lower energy, ^{201}Tl is the perfusion agent of choice among the single gamma emitters.[26-42] In an institution equipped with the cyclotron and the positron camera, a few short-lived positron-emitting tracers could be prepared *in situ* and used.

The determination of regional myocardial blood flow requires that the tracers be uniformly mixed in the blood, extracted, and retained in the perfusion bed of myocardium long enough for static imaging with a gamma camera. Two types of tracers are currently in use: soluble and insoluble.

The soluble tracers are administered intravenously and extracted by the myocytes on a single passage through the capillary bed. The total myocardial uptake is the cumulative effect of multiple extractions from vascular via the extracellular space. The upper limit of uptake is about 5% of the administered dose, the amount being determined by the fractional cardiac output (3 to 8%). The best image with a gamma camera can be obtained with ^{129}Cs,[25] but the delayed extraction and technical difficulty in production limit its usefulness (Table 2). Other tracers used in cardiovascular studies are shown in Table 3. The insoluble tracers (for example, ^{99m}Tc-labeled microsphere, ^{131}I-labeled macroaggregated albumin) are administered into the branches of the coronary artery through catheters (invasive procedure). The soluble tracers, for example,

Table 3
DECAY CHARACTERISTICS OF RADIONUCLIDES
OTHER THAN ALKALI METAL ION ANALOGS IN
CARDIOVASCULAR STUDIES

Tracer	Mode of decay	Photon energy (keV) and abundance	Physical half-life
^{11}C	β^+	511 (200%)	20.5 min
^{15}O	β^+	511 (200%)	2.0 min
^{52m}Mn	β^+	511 (56%)	21 min
^{67}Ga	Electron capture	93 (71%)	78 hr
		185 (24%)	
		300 (162%)	
		394 (4%)	
^{68}Ga	β^+, Electron capture	511 (178%)	68 min
		1077 (3%)	
^{99m}Tc	Isomeric transition	140 (90%)	6.0 hr
^{111}In	Electron capture	172 (90%)	67.9 hr
		247 (94%)	
^{113m}In	Isomeric transition	393 (63%)	100 min
^{123}I	Electron capture	160 (83%)	13 hr
^{131}I	β^-	364 (84%)	8.1 day
		284 (6%)	
		637 (7%)	
^{127}Xe	Electron capture	172 (21%)	36 day
		203 (61%)	
		375 (18%)	
^{133}Xe	β^-	81 (35%)	5.3 day

the potassium analogs (K^+, Cs^+, Tl^+, NH_4^+, and so forth) and the metabolic substrates (^{11}C-labeled or ^{123}I-labeled fatty acid, ^{11}C-labeled glucose or ^{18}F-labeled deoxyglucose), reach only the perfused myocyte and are retained in the perfused myocardium for a longer time than the inert gases and therefore appear as hot spots. However, blood flow and extraction are substantially reduced in ischemic or injured myocardium, and the area of ischemic and infarcted myocardium is seen as a cold spot. ^{201}Tl imaging always overestimates the size of the infarct because of poor uptake in the surrounding ischemic area. The labeled microspheres or macroaggregates (10 to 15 μm) are slightly larger than the diameter of the capillaries, and are completely extracted in the arteriolar capillary network and retained independent of cell function. Thus, they have been accepted as the reference standard for estimating regional myocardial blood flow. Because the capillary might be distorted or reduced in size by edema, an erroneous flow pattern may be observed. In the presence of an acute injury or a healing process, the microscopic vessels might appear in series; hence, labeled microsphere should be used judiciously as a regional blood flow reference in studies of myocardial infarction in patients.

Table 4 indicates the time course of biochemical and morphological changes in infarcting myocardium, so that the mechanisms of trace uptake could be directly interpreted in terms of pathological changes. The basic metabolic changes in ischemic myocardium could be imaged by labeled metabolic substrates, or pathological changes of dead myocyte by in vivo staining with metal-chelate-tracers and/or inflammatory response in infarcting myocardium could be measured by labeled white blood cells.

III. METABOLISM OF HEALTHY MYOCYTE AND UPTAKE OF DIFFUSIBLE TRACERS AND SUBSTRATE

Ling[35] studied the time course of uptake of Tl^+, Cs^+, and Na^+ ions and stoichiometric

Table 4

TIME COURSE OF BIOCHEMICAL AND MORPHOLOGICAL CHANGES IN INFARCTING MYOCARDIUM[a]

Time after ligation	Events in infarcting myocardium
5 sec	Cyanosis (oxygen-depleted hemoglobin)
5—30 sec	H[+] and lactic acid accumulation
30 sec	Swollen mitochondria; reduced glycogen (90% loss in 60 sec); loss of intracellular K[+] and increase in Na[+]
1—5 min	Nonspecific ECG changes; ST segment elevation; T-wave inversion
5 min	Aggregation of nuclear chromatin
20—25 min	Enlargement of mitochondria with calcium phosphate granule formation and loss of matrix density; ATP depletion and uncoupling of oxidative phosphorylation; lysosomal degradation; swelling of sarcoplasmic reticulum
60—90 min	Individual cell death commences at 30 min and completes at 90 min and no ECG formation occurs; loss of soluble protein (myoglobin) and various enzymes (CPK, LDH, SGPT) through leaky membranes; influx of Ca[2+] ion and plasma proteins; extracellular space communicating with intracellular space of dead cells
3—6 hr	Infiltration of neutrophils and other mediators of inflammation; increase in edema and irritability of ischemic cells
8—12hr	Disintegration of myofibrillar pattern (loss of Z and I bands)
24 hr	Increase in lymphocytic infiltration
4—5 days	Lymphocytes replace polymorphonuclear leukocytes; capillary sprouts; mobilization of fibroblasts; lymphocyte decreases; edema decreases
18—25 days	Necrotic tissue completely replaced by fibrous tissue

[a]   Data from Sobel.[159]

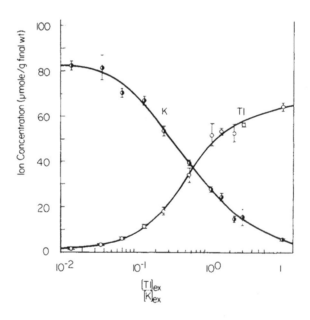

FIGURE 1.   Equilibrium distribution of Tl[+] and K[+] ions in frog muscle. (From Ling, G. N., *Physiol. Chem. Physics*, 9, 1977. With permission.)

displacement of K[+] with Tl[+] ion in the skeletal muscle of frog (Figure 1), and thus demonstrated the same adsorption site of these alkali metal ions. Britten and Blank[36] found that Tl[+] could effectively replace K[+] in activation of (Na[+] — K[+]) sensitive ATPase

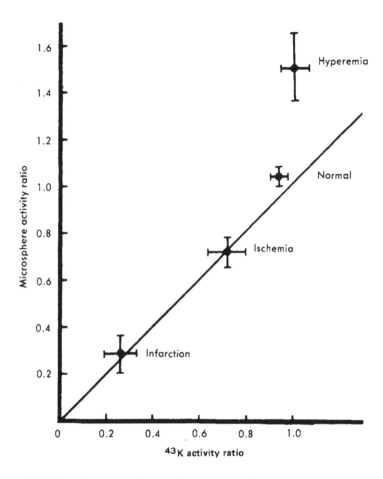

FIGURE 2. Correlation of blood flow as measured by microsphere versus potassium ion analogs, hyperemia. (From Prokop, E. K., Strauss, H. W., Shaw, J., Pitt, B., Wagner, H. N., Jr., *Circulation*, 50, 978, 1974. With permission.)

of rabbit kidney; the capacity of Tl^+ activation is ten times higher than that of K^+ ion. The affinity of Tl^+ and alkali-metal ions for ATPase enzyme is in the following order:

$$Tl^+ > K^+ > Rb^+ > NH_4^+ > Cs^+ > Li^+$$

It is not really known whether the same $K^+ - Na^+$ ATPase enzyme is also responsible for the uptake of Tl^+ ion. Figure 2 shows excellent correlation between uptake of K^+ ion and microsphere. Figure 3 shows the correlation of myocardial uptake between Tl^+ and K^+. Tl^+ could thus effectively replace K^+ ion in muscle cell. Figure 4 shows the time course of variation in biodistribution of different alkali metal ions in heart, blood, and liver of mice.

A. Extractable Diffusible Tracers
1. Alkali Metal Ions and Analogs

Because of the similarity of the biodistribution of K^+ and Tl^+, the use of ^{201}Tl for myocardial perfusion imaging was first suggested by Kawana et al.[26] in 1970. Lebowitz et al.[27] produced ^{201}Tl by proton bombardment of the naturally occurring ^{203}Tl target (abundance about 29.5%). The cross section at 30 MeV for the ^{203}Tl (p, 3n) ^{201}Pb reaction is about 800 mb. ^{201}Pb decays to ^{201}Tl (T½ = 9.4 hours) by an electron-

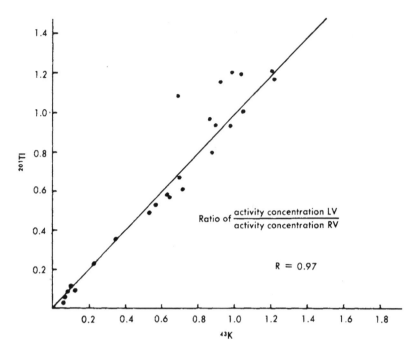

FIGURE 3.   Regional myocardial distribution of thallium-201 and potassium-43 in
dogs after ligation of left anterior descending artery. Data were expressed as the ratio
of activity of left ventricle to that of the right ventricle. (From Strauss, H. W., Langan,
J. K., Lebowitz, E., and Pitt, B., *Circulation*, 51, 641, 1975. With permission.)

capture process. The lead is complexed with EDTA, and the undesirable thallium (Tl^{3+})
in the target material is removed by a cation-exchange resin. The lead radioisotopes
are then allowed to remain overnight so as to permit the buildup of ^{201}Tl from ^{201}Pb.
^{201}Tl is then separated by cation resin from ^{201}Pb; $^{201}Tl^{3+}$ is eluted from the cation
exchanger, reduced to Tl^+, and neutralized and sterilized. The thallous chloride thus
obtained is carrier free. The biological distribution has been studied by several investigators.[28-31] The mechanism of Tl^+ uptake by the myocyte is not clearly understood.
Although thallium is a group IIIA element, thallous ion and K^+ have similar hydrated
ionic radii and demonstrate biological characteristics of K^+. The main difference between them appears in the comparatively prolonged cellular retention of thallous ion.[30]
For extraction by myocyte, the cations have to traverse the capillary wall and the interstitial space and be actively transported across the myocyte membrane.[31,32] The barrier
for the cations at the capillary wall is dependent on blood flow. From the interstitial
space, the cations are efficiently extracted by the active transport process via the Na^+
— K^+-ATPase on the myocyte membrane or a corresponding isoenzyme.

Differences in the fractional extraction during the first circulation and subsequent
extraction thereafter were observed among K^+, Rb^+, and Cs^+.[33,34] Extraction of both
the K^+ and Rb^+ was 70% during the first circulation,[31] but the relative extraction of
Rb^+ decreased thereafter. Comparative uptake of potassium ion and its analogs in
myocardium and liver and blood clearance in mice is shown in Figure 3.

In skeletal muscle, about 70% of the resistance of K^+ transport is located at the
capillary wall barrier, and about 30% is located at the myocyte membrane. This resistance is equally divided for Rb^+.[31] For Cs^+, both the fractional extraction and fractional
release rate are lower than those for K^+; only 20% of Cs^+ is extracted by dog heart on
each circulation. The maximal myocardial concentration of K^+ is reached in less than
5 min, while that of Cs^+ is reached in 2 to 3 hr. Because of this poor extraction and

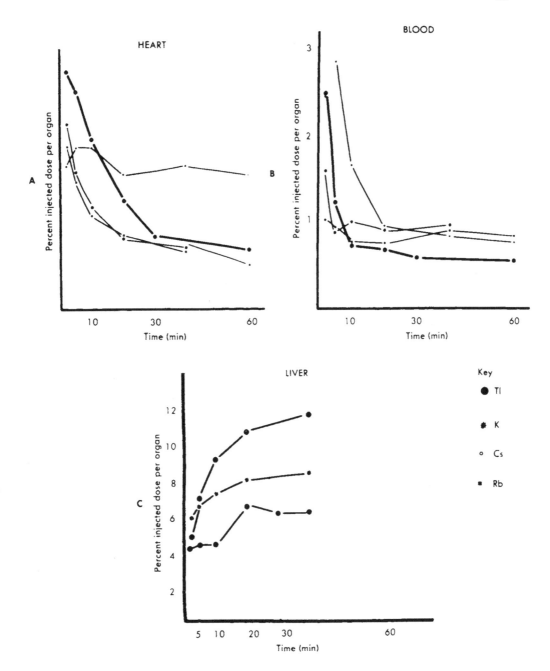

FIGURE 4. Time course of biodistribution of alkali metal ions of ^{201}Tl, ^{129}Cs, ^{81}Rb, and ^{43}K in heart, blood, and liver of mice. Variations in radiopharmacology of these four monovalent cations are obvious. Key: ● — Tl; * — K; O — Cs; and ■ — Rb. (From Strauss, H. W., Pitt, B., Rouleau, J., Bailey, I. K., Wagner, H. N., *Atlas of Cardiovascular Nuclear Medicine*, C. V. Mosby, St. Louis, 1977, 3. With permission.)

the slower kinetics of uptake, fibrotic tissue (low potassium pool) at the site of old myocardial infarct is difficult to delineate with cesium ion. The extraction efficiency of potassium analogs depends on the degree of regional blood flow and oxygenation of tissue. At low flow, the extraction efficiency increases to 100%, while at high flow, the extraction efficiency decreases. Similarly, the fractional escape rate is greater at high flow and less at low flow. Hypoxia alone, without change in flow, decreases the uptake of potassium analog, whereas increases in oxygen tension reverse the uptake.[25] The induction of hypoxia after cellular incorporation does not promote the loss of activity. Thus, in ischemic hypoxia, less cation is delivered to the myocardium because of reduction in flow and a lesser amount is extracted by the hypoxic cells; but, extraction efficiency increases because of prolonged residence of the potassium ion analog. Comparative studies of particulate and diffusible extractable tracers indicate that, under steady-state conditions, the initial distributions are the same. Only at high flow rates does the disparity between the two appear; lesser uptake of potassium ion analogs occurs because of a decrease in extraction efficiency.

Among the potassium analogs, $^{201}Tl^+$ is currently the agent of choice; usually 1.5 mCi is administered per study for the evaluation of patients with ischemia and infarct (Figure 5). When ^{201}Tl is injected into a patient at rest, only the muscle mass of the left ventricle is visualized. The right ventricle has one-third the thickness and 90% of the blood flow of the left ventricle and usually is not visualized. A focal decrease in activity appears to be due to previous myocardial infarct. The normal scan has a uniform distribution of tracer, with a decrease in tracer concentration in the apex. When the tracer is administered to a patient at stress, the myocardium is well-defined because of a decrease in lung background and loss of activity in the splanchnic bed. In coronary disease, transient myocardial ischemia can be observed as a perfusion defect only at stress; usually study at rest shows no abnormality. Patients who cannot exercise sufficiently to produce ischemia can be evaluated after the administration of a vasodilator, e.g., dipyridamole. Coronary blood flow through normal vessels will increase markedly, but flow to ischemic myocardium will increase only slightly, giving rise to a perfusion defect. The ischemic area can thus be well visualized without the patient being physically stressed.

Recent synthesis by Deutsch et al.[43] of cationic complexes of $[^{99m}Tc(III) (diars)_2 Br_2]^+$, (diars = orthophenylene bis-[dimethyl-arsine]) and evaluation in the canine model demonstrate a new generation of potential myocardial perfusion tracer that probably mimics ^{201}thallous ion. The chelating agent upon heating with $^{99m}TcO_4^-$ forms this complex and, thus, works also as a reducing agent. The myocardial uptake at 20 min after intravenous administration is half of that of ^{201}Tl; the major fraction of the $[^{99m}Tc (diars)_2 Br_2]^+$ complex is cleared by the hepatobiliary system. This preliminary finding holds promise for the future development of other ^{99m}Tc-complexes which might replace ^{201}thallous ion as well as development of labeled phospholipids.

In the future, other less toxic elements could replace arsenic in this category of perfusion agents. Stannous chloride has been extensively used for reduction of ^{99m}Tc-pertechnetate prior to chelation. The drawback of this system is that the excess stannous ion also must be kept in solution with 5 to 200 mg of excess chelating agents, also Tc is formed in oxidation states not suitable for preparation of bi- or polynuclear organometallic complexes which might serve as a perfusion agent. Several other reducing agents, e.g., dithionite, borohydride or use of chelating agent that also functions as a reducing agent, might be used in the future.[44,45]

The other approach taken by several investigators is to develop iodine-labeled receptor binding tracers on myocardial cell membranes. Most of these tracers fall into the group of β-adrenoreceptors or muscarinic cholinergic blockers. For the success of these tracers to be used as a myocardial perfusion agent, the specific activity should be very high; since the concentration of receptors is low, nonspecific binding with serum and other proteins and other cell membrane receptors should be low. Up till now, no such agent is available as a perfusion agent,[46] although they are useful for in vitro assay. Further investigation is necessary in this area to see if labeled receptors could play a role in imaging.

The possibility of using $^{13}NH_4^+$ ion for myocardial perfusion imaging was first sug-

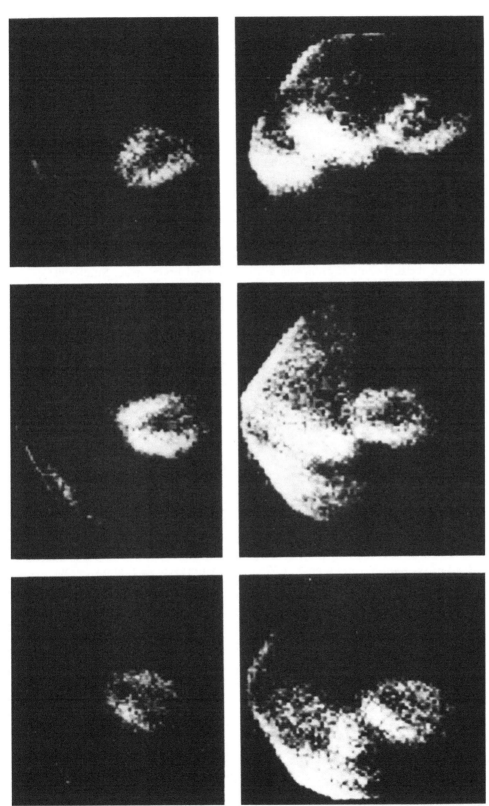

FIGURE 5. Myocardial perfusion image after injection of ²⁰¹Tl at rest (upper panel) and at stress (lower panel). Views were obtained diminution of splanchnic activity is observed with injection at stress. (From Strauss, H. W. and Pitt, B., *Semin. Nucl. Med.*, 7, 49, 1977. By permission of Grune & Stratton).

gested by Hunter, Monahan, and co-workers.[47-58] Harper and co-workers,[49] confirmed their findings and demonstrated localization of NH_4^+ ion in viable human myocardium. The present method of production involves a ^{16}O (p, α) ^{13}N reaction on a water target with 15 MeV protons. The ^{13}N-labeled nitrous oxide produced in the target is reduced with titanous hydroxide in basic solution and is isolated from the mixture by distillation. Very pure $^{13}NH_4^+$ (99.6%) ion containing only a few micrograms of carrier ammonium ion could be produced by this method.[50] Preparation, purification, sterilization, and tube transport to the patient take 20 to 30 min.

Intravenously administered $^{13}NH_4^+$ ion is rapidly cleared from the circulation, being extracted by the liver (15%), lungs, myocardium (2 to 4%), brain, kidneys, and bladder. Myocardial uptake is rapid; about 90% extraction occurs in a single pass. The myocardial uptake remains constant for 30 min.[50] The biological half-life in canine heart is about 72 min. NH_4^+ ion in myocardium is metabolized mainly to glutamine via the glutamine synthetase pathway.[51-54] The neutral NH_3 gas is freely permeable to all cell membranes, but at a physiologic pH of 7.4, the gas is in cationic form that is impermeable to cell membrane. The reaction $NH_3 + H^+/\rightleftharpoons NH_4^+$ is extremely rapid. This may explain the diffusion of ammonia gas into normal brain tissue. Studies in human red cells indicate that NH_4^+ ion may enter the red cell by the substitution of the K^+ and the utilization of the $Na^+ - K^+-ATPase$ enzyme in the membrane, but this mechanism of uptake has not yet been demonstrated in myocardial tissue.

Pulmonary uptake is transient in nonsmokers. In smoker's lung, uptake is substantial, and delayed lung clearance makes myocardial imaging difficult. The second metabolic pathway of NH_4^+ ion is its conversion to carbamoyl phosphate, which is the initial step in the synthesis of urea. This pathway may be important in patients with myocardial disease. Coincidence imaging with a positron camera provides high-resolution images. Three-dimensional reconstruction of the myocardial radionuclide distribution permits visualization of the transaxial images from the apex to the base of the heart.

2. Manganese Ion

Manganese is considered a tracer metal with a body burden of 12 to 20 mg. In the plasma, manganese is bound to a specific globulin, transmanganin. It is primarily an intracellular ion located in the mitochondria of liver, kidney, and muscle cells. Several investigators[59-63] studied the biodistribution of manganese with the tracer ^{54}Mn in rats and dogs. About 25% of administered $^{54}Mn_2^+$ ion localized in myocardium, 2% in lung, 42% in liver, and 8% in kidneys at 30 min after intravenous administration. Myocardial uptake reached a plateau value at 0.5 to 1 hr after administration. Although higher values of myocardial/blood ratios ($\simeq 50$) were obtained, the myocardium to lung and liver ratios were significantly low.

^{52}Fe-^{52m}Mn generator was developed by Ku et al.[61] ^{52m}Fe as anionic chlorocomplex is absorbed on anion exchange resin and eluted with $8N$ HCl. HCl is removed by evaporation and ^{52m}Mn ion is dissolved in buffer. Ischemic area of the size of 2.5 cm^2 with 50% reduction in myocardial blood flow was imaged with a positron emission transaxial tomography in a canine model (PETT III).

3. Inert Gases for Measurements of Blood Flow

Inert gases diffuse freely across endothelial and myocardial cells-membranes.[64-67] Therefore, clearance or washout of radionuclides of inert gases from the injected depot depends on blood flow. ^{133}Xe tracer is a lipophilic substance which maintains a diffusion equilibrium between interstitium and capillary blood regardless of flow rate. After intracoronary injection of ^{133}Xe gas dissolved in saline, the disappearance of tracer from myocardium is recorded with a gamma camera interfaced with a computer. Disappearance of tracer from different regions of myocardium is determined with a light pen. Regional blood flow is proportional to the slope of the disappearance curve. Higher blood flow at stress could be visualized by a steeper disappearance curve.

The major fraction of ^{133}Xe is eliminated through the lung, only a small fraction recirculates. This small amount of circulating ^{133}Xe influences the washout or disappearance and leads to systematic underestimation of washout rate which becomes considerable at high flow rates.

B. Substrates and Substrate Analogs as Tracers
1. Metabolism of Fatty Acids in Myocardium, Labeled Fatty Acids and Analogs

In the blood, free fatty acids are mainly carried by serum albumin (one molecule of serum albumin could carry as many as 20 molecules of fatty acid). The concentration of fatty acid in plasma is (8 to 31 mg)/(100 mℓ of plasma); this concentration is susceptible to change according to physical work, excitement, or physical stress and level of blood gases. The fatty acids are taken up by most tissues, specifically liver and muscle, for satisfying energy requirements.

Fatty acid oxidation normally accounts for 60 to 80% of energy production by the heart. In moderate hypoxia or ischemia, exogenous fatty acid and endogenous fatty acid liberated by the hydrolysis of triglycerides with myocardial enzyme lipase are utilized in preference to glucose. With marked hypoxia (O_2 delivery less than 20% of normal value) after the onset of severe ischemia, anaerobic metabolism provides a substantial proportion of energy by glucose metabolism. In blood, fatty acids are mainly bound to albumin. Myocardial extraction of fatty acid depends on chain length, molarity of fatty acid in albumin, metabolic integrity of the myocyte, perfusion and myocardial oxygen requirement. Ischemia and hypoxia both lead to decreased extraction of fatty acid.[68-102]

The oxidation of a gram-molecule of palmitic acid produces carbon dioxide, water, and energy as shown in the following equation:

$$C_{15} H_{31} COOH + 23 O_2 \rightarrow 16 CO_2 + 16 H_2O + 2,330 \cdot 5 \, kcal \qquad (2)$$

Nearly 40% of this energy is stored in ATP molecules which can be metabolized by the ATPase enzyme for energy requirements.

Long-chain fatty acids in the interstitial or intracellular fluid are bound to soluble proteins and are transported across mitochondrial membrane by acyl coenzyme A carnitine transferases specific for chain length.[85,86] They are oxidized for the production of energy in the mitochondria of cells by a series of reactions that operate in repetitive manner to shorten the chains by two carbon atoms at a time. This process is called β-oxidation. As a result, each fatty acid molecule with 16 carbon atoms is converted to eight molecules of an intermediate known as acetyl coenzyme A (acetyl CoA). Acetyl CoA does not normally accumulate in the cell but becomes enzymatically condensed with "oxaloacetate," a substance derived largely from carbohydrate metabolism. The condensation product is citrate, which is a major component of another cyclic series of reactions known as Krebs cycle. The Krebs cycle serves as a common pathway for the final oxidation of nearly all food material, whether derived from carbohydrate, fat, or protein (Figure 6). The minor pathway for fatty acid metabolism is through α- and ω- oxidation.

In repetitive ischemia, triglycerides accumulate due to the combination of glycerol obtained from anaerobic glycosis and unused excess fatty acid. Thus positive uptake in myocardium reflects aerobic metabolism. On the other hand, under poor flow and hence, poor delivery, the lack of accumulation of fatty acid is not really an indicator of the lack of β-oxidation.

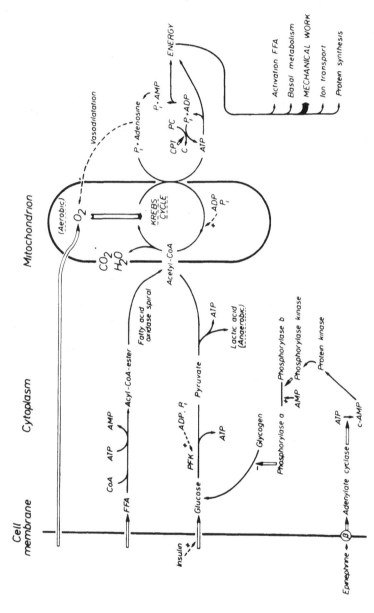

FIGURE 6.   Major steps in the production of energy in muscle cells and of the factors modulating it. Myocardial muscle cells preferentially use free fatty acids (FFA) as substrate for oxidative metabolism; the fatty acids have to be combined with coenzyme A (CoA) before they can enter the mitochondria and be transformed to acetyl coenzyme A (acetyl CoA). The vascular smooth muscle cells preferentially use carbohydrates as substrate and degrade them aerobically and anaerobically. The aerobic metabolism of a fatty acid with 11 C-atoms yields the net production of 95 molecules of adenosine triphosphate (ATP); that of glucose yields 38 molecules of ATP. The breakdown products of ATP have a positive feedback on the metabolism chain, which continuously adjusts energy production to the needs of the cell. AMP = adenosine monophosphate. ADP = adenosine diphosphate. c-AMP = cyclic AMP. P_i = inorganic phosphate. C = creatine. PC = phosphocreatine. PFK = phosphofructokinase. CPT = creatine phosphotransferase, β = beta-adrenergic receptor. (From Shepherd, J. T. and Vanhoutte, P. M., *The Human Cardiovascular System. Facts and Concepts*, Raven Press, New York, 1979, 46. With permission.)

Evans et al.[69] first attempted myocardial imaging with ^{131}I-labeled oleic acid. Because iodination saturates the double bond in the fatty acid molecule, myocardial extraction is substantially reduced and blood clearance is slowed down. This agent has never become clinically useful. Recently, Robinson et al.[70] prepared terminally radioiodinated 16-hexadecenoic acid with [123]I. This tracer was more efficiently extracted by the myocardium (2 to 4%) than ^{131}I-labeled fatty acid prepared by double-bond saturation.

Poe et al.[71] performed successful clinical trials with a limited number of patients. A dose of 5 mCi of ^{123}I-hexadecenoic acid was adequate for imaging. The initial distribution and uptake parallel those of potassium analogs and ^{11}C-labeled oleic and stearic acids. The half-times of biphasic clearance from the blood and myocardium are 1.4 and 20 min, respectively. Unfortunately, the iodine-labeled fatty acids are rapidly metabolized, and the free iodine that is liberated subsequently increases the nonmyocardial background and reduces the reliability of quantitative measurements. The breakdown products of 16-iodo-9-hexadecenoic acid and 16-bromo-9-hexadecenoic acid (starting material), namely, iodoacetate and bromoacetate, are toxic. Iodoacetate is present only in tracer amounts, but the bromoacetate could cause substantial toxicity, particularly when administration of multiple doses is anticipated.

There is a renewed interest in the use of radioiodinated fatty acid for imaging viable myocardium. The terminally radioiodinated saturated fatty acids were recently evaluated for myocardial imaging by several investigators.[72-75] The radioiodine in the para position of the benzene ring of conjugated fatty acid localizes in the normally perfused myocardium by diffusion and appears more stable in vivo than iodinated aliphatic fatty acid; the para-iodo-derivative has less hindrance than the ortho-derivative and has higher extraction and slower rate of clearance from myocardium. The degree of in vivo deiodination is less, due to the presence of radioiodine in the benzene ring. The three components of clearance of radioactive fatty acid from myocardium represents initial distribution, metabolism of free fatty acid and oxidation to CO_2, and slowest clearance due to metabolism of triglyceride of fatty acid. These metabolic data from the clearance curve of ^{11}C-labeled palmitic acid from normal, ischemic, and infarcted myocardium could be easily interpreted. On the other hand, several species of radioactive metabolites, e.g., free radioiodine, free iodobenzoic acid, iodoacetic acid, iodopropionic acid are formed in normal, ischemic, and infarcted myocardium due to the presence of these radiometabolites.

The metabolic rate of fatty acid is difficult to interpret. Since radioiodide accounts for the major portion ($\simeq 40\%$) of these metabolites, several investigators make a second administration of radioiodide for background subtraction in myocardium and increase contrast enhancement of myocardial images.

The palmitic acid was synthesized by the carboxylation of reagent, pentadecyl magnesium bromide with [11]CO_2. The ^{11}C-labeled palmitic acid is solubilized by complexing with albumin. Weiss et al.[94,95] used ^{11}C-labeled palmitate and a positron camera to perform quantitative studies of myocardial metabolism in experimental animals and patients. The half-life of disappearance from blood was 2 to 3 min and that of myocardium was 5 hr. In normally perfused myocardium, fatty acid metabolism by the aerobic process is the preferred pathway for production of high-energy phosphate compounds. Reduced blood flow in ischemic myocardium decreases ^{11}C-palmitate uptake. Thus, the use of physiologic substrates labeled with short-lived radionuclides permits the noninvasive detection of the regional metabolic alterations accompanying myocardial ischemia.

Machulla et al.[90,91] studied the biodistribution of ^{11}C, [34m]Cl, [77]Br, and ^{123}I-labeled ω-heptadecanoic acid in mice. Comparative studies of myocardial uptake and blood clearance indicate that myocardial uptake of labeled heptadecanoic acid has the following order:

$$CH_3(CH_2)_{14}\ ^{11}COOH > \ ^{123}I\ CH_2\ (CH_2)_{15}\ COOH > \ ^{77}BrCH_2(CH)_{15}COOH >$$

$$^{34\ m}ClCH_2(CH_2)_{15}COOH$$

On the other hand, for labeled α-stearic acid, the uptake was in the following order:

$$CH_3(CH_2)_{14}\ ^{11}COOH > CH_3(CH_2)_{15}CH^{34\ m}Cl-COOH > CH_3(CH_2)_{15}-CH-$$

$$^{123}I-COOH > CH_3-(CH_2)_{15}CH^{77}Br-COOH$$

Extraction of ω-halo fatty acid is higher than that of α-halo fatty acids; among the ω-halo fatty acids, the highest uptake was observed for the 17-iodoheptadecanoic acid whose extraction is slightly lower than that of ^{11}C-labeled palmitic acid. The observed pattern of myocardial uptake was explained by steric hindrance and inductive effects of halogen atom on the metabolism of fatty acid.

Knust et al.[92] recently evaluated the myocardial and hepatic extraction of ^{18}F-labeled fatty acids (^{18}F-label on different C-atoms) in mice. Among the ^{18}F-labeled fatty acids the myocardial uptake in mice was in the following order:

$$17\text{-}(^{18}F)\text{-heptadecanoic acid} > 9\text{-}10(^{18}F)\text{-stearic acid} >$$

$$16\text{-}(^{18}F)\text{-hexadecanoic acid} > 2(^{18}F)\text{-stearic acid}$$

2-^{18}F fluorostearic acid shows poor myocardial extraction and higher hepatic extraction; the fatty acids with ^{18}F-label in the middle or at the end of carbon chain exhibit extraction and elimination behavior similar to that of analogous ^{11}C-labeled compounds. Also differences had been observed in the metabolism of 16-^{18}F-fluorohexadecanoic acid (even number of C-atoms) and 17-^{18}F-fluoroheptadecanoic acid (odd number of C-atoms) only. For the former, free fluoride ion was found in the metabolites. These results indicate that β-oxidation of even-numbered fatty acids forms (^{18}F) fluoroacetic acid which participates in the Krebs' cycle, whereas odd-number fatty acids give rise to β-(^{18}F) fluoropropionic acid which undergoes further dehalogenation giving rise to free fluoride ion.

A variety of approaches to preparation of fatty acid derivatives labeled with Tc-99m have been proposed and tested. These include chelation of reduced Tc-99m ion by EDTA, DTPA, and IDA derivatives[88,89] as well as by fatty acid derivatives containing $-OH$ and $-SH$ groups. Unfortunately, none of these agents have been successful for myocardial imaging. This failure is due to modification of structure which is required to produce chelating fatty acid derivatives, the presence of a metal ion; and, also, because of the participation of the required fatty acid carboxyl group in the chelation itself. Only labeled fatty acids which closely approximate the stereochemical and electronic configurations of the physiologic molecules have shown promise for in vivo measurements of myocardial perfusion and/or metabolism.

Fatty acids retain their original biological properties when selected double bonds are isosterically substituted with S, Se, or Te.[76,77] Knapp et al.[78,79] prepared ^{123m}Tellurium-labeled 9-tellurapentadecanoic acid, $CH_3(CH_2)_5$-^{123m}Te-$(CH_2)_7$ COOH, 9-tellurapentadecanoic acid $CH_3(CH_2)_7$-^{123m}Te-$(CH_2)_7$ COOH as isosteres of palmitoleic and several other derivatives and studies the effect of total chain length and the position of tellurium heteroatom on myocardial uptake and retention in rodents and monkeys.

Klein et al.[80] demonstrated that isolated beating rabbit heart perfused with ^{14}C-labeled palmitate (0.4 mM) at high flow $>$ 15 mℓ/min) incorporated 40% of the ^{14}C-palmitate as neutral lipids, 5% as polar lipids, and 39% as acyl CoA, fatty acid, and short chain CoA derivatives. On the other hand, hearts perfused at low flow ($<$5 mℓ/min) exhibited reduced fatty acid uptake and 80% of the palmitate was incorporated into neutral lipids. This increase in ischemic rabbit myocardium at low flow in the content of triglyceride occurred due to reduction of oxidation of fatty acid and conversion to triglyceride. Knapp et al.[78] also demonstrated that ^{123m}Te-analog has higher extraction than the corresponding ^{73m}Se-analog and myocardium metabolism and retention depend on chain length and not on the position of Te. Unlike ^{11}C-labeled fatty acids, ^{123m}Te-9-telluraheptadecanoic acid is degraded into tellurium dioxide and retained into myocardium for a longer period of time. Presently, synthesis of fatty acid analogs with 9-telluraheptanoic acid moiety and tracers of ^{77}Br- or ^{123}I-labeled iodophenyl or ^{123}I label at the ω-terminal is being attempted, with the idea that nonradioactive Te-moiety will prevent further oxidation leading to the prolonged retention and imaging of perfused myocardium.

2. Glucose Metabolism in Myocardium, Labeled Glucose and Analogs

Glucose transport and uptake in myocardium are passive carrier-mediated processes involving an equilibration of extra- and intracellular glucose concentrations. Glucose enters into biochemical reaction sequences by which the potential chemical energy is employed for ATP synthesis. The maximum yield of ATP is obtained by complete oxidation to CO_2. The first major fate of glucose in mammalian cells is its conversion to glucose-6-phosphate with the enzyme hexokinase as shown by the following equation:

$$\text{Glucose} + \text{ATP} \xrightarrow[\text{hexokinase}]{\text{Mg }2^+} \text{glucose-6-phosphate} + \text{ADP} \qquad (3)$$

In the intermediate sequence, glucose is converted into pyruvic acid in the cytoplasm. The pyruvic acid is then oxidized in mitochondria to CO_2 via acetyl CoA and the tricarboxylic acid cycle. All cells could obtain a limited amount of energy in absence of oxygen (anaerobic glycolysis). In this reaction only two molecules of ATP are obtained and glucose is converted into lactic acid as shown below:

$$\text{Glucose} + 2\text{ ADP} + 2\text{ PO}_4^{3-} \rightarrow 2\text{ lactic acid} + 2\text{ ATP} \qquad (4)$$

The only difference between 2-deoxyglucose and glucose is the replacement of $-$OH group on the second carbon by $-$H atom. The structures of glucose, its derivatives, and 2-deoxyfluoroglucose are shown in Figure 7.

Although the primary energy source for the normal myocardium under physiologic conditions is β-oxidation of fatty acids, under stress, the ischemic or anoxic myocardium receives energy by glycolysis. Thus, combining the metabolism of labeled fatty acid and glucose, we could follow the metabolic status of myocardium by in vivo imaging.

Gallagher et al.[96] showed that ^{18}F-labeled deoxyglucose is rapidly extracted by the myocardium of animals and humans, and that the radioactivity in heart (3 to 4%) and brain (2 to 3% of the administered dose) remains relatively constant for 2 hr after injection. The radioactivity from liver, lungs, kidneys, and blood clears rapidly. At 120 min, the heart-to-lung ratio is 12 and the heart-to-liver ratio is 32. Most of the radioactivity in the myocardium is in the left ventricular wall and the intraventricular septum. Cross-sectional images of these regions were well visualized using emission tomography.

Because ^{11}C-labeled glucose is poorly extracted in the myocardium by a passive car-

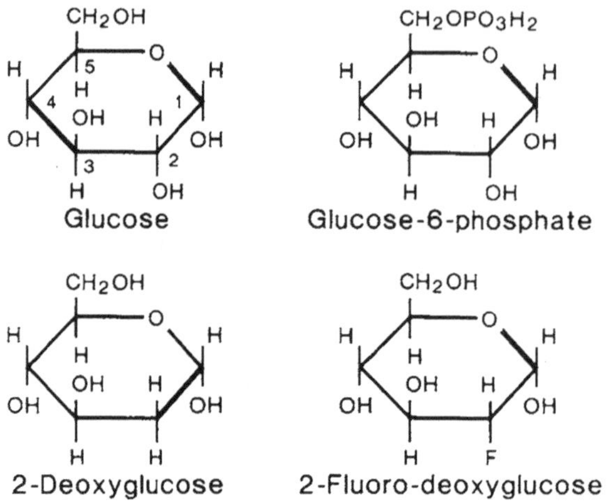

FIGURE 7.   Structures of glucose and analogs.

rier-facilitated process, and is subsequently lost from the myocyte, emission tomography with ^{11}C-labeled glucose was not successful. The metabolic pathway of glucose as shown below is different from that of ^{18}F-labeled deoxyglucose.

$$\text{Glucose} \underset{}{\overset{\text{Hexokinase}}{\rightleftharpoons}} \text{Glucose-6-phosphate} \rightleftharpoons$$

Glycolysis or Glycogen Storage   (5)

$$^{18}\text{F-deoxyglucose} \rightleftharpoons {}^{18}\text{F-deoxyglucose-6-phosphate} \rightarrow$$

Metabolic Trap   (6)

Although ^{18}F-labeled deoxyglucose like glucose is a suitable substrate for hexokinase, ^{18}F-labeled deoxyglucose phosphate is a poor substrate for subsequent metabolic steps, and the permeability of ^{18}F-labeled deoxyglucose phosphate is low. Prolonged retention in myocardium makes this agent a suitable marker for myocardial metabolism and imaging, especially in an ischemic or a hypoxic condition in which this glucose derivative could be used as a major source of energy. ^{18}F-labeled deoxyglucose, in combination with a positron camera, thus may provide the basis for the measurement of regional myocardial glucose metabolism.

[11]C-labeled orthomethyl (^{11}C)-D-glucose[99] is a nonmetabolizable analog of glucose and appears as a suitable indicator for investigating the sugar transport across the cardiac cell membranes. Due to longer time of retention in myocyte, imaging with this marker will be more likely than ^{11}C-labeled glucose. The combination of ^{18}F-deoxyglucose and ^{11}C-labeled palmitic acid appears ideal for studying the complementary metabolic role of glucose and fatty acid in ischemic and normal myocardium.[97]

Recently, ^{11}C-labeled acetate[100] and ^{11}C-labeled lactate[101] have been used for studying intermediary metabolism of myocyte. These tracers also appear suitable for regional myocardial metabolism. Also [30]P-phosphate[102] appears suitable for imaging, phosphate incorporation in ADP and ATP in mitochondria of normal myocyte.

3. Labeled Amino Acids and Analogs

The presence of nitrogen in important biological molecules, for example, proteins and nucleic acids, has stimulated research in the rapid production of ^{13}N-labeled compounds that are precursors in the synthesis of these macromolecules. With the availa-

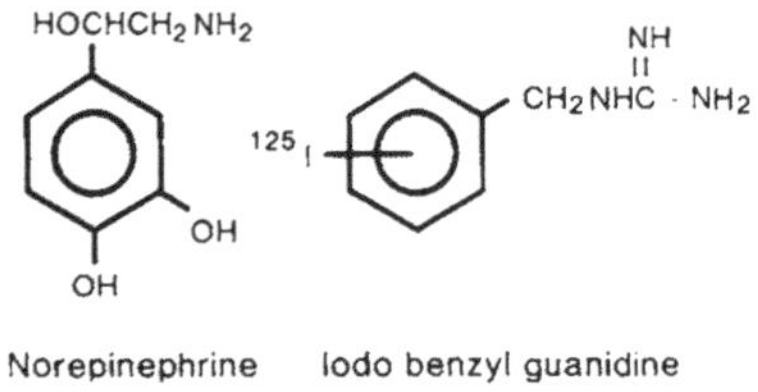

FIGURE 8. Structures of norepinephrine and ^{125}I-labeled benzyl guanidine.

bility of ^{11}C- and ^{13}N-labeled amino acids, there is a renewed interest in ascertaining whether myocardial amino acid uptake might provide an in vivo assessment of protein synthesis and degradation. The branched chain essential amino acids, e.g., leucine, isoleucine, valine, etc., are oxidized by the skeletal and cardiac muscle.[109,110] These amino acids provide a carbon chain for intermediate substrates in the tricarboxylic acid cycle and the amino acid group released is transferred to pyruvate derived from glycolysis to form alanine and perhaps glutamine.

Several amino acids have been synthesized[103,109] enzymatically from ^{13}N-labeled ammonia, and have been evaluated for uptake in myocardium, pancreas, and liver. The myocardial extraction of ^{13}N-labeled glutamine and glutamic acid was lower than that of ^{13}N-labeled ammonium ion. Gelbard et al.[106] synthetized ^{13}N-asparagine by the enzyme asparagine-synthetase and observed higher myocardial uptake of ^{13}N-labeled asparagine in dogs, but observed no uptake in rabbits and human myocardium. Myocardial uptake of different amino acids also was found to be dependent on the species of animals. On the other hand, ^{13}N-labeled glutamate was highly extracted by human myocardium and myocardium of patients with coronary artery disease extracts more glutamate than normal myocardium. Scintigraphic studies with a positron camera could provide useful information about the metabolic fate of different nitrogen-labeled amino acids.

C. Tracers for Myocardial Catecholamine Pool and Analogs

The catecholamines, norepinephrine, and epinephrine are naturally occurring neurohormones in human lungs and are abundant in blood and myocardium;[110-116] their levels in blood rise when appropriate stimuli are applied. After intravenous administration, the organs with extensive sympathetic innervation accumulate epinephrine or norepinephrine and store most of it in a chemically unchanged form in the sympathetic nerve endings. The organ uptake is dependent on blood flow and the density of sympathetic nerve endings in the tissue. Using ^3H-labeled norepinephrine, Whitby and coworkers[112] demonstrated that after intravenous administration in animals, spleen, heart, adrenal gland, lung, and liver were the major sites of localization. Fowler et al.[113] used ^{11}C-labeled norepinephrine in mice and dogs for biodistribution and successful myocardial imaging. At 30 min after intravenous administration, 0.084, 0.037, 0.026, 0.026, 0.024, and 0.010 and 0.004 percentage of injected dose per gram were found in adrenal medulla, kidney, spleen, myocardium, liver, lung, and blood, respectively.

Recently, Wieland et al.[115,116] reported the synthesis and biodistribution of ^{131}I-labeled benzyl guanidine derivative and demonstrated higher uptake in adrenal medulla and myocardium. Structures of norepinephrine and iodobenzyl guanidine are shown in Figure 8. About 0.6-0.8% of administered dose localizes in myocardium of dogs, monkeys, and humans at 2 to 3 hr after intravenous administration.

The in vivo deiodination of the meta-derivative is small. The half-life of efflux curve of iodobenzylguanidine from the myocardium approximates 5 hr. The results of a reserpine blocking study show myocardial uptake reduced by 50%. This indicates that myocardial uptake of M-IBG is by entrapment within the neuronal storage vesicles. This tracer is suggested to share the same uptake by active transport, storage, and release mechanism as norepinephrine, but is not metabolized by monoamine-oxidase or catechol-o-methyl transferase. The prolonged retention facilitates myocardial imaging. This radioiodinated tracer may be used for the evaluation of human myocardial catecholamine turnover and pool under different stimuli.

IV. LOCALIZATION OF TRACERS IN INFLAMED AND DEAD MYOCYTE

In the absence of integrity of membranes of capillary endothelial cells and myocyte and subcellular organelles in infarcted myocardium, a host of large and small molecules and colloidal particles enters and leaves the dead myocyte, leading to participation of tracers in precipitate formation, e.g., calcium phosphate, CaF_2, $^{113}Sn^{2+}$−phosphate (in presence of Ca^{2+} or phosphate analogs), transchelation of organomercurial compounds, binding of ^{99m}Tc-chelates with soluble proteins and enzymes, and some absorption of ^{99m}Tc-complexes in amorphous and crystalline hydroxyapatite and the binding of labeled antibody with myosin, myoglobin, or antiorganelle antibody.[117-184] The time course of biochemical and morphological changes in infarcting myocardium is shown in Table 4. The pathological changes that occur in ischemic and infarcting myocardiumaffecting ion pump permit imaging with ^{201}Tl ion as a cold spot. Disruption of endothelial cell junction and myocyte membrane rupture lead to the influx of small ions and serum proteins. Anionic ^{99m}Tc-complexes which form weak association complexes with soluble proteins also localize into this region of pathological change and give rise to hot spots. Circulating white blood (PMN and lymphocyte) cells labeled with $^{111}Indium$-oxine could be used to determine inflammatory response. ^{111}In-labeled platelets might be used to delineate the platelet deposition on ulcerating lesion and thrombus formation on the wall of the damaged vessel. Formation of platelet microembolism subsequent to implantation of prosthetic valve could be quantitated by tissue/blood ratio; these microembolisms will be trapped in cerebral and visceral organs according to cardiac output. Table 5 computes the mean ratio of radioactivity of infarcted myocardium to normal myocardium and blood of a host of tracers in experimental infarct model obtained in the author's laboratory and elsewhere.

A. Divalent and Trivalent Cations and Anions

Due to the breakdown of myocyte membrane and inactivation of ion-pumps in infarcting myocardium, influxes of sodium, calcium, divalent, trivalent cations and anions occur. Since the soluble proteins and enzymes are lost, which keeps these cations in solution by protein binding, the ionic concentration reaches solubility product value of Ca-phosphate and precipitation of these insoluble compounds occurs. If radioactive cations and anions (e.g., $^{169}Yb^{3+}$, $^{113}Sn^{2+}$, ^{45}Ca, $^{18}F^-$, $^{32}P_2O_7^{4-}$, etc.) are made available by intravenous administration, they should localize in infarcting myocardium at sufficient concentrations to delineate infarcting myocardium. This is more likely at intermediate flow regions since at both regions of high and low flow, the possibility of washout and inaccessibility is high. Due to the loss of tight junctions of endothelial cells in the capillary bed, localization of ^{99m}Tc-labeled sulfide colloid on the basement membrane of capillary vessel in infarcting myocardium is also reported.[120] With the use of a positron camera, Cochavi et al.[125] delineated the region of infarcting myocardium with $Na^{18}F$; one hour after injection and 24 to 72 hr after coronary artery ligation. The time course of biodistribution is similar to that of ^{99m}Tc-pyrophosphate, another bone-seeking pharmaceutical.

Table 5

MEAN RATIO OF RADIOACTIVITY OF INFARCTED MYOCARDIUM-TO-NORMAL MYOCARDIUM AND BLOOD IN EXPERIMENTAL MYOCARDIAL INFARCT MODEL IN RABBITS, 48 HR AFTER LIGATION OF LAD ARTERY

Ratio	Tc-PP[a]	Tc-HEDP[a]	Tc-DMSA[a]	Tc-GH[a]	Tc-Tet[a]	^{3}H-Tet[a]	Tc-DTPA[a]	^{111}In-WBC[a]	I-HSA[a]	I-HU[a]	I-Myosin antibody[a]	I-Myosin Fab'$_2$[a]
IM/NM	40	50	15	8	20	8	3	10	2	2	3	5
IM/Blood	12	14	4	4	6	3	4	6	0.4	0.5	—	—

Ratio	^{85}Sr-[b] microsphere	^{82}Br-[a]	Hg^{2+}[a]	^{203}Hg-fluorescein[a]	^{3}H-fluorescein[a]	^{45}Ca^{2+}[a]	^{113}Sn^{2+}[a]	^{169}Yb^{3+}[a]	^{99m}Tc$_2$S$_7$[a]
IM/NM	0.3	3	15	25	3	9	6	6	3
IM/Blood	—	—	3	5	—	4	3	3	0.6

Ratio	^{18}F-	^{33}P$_2$O$_7$$^{4-}$[a]	^{54}Mn^{2+}	[^{99m}Tc-(diars)$_2$Br$_2$]$^+$	^{201}Tl[c]	^{11}C-palmitate	^{123}I-IDA	^{18}F-FDG	^{123}I-MBG
IM/NM	8	8	0.3	0.2	0.2	0.2	0.2	0.2	0.2
IM/Blood	4	3	50	0.6	0.6	0.9	0.6	0.6	0.6

a Twenty-four hours after intravenous administration.

b Five minutes after intra-auricular injection.

c For cold spot agents, these ratios were obtained at 30 min after intravenous administration, the values were computed from literature.

B. Organometallic Compounds of Mercury

Using [203]Hg-chloromerodrin, Carr and co-workers[127] first demonstrated that the acutely infarcted myocardium is an area of increased radioactivity, after imaging both the living dog and the excised heart, and an infarcted myocardium-to-normal myocardium (IM/NM) ratio of about 15 was obtained. Unfortunately, the clinical trials of [203]Hg-chlormerodrin were not successful. Malek et al.[128] and Hubner[129] showed that [203]Hg-labeled hydroxymercurifluorescein also localized into dead muscle cells in the acute phase. In a reperfusion model obtained by the temporary ligation of the left anterior descending coronary artery, followed by reperfusion, higher ratios of IM/NM, 15 to 100, were obtained. Without reperfusion, a lower ratio (about 10) was obtained, and external imaging was possible in all cases.

When the time course of the IM/NM ratio was followed in a canine model, the highest ratio was obtained three days after ligation. The ratio decreased to unity by the end of the third week. The clinical trial with [203]Hg-fluorescein was not successful, either. Long physical half-life, the nonpenetrating radiation, and prolonged renal retention make the [203]Hg– and [197]Hg-labeled agents undesirable for infarct imaging in human patients. Studies with ^{3}H-labeled fluorescein indicate that IM/NM ratio is $\simeq 2$ for fluorescein. This indicates that fluorescein is only a carrier molecule and mercuric ion from Hg-fluorescein undergoes transchelation reaction, whereby Hg^{2+} ion binds with myocyte enzymes and proteins in dead myocyte. Because the absolute uptake of Hg-labeled fluorescein in infarcted myocardium is high, attempts were made to conjugate ^{99m}Tc to the Hg-fluorescein derivatives. Preliminary results indicate that ^{99m}Tc-chelates of nonradioactive Hg-labeled fluorescein give an IM/NM ratio of about ten with a lower value of absolute uptake into infarcted muscle.

C. Labeled Metal Chelates for Imaging Acute Myocardial Infarct
1. ^{99m}Tc-Tetracycline

Malek and his colleagues[130] showed that the antibiotic tetracycline and its analogs localize into infarcted myocardium in the acute phase. The kinetics of the accumulation of tetracycline studied by the fluorescence of this antibiotic in the canine model can be divided into four phases: in phase 1, immediately after administration, only the normal tissue showed fluorescence and no fluorescence was observed in the infarcted muscle; in phase 2, three hours after injection, fluorescence disappeared from the normal myocardium and intensified along the borders of the infarct zone; in phase 3, four days after infarct, intense fluorescence was limited to the margins of the infarct zone, and after four days the continuous fluorescence along the margin broke into clumps; and in phase 4, seven days after infarction, tetracycline diffused through the whole of the infarcted myocardium, and histologic study in the same canine model indicated that the kinetics of tetracycline concentration in the infarct zone correlated well with that of calcium ion (von Kossa stain).

Early attempts to label tetracycline with gamma-emitting radioisotopes (specifically ^{131}I) for imaging myocardial infarct were not successful. Iodinated tetracycline was not stable. Exchange labeling at low pH leads to the opening of the ring structure of tetracycline,[132] and slow blood clearance and low IM/NM ratio of iodinated tetracycline did not lead to successful infarct imaging.

In 1972, Dewanjee et al.[133] synthesized ^{99m}Tc-tetracycline as well as its analogs, and Holman and Dewanjee successfully imaged infarcted myocardium in the canine model.[134] This ^{99m}Tc chelate became a new tool in cardiovascular nuclear medicine for the accurate detection, localization, and sizing of experimental infarcts. Holman et al.[135] performed clinical trials 24 hr after the injection of 20 mCi of ^{99m}Tc-tetracycline. The optimal time for imaging was within the first three days after onset of symptoms. Abnormal scans in patients with myocardial infarct returned to normal between one and two weeks after the onset of chest pain. Excellent correlation was obtained between scintigraphic and electrocardiographic studies and serum creatine-phosphokinase levels. Because ^{99m}Tc-tetracycline does not localize in ischemic tissue, viable tissue could not be differentiated from irreversibly damaged tissue. The major limitation of ^{99m}Tc-tetracycline for myocardial infarct imaging is the slow clearance from the blood pool, and clear visualization of the infarcted zone by imaging was possible only at 24 hr after injection.

2. ^{99m}Tc-Pyrophosphate and ^{99m}Tc-Diphosphonate

Bonte et al.[136] first used a ^{99m}Tc-labeled pyrophosphate for imaging myocardial infarct. The ^{99m}Tc-labeled phosphate derivatives were first synthesized by Subramanian and McAfee[137] for skeletal imaging. Among the ^{99m}Tc-labeled phosphate and phosphonate derivatives, ^{99m}Tc-labeled pyrophosphate[138] is the most widely used tracer for imaging infarcted myocardium. ^{99m}Tc-labeled hydroxyethylene diphosphonate (HEDP), methylene diphosphonate (MDP), and imidodiphosphate (IDP) also have been evaluated for the same purpose; the absolute uptake of ^{99m}Tc-pyrophosphate is higher than that of ^{99m}Tc-HEDP.[139] Because of the presence of two physiologic sinks (skeletal uptake and renal clearance), the imaging of infarcted tissue is possible at 1 to 2 hr after injection. In an experimental infarct model,[136] myocardial damage in the acute phase was well visualized at 10 to 12 hr after ligation of the left anterior descending artery. The intensity of ^{99m}Tc-pyrophosphate uptake in infarct increases to maximal value (IM/NM ratios between 20 and 100) at 48 to 72 hr, fades after 6 to 7 days, and is absent 14 days after the onset of acute myocardial infarct. Usually 15 to 20 mCi of ^{99m}Tc-pyrophosphate is adequate per study. ^{99m}Tc-pyrophosphate when kept at 0° C is stable 3 to 4 hr after preparation. The amount of [99m]TcO$_4^-$ in the preparation of ^{99m}Tc-pyrophosphate could degrade the quality of the image; in the presence of a large amount of stannous ion, the free [99m]TcO$_4^-$ will label the red blood cells, thus delaying blood clearance, and the stomach uptake will make the delineation of infarct difficult during interpretation.

Unlike the localization in the canine infarct model, different patterns of uptake were observed on clinical trials;[140] some infarcted tissues maintained low levels of increased activity for several weeks, and in some patients for months. The procedure is sensitive but not specific for diagnosis of myocardial infarct. The uptake of ^{99m}Tc-pyrophosphate[141-144] may be traced to limited cell death in the myocardium; dystrophic calcification in the valve, pericardium, or ventricular wall; or abnormal glycoprotein deposition (amyloid, etc.). False-positive diagnosis might also occur owing to breast tumors, inflammation, chest wall diseases of ribs or muscles, metastatic carcinoma, cardioversion, and coronary spasm. Repeated studies performed at 24 hr and 72 hr in the presence of diffuse or low uptake increase the diagnostic accuracy of infarct imaging. Myocardial dosimetry was performed by Lancaster et al.[145] The uptake of ^{99m}Tc-labeled phosphates and phosphonates has been evaluated by several investigators.[146-150] Unlike Hg-labeled organometallic compounds, the ^{99m}Tc-labeled compounds do not undergo transchelation reaction. Reconstructed images thus could provide quantitative information about the extent of muscle damage and the detection of small and posterior wall lesions.[151]

In spite of several studies,[152-158] the factors that determine the uptake of these different tracers are not definitely known. Sobel et al.[159-162] have provided the time course of biochemical and morphologic changes in infarcting myocardium (Table 4) which can be useful in understanding the mechanisms of localization. Buja and his colleagues[156] have ascribed the increased uptake of ^{99m}Tc-pyrophosphate to the deposition of calcium phosphate granules in mitochondria. Dewanjee and Kahn[153] showed that, depending on the type of pathophysiologic change, soluble proteins and calcium phosphate could both be responsible for the localization of ^{99m}Tc-pyrophosphate in infarcted myocardium, while the nonspecific binding of Tc-chelates with denatured proteins might have a major role. This might also explain why other non-bone-seeking ^{99m}Tc-chelates, for example, Tc-glucoheptonate, Tc-tetracycline, Tc-heparin, could concentrate into infarcted tissue.

Preliminary results in our laboratory indicate that metal ions (except the alkali metals, Mg^{2+} and Mn^{2+}) or metal chelates administered intravenously localize into ischemic and infarcted tissue from the vascular space either as a metal protein complex or free metal ion. The free ions, for example Ca^{2+}, give rise to higher IM/NM ratios (greater than 4). However, a metal protein complex, for example, $^{59}Fe^{3+}$-transferrin or $^{64}Cu^{2+}$-ceruloplasmin gives rise to lower ratios (less than 4). In general, the latter represents a larger pool of protein in the interstitial space communicating with leaky cells in ischemic and infarcted myocardium. Shen and Jennings[161] showed that a large fraction of Ca^{2+} participates in calcium phosphate formation in the mitochondria of ischemic and infarcted muscle cells, leading to the continuous influx of Ca^{2+}. Ca-phosphate only, and not Ca^{2+} ion, traps Tc-phosphates.

Any metal ion or metal chelates administered intravenously can bind to any of these several components in ischemic and infarcted myocardia:[163] extracellular protein, intracellular proteins and enzymes, and mitochondrial or cytoplasmic particles of calcium phosphate or fibrin deposits. Recently, the extracellular pool (bromide and soluble protein space) has been measured in the experimental myocardial infarct using ^{82}Br, ^{131}I-human serum albumin, and ^{131}I-labeled hyaluronidase. From the onset of ligation to the healing phase of acute infarct (7 to 15 days after ligation), the IM/NM ratios for ^{82}Br, ^{131}I-HSA, and ^{131}I-hyaluronidase were between 2 and 3. This indicates a larger extracellular fluid space (almost two to three times the normal value) in the infarcted myocardium. Because of reduction of wall motion after myocardial infarct, the flow of extracellular fluid in this region is reduced. Intravenously administered radioactive material reaches this extracellular space by collateral circulation. Because of the stagnation of extracellular fluid, there is a prolonged retention of radioactive material, facilitating binding with cytoplasmic organelles, proteins, and particles. Depending on the abundance and intensity of binding of each of these cytoplasmic proteins and particulates, the radioactive material will be redistributed in the extracellular and intracellular spaces of dead cells.

By the use of the modified Hummel-Dreyer technique,[164] equilibrium dialysis (Figure 9) and gel filtration, the degree of serum protein binding of different ^{99m}Tc-labeled chelates has been determined (Tables 6 and 7). The degree of protein binding decreases in the following order:

Tc-DMSA > Tc-pyrophosphate > Tc-MDP > Tc-glucoheptonate > Tc-DTPA

For ^{99m}Tc-pyrophosphate, the competitive binding with calcium phosphate, human serum albumin, bovine γ-globulin, cellulose, and dextran was determined. The degree on nonspecific binding of Tc-pyrophosphate decreases in the following order:

calcium phosphate > soluble proteins and

enzymes > dextran > myosin > cellulose

The binding of Tc-pyrophospyate with calcium phosphate is eight or nine times higher than that of soluble protein, but the variety and abundance of proteins in the infarcted tissue far outweigh the amount of calcium phosphate particulate that might be formed in the mitochondria, myoplasm, or even extracellular space. Recent results of Reimer et al.[157] have indicated that damaged reperfused myocardium contains about

SEQUENTIAL DIALYSIS OF RADIOPHARMACENTICALS IN PRESENCE OF HUMAN SERUM ALBUMIN

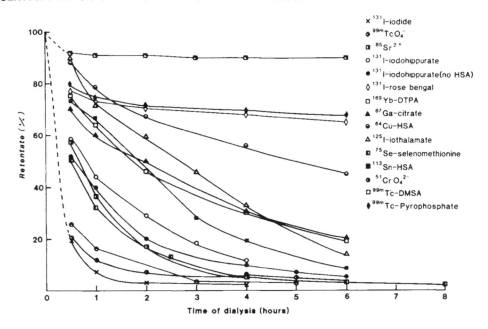

FIGURE 9. Protein binding of radiopharmaceuticals by sequential dialysis. Higher amount of tracers is retained inside cellulose tubing due to the formation of reversible association complex between tracer and albumin.

1 μmol/g of Ca^{2+} (two to three times higher than that in normal muscle). On the assumption that all the Ca^{2+} (upper limit) participates in the calcium phosphate formation, only 84 μg of calcium phosphate could be found. However, there were about 90 mg of soluble proteins and enzymes per gram of muscle, a value that outweighs the calcium phosphate by a factor of 963 (Table 8), and total calcium phosphate in infarcted myocardium could account for less than 1% of the total ^{99m}Tc-pyrophosphate in infarcted myocardium.

Calcium phosphate, as such, could not account for all the binding of ^{99m}Tc-pyrophosphate and might have a minor role in the retention of Tc-pyrophosphate. Insoluble protein (myosin, etc.) does not bind Tc-pyrophosphate (Table 9). Heat-denatured albumin (60 min heat at 45°C) binds more strongly than does native protein (Figure 10). Dewanjee recently performed competitive equilibrium dialysis of ^{99m}Tc-complexes in presence of serum albumin, gamma globules, myosin, and calcium phosphate. Most of the ^{99m}Tc chelates were found to bind nonspecifically with soluble proteins and enzymes and absorbed to a certain extent in calcium phosphate. Recent results of the distribution of ^{131}I-HSA, ^{85}Sr microspheres, and ^{99m}Tc-pyrophosphate in our laboratory indicate a good correlation between intracellular and extracellular protein pools and the IM/NM ratio of ^{99m}Tc-pyrophosphate but no correlation with regional myocardial blood flow (Figure 11).

Among the ^{99m}Tc-chelates, ^{99m}Tc-pyrophosphate has the highest IM/NM ratio and ^{99m}Tc-DTPA has the lowest (Table 5). These results in Tables 8, 9, and 10 were obtained in the experimental myocardial infarct in New Zealand albino rabbits obtained by ligation of left anterior descending coronary artery. There is also a correlation of relative binding of soluble protein and intensity of retention into infarcted myocardium. Hence, ^{99m}Tc-pyrophosphate that is bound to serum protein probably undergoes an exchange reaction with denatured soluble cytoplasmic and extracellular protein and some chemisorption into calcium phosphate in the myoplasm (Figure 12).

Table 6

Mean ±S.E. [a] OF PERCENTILE RADIOACTIVITY IN RETENTATE AND CELLULOSE TUBE EQUILIBRIUM DIALYSIS OF Tc-99m-PERTECHNETATE AND Tc-99m-CHELATES WITH HUMAN SERUM ALBUMIN [b]

Dialysis	^{99m}TcO$_4^-$	^{99m}Tc-DMSA	^{99m}Tc-PP	^{99m}Tc-MDP	^{99m}Tc-GH	^{99m}Tc-DTPA	^{99m}Tc-HIDA
Control							
Retentate	1.13 ± 0.34	3.94 ± 1.91	2.15 ± 0.76	2.63 ± 1.14	1.42 ± 0.56	1.67 ± 0.71	1.46 ± 0.61
Cellulose tubing	1.82 ± 0.42	45.63 ± 15.61	35.76 ± 9.68	21.24 ± 8.63	29.42 ± 10.46	6.41 ± 2.16	14.23 ± 4.18
Albumin							
Retentate	4.59 ± 1.89	82.59 ± 7.63	55.71 ± 6.73	41.38 ± 8.45	22.75 ± 7.25	10.86 ± 3.42	8.76 ± 2.63
Cellulose tubing	1.02 ± 0.64	12.54 ± 1.63	11.67 ± 2.64	3.32 ± 1.19	6.29 ± 2.31	5.46 ± 1.94	5.73 ± 2.14

Note: ^{99m}Tc-DMSA = ^{99m}technetium dimercaptosuccinate; ^{99m}Tc-PP = ^{99m}technetium pyrophosphate; ^{99m}Tc-MDP = ^{99m}technetium methylene diphosphonate; ^{99m}Tc-GH = ^{99m}technetium glucoheptonate; ^{99m}Tc-DTPA = ^{99m}technetium diethylenetriaminepentaacetate.

[a] 4 sets of experiments.
[b] 24 hours at 4°C.

Table 7

AFFINITY CONSTANTS OF TcO$_4^-$ ION, Tc-CHELATES AND HUMAN SERUM ALBUMIN

AC	Tc-HIDA[a]	Tc-DMSA	Tc-PP	Tc-HEDP	Tc-GH	Tc-DTPA	TcO$_4^-$
LAC × 10^3	4.07 ± 1.21	22.86 ± 5.42	5.50 ± 1.24	—	6.36 ± 1.21	1.88 ± 0.93	1.96 ± 0.47
HAC × 10^5	1.38 ± 0.35	1.24 ± 0.25	0.91 ± 0.17	0.78 ± 0.25	1.13 ± 0.43	0.56 ± 0.34	—

[a] Tc-HIDA = ^{99m}technetium hepatobiliary imindiacetate.

Table 8

CONTRIBUTION OF SOLUBLE PROTEINS AND ENZYMES AND CALCIUM PHOSPHATE FOR THE UPTAKE OF ^{99m}Tc-PYROPHOSPHATE IN INFARCTED MYOCARDIUM[a]

	Normal myocardium	Infarcted myocardium
Calcium ion (μmol/g of muscle)	0.6 ± 0.03	0.8 ± 0.04
Calcium phosphate (μg)	62.58	83.44
Soluble proteins and enzymes (mg)	45	90
Soluble proteins and enzymes/calcium phosphate	719	963
Percent uptake by calcium phosphate in infarcted myocardium	—	$\leqslant 0.8$[a]

[a] These values were determined by competitive equilibrium dialysis of ^{99m}Tc-pyrophosphate in presence of calcium phosphate, serum albumin or myosin.

Table 9

MEAN $\pm$ S.E. OF PERCENTILE ACTIVITY DISTRIBUTION OF ^{99m}Tc-PYROPHOSPHATE AND ^{131}I-HUMAN SERUM ALBUMIN IN FRACTIONS OF SERUM PROTEINS AND FIBROUS PROTEINS, ACTIN, AND MYOSIN FROM THE NORMAL AND INFARCTED MYOCARDIA IN THE RABBIT MODEL[a]

	Normal myocardium		Infarcted myocardium	
	^{99m}Tc-PP[b]	^{131}I-HSA[b]	^{99m}Tc-PP	^{131}I-HSA
Serum protein	39.23 ± 4.65	49.67 ± 5.89	36.05 ± 4.75	49.36 ± 5.13
Actin and fibrous proteins	34.42 ± 4.12	17.38 ± 1.86	32.96 ± 3.42	30.89 ± 3.11
Supernatant	24.02 ± 2.84	31.61 ± 3.67	29.10 ± 3.12	19.09 ± 2.11
Myosin	2.33 ± 0.28	1.34 ± 0.19	1.89 ± 0.21	0.66 ± 0.05

[a] Twenty-four hours after ligation and 3 hours after administration of the tracers.
[b] ^{99m}Tc-PP = ^{99m}technetium pyrophosphate; ^{131}I-HSA = ^{131}I-human serum albumin.

The fragile nature of mitochondria and the problem of separating the subcellular organelles in the infarcted myocardium make it difficult to quantitatively estimate the distribution of ^{99m}Tc-pyrophosphate in nuclei, mitochondria, microsome, and soluble protein fractions. However, because of the weak bonds formed between ^{99m}Tc-pyrophosphate and the protein, autoradiography of Tc-pyrophosphate in infarcted tissue is very difficult. Most of the radioactivity washes off during fixation, staining, and the washing procedure. On the assumption that about 30% is the upper level of binding for Tc-pyrophosphate in the particulate material (that is, calcium phosphate) in the mitochondria and microsome (Table 10) as observed by subcellular distribution study, the rest of the radioactivity could only be accounted for by binding with extracellular and intracellular-soluble protein in the infarcted tissue. The results of gel filtration of soluble protein after separation of nuclei, mitochondria, and microsomes from normal and infarcted rabbit myocardia are shown in Figure 13. A mixture of ^{99m}Tc-pyrophosphate and ^{113}Sn(II)-pyrophosphate was injected into the rabbits. Most of the ^{99m}Tc activity is bound to macromolecules. ^{99m}Tc-pyrophosphate absorbed in calcium phosphate granules will not be eluted from the column and will settle down during centrifugation in the organelle fraction. Using a mixture of ^{99m}Tc-pyrophosphate and ^{32}P-pyrophosphate, Dewanjee and Kahn[153] demonstrated that chelation of ^{99m}Tc ion with pyrophosphate increases protein binding affinity (Figure 14).

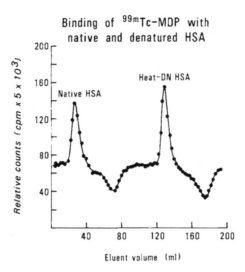

FIGURE 10.   Enhanced binding of denatured
human serum albumin (DN HSA) with ^{99m}Tc-
methylene diphosphonate with respect to that of
native human serum albumin (HSA).

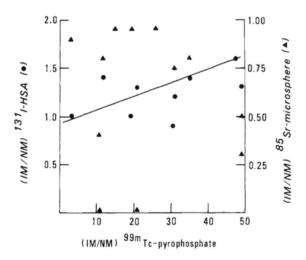

FIGURE 11.   Correlation between IM/NM ratio of ^{99m}Tc-py-
rophosphate in myocardial infarct and iodine-labeled human
serum albumin (HSA) is positive. Correlation between IM/NM
ratio of ^{85}Sr-microsphere and IM/NM ratio of ^{99m}Tc-pyrophos-
phate is lacking.

For Tc-glucoheptonate, which does not localize into bone, the degree of binding
decreases in the following order:

soluble proteins and enzymes > calcium phosphate > cellulose

The strength of protein binding of ^{99m}Tc-glucoheptonate is two to three times higher
than that of calcium phosphate. For ^{99m}Tc-labeled tetracycline, ^{99m}Tc-glucoheptonate
and ^{99m}Tc-labeled DMSA, the binding with denatured macromolecules (protein being
the basic ingredient) is the principal reason for the localization into infarcted myocar-

Table 10
SUBCELLULAR DISTRIBUTIONS (M ± SE) of ^{131}I-HSA, ^{45}Ca, ^{32}P-PYROPHOSPHATE, AND ^{99m}Tc-PYROPHOSPHATE IN NORMAL AND INFARCTED MYOCARDIA

	^{131}I-HSA[a] (%)	^{45}Ca[a] (%)	^{32}P (%)	^{99m}Tc-PP (%)
Normal myocardium				
Nuclei	0.8 ± 0.2	11.2 ± 1.8	5.6 ± 0.6	14.4 ± 1.8
Mitochondria	15.2 ± 2.3	18.3 ± 2.5	6.9 ± 0.7	16.5 ± 2.3
Microsomes	4.4 ± 0.5	7.0 ± 1.1	2.9 ± 0.4	5.2 ± 0.6
Soluble proteins	79.6 ± 9.5	63.5 ± 7.1	84.5 ± 9.9	63.9 ± 7.3
Infarcted myocardium				
Nuclei	2.4 ± 0.35	10.5 ± 1.8	6.5 ± 0.7	19.5 ± 2.1
Mitochondria	2.9 ± 3.1	18.2 ± 2.2	5.5 ± 0.6	21.4 ± 2.2
Microsomes	4.1 ± 0.41	10.3 ± 1.6	2.8 ± 0.3	5.3 ± 0.5
Soluble proteins	90.6 ± 11.5	61.0 ± 7.5	85.2 ± 8.5	53.9 ± 5.4

[a] ^{131}I-HSA = ^{131}I-human serum albumin; ^{99m}Tc-PP = ^{99m}technetium pyrophosphate.

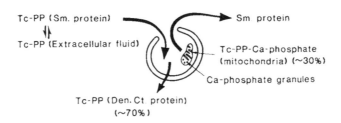

FIGURE 12. Schematic representation of interaction among ^{99m}Tc-pyrophosphate-serum protein complex: {Tc-PP} {Sm. protein}; serum protein (Sm. protein), and ^{99m}Tc-pyrophosphate-denatured cytoplasmic protein-complex (Tc-PP-{Den. Ct.} protein) of infarcted myocardium.

dium. The role of calcium phosphate binding will be even lower than that of Tc-pyrophosphate. Recently, it was also demonstrated that denatured protein binds Tc-chelates more strongly than does native protein. At low pH in the infarcted myocardium, most of the macromolecules remain in the denatured state. Unfolding of the tertiary structure of these macromolecules further enhances nonspecific binding of materials available in the extracellular space. However, a strong protein binder like Tc-DMSA will delay the transition from vascular to extracellular space, and from extracellular to intracellular space, and will increase the optimal time for myocardial infarct imaging. Most of the ^{99m}Tc-chelates that we are presently using are negatively charged and do not localize into live cells to a great extent. Dewanjee and Prince[152] showed that uptake by live cells in tissue culture, as well as in fresh papillary muscle, is diffusion dependent. However, because of breakdown in membrane, certain ^{99m}Tc chelates bind irreversibly to cytoplasmic contents. The uptake of ^{99m}Tc-pyrophosphate is 15 to 20 times higher in dead cells than in live cells (Figure 15). In general, in normal bone, the extraction[154] of ^{99m}Tc-labeled diphosphonate and pyrophosphate is lower than that of Sr^{2+}. This has been ascribed to the larger molecular size of the Tc-chelates. The extraction of Sr^{2+} is similar to that of F$^-$ and is about 65%. However, the corresponding values for Tc-HEDP and Tc-pyrophosphate are 27% and 43%, respectively. Poor ex-

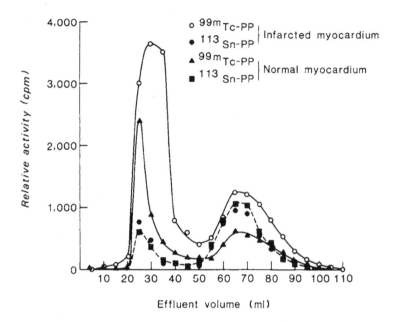

FIGURE 13.   Gel filtration of soluble cytoplasmic protein obtained after separation of nuclei, mitochondria, and microsomes from normal and infarcted rabbit myocardium. A mixture of ^{99m}Tc-pyrophosphate (Tc-PP) and ^{113}Sn-pyrophosphate (Sn-PP) was injected into the rabbits. Most of the ^{99m}Tc activity is protein bound.

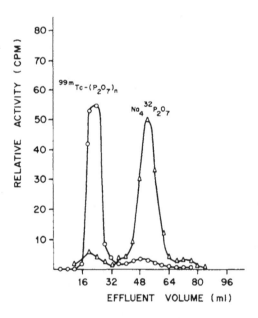

FIGURE 14.   Sephadex gel filtration of rabbit serum containing ^{99m}Tc-pyrophosphate and ^{32}P-pyrophosphate. Pyrophosphate only when chelated with ^{99m}TcO^{2+} ion strongly binds serum protein.

traction accompanied by poor protein binding of Tc-HEDP makes Tc-pyrophosphate better in infarct imaging.[139] Presently, the belief is that both calcium phosphate and

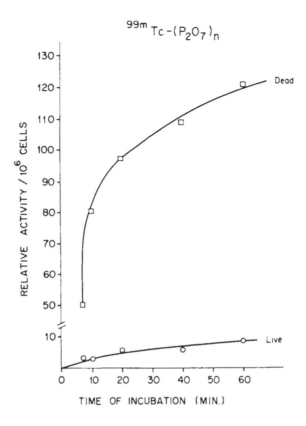

$$^{99m}Tc-(P_2O_7)_n$$

FIGURE 15. Uptake of ^{99m}Tc-pyrophosphate by live and dead cells in culture medium.

the denatured proteins in the dead cells are responsible for the localization of ^{99m}Tc chelates into myocardial infarct.

3. ^{99m}Tc-Glucoheptonate and ^{99m}Tc-Heparin

Several investigators[166,167] showed that ^{99m}Tc-glucoheptonate also localizes into infarcted myocardium. After myocardial infarct in the dog by the ligation of the left anterior descending artery, an IM/NM ratio of 10 to 20 was obtained at 24 hr. Localization of ^{99m}Tc-glucoheptonate occurred rapidly, and an IM/NM ratio of 6 to 7 could be obtained as early as 3 to 4 hr after occlusion of the left anterior descending artery and 4 to 6 hr after injection. Because the binding of ^{99m}Tc-glucoheptonate with soluble protein and calcium phosphate is lower than that of ^{99m}Tc-pyrophosphate, the uptake in infarcted myocardium is low. Although satisfactory images were obtained in the experimental animal models, results of clinical trials of this agent were not encouraging. Recently, Kulkarni et al.[149] synthesized ^{99m}Tc-heparin and used it in experimental infarct model; uptake might be due to collagen and other protein binding mechanisms of tracer in infarcting myocardium.

4. ^{67}Ga-Citrate, ^{131m}In- and ^{68}Ga-Labeled EDTMP and DTPMP

Kramer et al.[168] showed that ^{67}Ga-citrate also localized into acutely infarcted myocardium. Unlike ^{99m}Tc-pyrophosphate, ^{67}Ga does not preferentially localize into dead cells.[154] After intravenous injection, a fraction of ^{67}Ga binds with white blood cells and the ^{67}Ga-labeled white blood cells infiltrate the damaged muscle along with ^{67}Ga-transferrin. The absolute uptake and the IM/NM ratio are low. Imaging with ^{67}Ga-

or ^{111}In-labeled white blood cells might be useful for the diagnosis of inflammatory disease of the myocardium.

Dewanjee et al.[169-172] synthesized the [113m]In- and ^{67}Ga-labeled ethylenediaminetetramethylenephosphonate (EDTMP) and diethylenetriaminepentamethylenephosphonate (DTPMP) chelates and evaluated their uptakes in experimental myocardial infarct. These agents behave like ^{99m}Tc-pyrophosphate (with the exception of poor bone uptake and protein binding).

The predominant concentration was in the bones, bladder, kidneys, and infarcted myocardium. For [113m]In-EDTMP, an average ratio of IM/NM of 30 was obtained 24 hr after ligation. This ratio peaked to 50 at 48 hr after ligation. For ^{68}Ga-EDTMP , the IM/NM ratio was 7 to 8. The two pairs of [113m]In-EDTMP and [201]Tl[+] or ^{68}Ga-EDTMP and [13]NH$_4^+$ ion might be used for multiple imaging of myocardial infarct with gamma or positron cameras, respectively.

5. Protein Binding and Affinity Constants of Metal Chelates

Most of the radiopharmaceuticals after intravenous injection bind nonspecifically with serum protein in a reversible way and thus stabilize the metal complex. Both hydrolysis and oxidation of Tc-chelates are inhibited over a period of one to two days, once the radiopharmaceuticals are mixed even with only small amounts of serum protein. We have observed that even in the absence of ascorbic acid, Tc-MDP is stable over a period of 24 hr (Tc-99m-pertechnetate = 2.3% of $TcO(OH)_2$, $Sn(OH)_2$ colloid $\simeq$ 1 to 2%) in the presence of rabbit or human serum. Biodistribution experiments showed no evidence of transchelation after 24-hr incubation with HSA. This type of weak bond formation may involve H-bond, induced dipole and other electrostatic interactions, and the bond energies may be of the order of a few kilocalories per mole. Other types of bond formations are possible.

Dewanjee et al.[171] performed sequential and equilibrium dialysis of several radiopharmaceuticals in presence and absence of human serum albumin and determined the equilibrium time of dialysis at 24 hr. With the addition of excess of [99]technetium carrier, and equilibrium dialysis, Dewanjee et al.[171] also determined affinity constants of ^{99m}Tc-chelates (Table 7) and [113]Sn-chelates in presence of serum albumin, bovine gamma globulin, and myosin. The biphasic Scatchard plot indicates high and low affinity binding of ^{99m}Tc-chelates with soluble protein (Figure 16) and one affinity constant of lower value for ^{99m}Tc-pertechnetate. The two values of affinity constant of Tc chelates indicate two predominant Tc-species or two binding sites on serum protein.

To determine the role of stannous ion in the formation and coexistence in reconstituted kits of [99m]Tc- and Sn-chelates, Dewanjee et al.[173,174] performed biochemical and biodistribution studies. The sequential and equilibrium dialysis studies (Figures 17-19) indicate that hemoglobin binding affinity of hemoglobin is higher than that of serum albumin. Unlike ^{99m}Tc-chelates, Sn(II)-chelates have one affinity constant value for the binding with serum albumin (Figure 18). These studies also indicate that ^{99m}Tc-chelates and tin (II)-chelates coexist as separate chemical entities and the only role of soluble stannous complexes is the reduction of ^{99m}Tc-pertechnetate ion. The structural studies of ^{99m}Tc-compounds[44] and above-mentioned biochemical studies help us in understanding the role of ^{99m}Tc-chelates in protein binding and subsequent distribution in various organs in health and disease.

6. Tracer Equilibration in Blood and Lymph Pool

Although the blood clearance of a host of radiopharmaceuticals has been extensively studied in different animals and humans, the time of highest concentration of these tracers in lymph of extracellular space has not been determined. By cannulation of mesenteric lymph duct, intravenous administration of tracers and serial lymph collec-

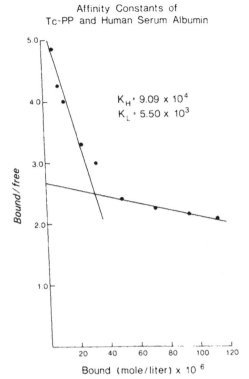

Affinity Constants of
Tc-PP and Human Serum Albumin

$K_H = 9.09 \times 10^4$
$K_L = 5.50 \times 10^3$

Bound (mole/liter) x 10^6

FIGURE 16. Scatchard plot of the binding of
^{99m}Tc-pyrophosphate with human serum albumin. Biphasic curve indicates the presence of two
slopes representing high and low affinity constants.

DIALYSIS OF TIN (II) CHELATES

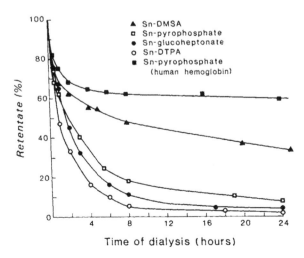

▲ Sn-DMSA
▫ Sn-pyrophosphate
● Sn-glucoheptonate
○ Sn-DTPA
■ Sn-pyrophosphate
(human hemoglobin)

FIGURE 17. Protein binding of tin(II) complexes with serum
albumin and hemoglobin by sequential dialysis. Retention of
stannous ion with hemoglobin is higher than that of albumin.

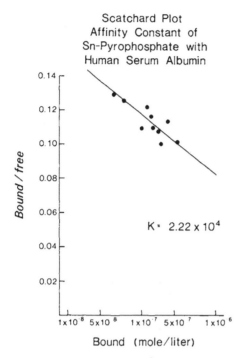

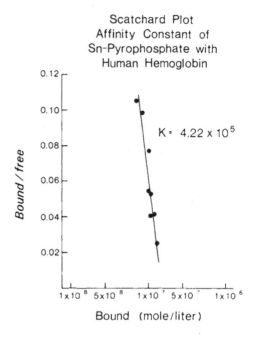

FIGURE 18.   Scatchard plot of the binding of tin(II) pyrophosphate with human serum albumin. Straight line indicates one affinity constant value of one homogeneous binding substance and one ligand.

FIGURE 19.   Scatchard plot of the binding of tin(II) pyrophosphate with hemoglobin. The steeper slope of tin(II) pyrophosphate binding with hemoglobin with respect to that of tin(II) pyrophosphate with albumin indicates higher affinity of stannous ion for hemoglobin.

tion in the Sprague-Dawley rats, Dewanjee determined the time-activity curve of a host of radiopharmaceuticals in lymph (Figure 20). For replenishing fluid loss during lymph collection, isotonic saline was intravenously administered with a constant infusion pump. For tracers of low molecular weight, the peak time of appearance in lymph is 12 to 15 min; on the other hand, for human serum albumin the buildup of lymph activity occurs slowly. Subsequent fall in activity of low molecular weight tracers parallels that of blood clearance. The slightly delayed appearance of ^{67}Ga-citrate indicates tranchelation within proteins. The lymph-to-blood radioactivity ratio has been 1:6-15.

7. Effect of Excess Stannous Ion on Oxidative Phosphorylation in Mitochondria

Since 0.2 to 0.5% of the administered tin localized into infarcted and ischemic tissue which might aggravate the hypoxic tissue by further oxygen depletion, we investigated the effect of stannous pyrophosphate directly on oxidative phosphorylation of isolated mitochondria. Mitochondria were isolated from cat heart by subcellular distribution.

Chance and Williams[176] have described five steady states of respiratory activity in mitochondria, of which state III (rapid state of respiration in presence of oxygen, substrate, and ADP) and state IV respiration (slower rate when all ADP is phosphorylated indicating oxygen use without ATP production) were specifically investigated in presence of Sn(II)-pyrophosphate. When respiration rate of mitochondria is controlled by phosphorylation, the system is tightly controlled. The ratio of state III and state IV respiration is called the respiratory control index (RCI) and high RCI value is an indication of intactness of mitochondrial function.

Figure 21 shows that there is no major change in the oxygen electrode tracings of cat heart mitochondria in presence of glutamate as substrate and glutamate and 6.11 × $10^{-8} M$ of Sn(II)-pyrophosphate. The amount of oxygen consumption during state III

Peak time and lymph clearance of radiopharmaceuticals

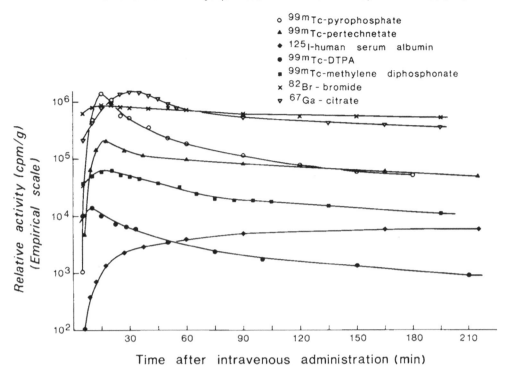

FIGURE 20. Time course of tracer appearance and clearance in rat lymph after intravenous administration. Three rats were used per tracer study.

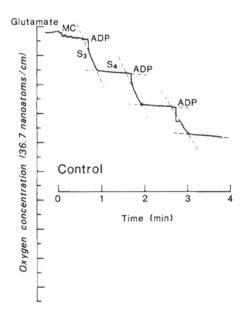

FIGURE 21. Oxygen electrode tracings of cat heart mitochondria in the presence of glutamate as substrate (control). Mitochondria (MC); adenosine diphosphate (ADP); stage III respiration (S_3); stage IV respiration (S_4).

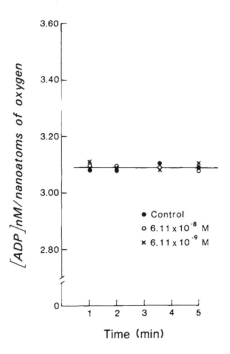

FIGURE 22. ADP:O ratio in presence of Sn(II)-pyrophosphate.

respiration is determined after addition of a known amount of ADP and from this the ADP/O ratio was calculated. There is no major change of ADP/O ratio as the amount of Sn(II)-pyrophosphate is increased (Figure 22). The rate of oxygen consumption for state III respiration per mg of mitochondrial protein is an expression of the enzymatic activity of the protein. No major change in state III respiration is observed between control and $6.11 \times 10^{-9} M$ of Sn(II)-pyrophosphate but the rate of respiration decreases faster for $6.11 \times 10^{-8} M$ of Sn(II)-pyrophosphate. A drastic change is observed in state IV respiration specifically in presence of $6.11 \times 10^{-8} M$ of Sn(II)-pyrophosphate (Figure 23). Since RCI value was obtained from the ratios of slopes of state III and state IV respiration, the RCI value in presence of $6.11 \times 10^{-8} M$ of Sn-pyrophosphate was reduced from 63% to 48% of normal value (Figure 24). In ischemic myocardium, the stage III respiration is $\simeq$ 50% reduced bringing down the RCI value to a value of 24% of normal value.

Kappas and Maines[177] indicated the effect of Sn^{2+} ion on the induction of heme-oxygenase to heme breakdown from the conjugate in cytochrome thus affecting drug detoxification, specifically in liver and kidney tissue. Similar effect was also noted for other divalent metal ions of Cd^{2+} and Pd^{2+} ions. Further study is necessary to critically evaluate the subtle and long-term effect of the Sn(II) compounds on human physiological system. We have added 100 times the amount of human dose of Sn(II)-pyrophosphate to hemoglobin to see any shift in the hemoglobin dissociation curve. It is gratifying to note that even at that concentration, no shift in the hemoglobin-oxygen dissociation curve was observed.

D. Labeled Antibodies Against Myosin, Myoglobin, and Organelles

Khaw et al.[179-182] developed a method for the localization and sizing of myocardial infarcts which is based on selective binding in areas of ischemic injury and infarct of ^{131}I-labeled antibody (or antibody fragments) that is specific for cardiac myosin. An antibody to purified canine cardiac myosin was isolated from antimyosin antiserum

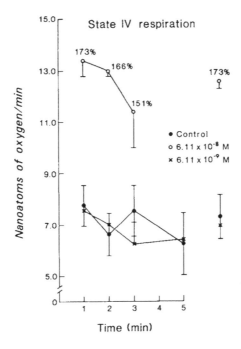

FIGURE 23. Rate of oxygen consumption by cat heart mitochondria, state IV respiration.

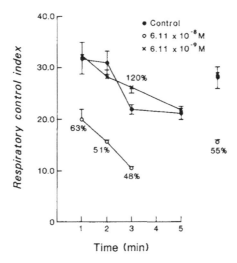

FIGURE 24. Respiratory Control Index in presence of Sn(II)-pyrophosphate.

of the rabbit by affinity chromatography with a myosin-sepharose conjugate. When ^{131}I-labeled antibody was administered intravenously, 24 hr after ligation of the left anterior descending artery, IM/NM ratios of 4.2 for endocardial and 2.9 for epicardial layers were obtained at 24 hr. When Fab'$_2$ fragments of antimyosin antibody obtained by pepsin digestion were used, the IM/NM ratio increased to 6.1 in endocardial and to 3.3 in epicardial layers 24 hr after injection. The concentration of labeled antibody or Fab'$_2$ fragment was greatest in areas where flow was reduced to less than 5% of normal flow. Recently, Khaw et al. labeled Fab$'_2$ fragment with reduced ^{99m}Tc cation and imaged infarcted myocardium in the canine model.

Recently, Parkey et al.[183,184] reported the selective concentration of ^{131}I-labeled antibody against cardiac myoglobin in the area of gross infarction 48 to 72 hr after ligation of the left anterior descending artery. Although the feasibility of imaging infarcted myocardium had been demonstrated with ^{131}I-labeled antibody against cardiac myosin or myoglobin, the difficulty of separation, purification, and labeling, and slow blood clearance leading to delayed imaging will be limiting factors for use in diagnostic nuclear medicine.

V. IMAGING OF REGIONAL MYOCARDIAL BLOOD FLOW: LABELED MICROSPHERES AND MACROAGGREGATES

Myocardial perfusion scintigraphy with particulate indicators[185-193] and diffusible indicators attempts to evaluate myocardial viability by determining regional myocardial blood flow. It is assumed that perfused myocardium is viable. The requirements for particulate indicators include complete mixing with the afferent blood supply and redistribution according to the cellular elements of blood and complete extractability at the capillary or precapillary network. Also the deposition and retention of the particles in the capillary bed should not perturb the circulation. Large microspheres (about 50 μm) are completely extracted; 95 to 97% of smaller microspheres (about 7 μm) are also extracted. Use of larger microspheres is discouraged because they can disrupt capillary flow drastically.

In general, the biodegradable-labeled particles of albumin satisfy these requirements. The carbonized plastic or ceramic microspheres prepared in controlled sizes have been used in animal experiments[186] for quantitative circulation studies; the leaching of radionuclide from the plastic microsphere can result in errors in these studies.

The disadvantage is that coronary catheterization is necessary. In labeling particles, the use of two radioisotopes with reasonable separation of photon energy in the left and right coronary arteries permits visualization of perfused areas of myocardium by each artery and identification of collateral circulation. Two to four mCi of ^{99m}Tc-human albumin microspheres containing 10,000 to 50,000 particles (size range = 10 to 45 μm) could be administered in one artery. The other artery could be perfused with 150 μCi of ^{131}I-macroaggregated albumin (500,000 to 800,000 particles in the size range of 10 to 90 μm) or ^{113}In-labeled microspheres or macroaggregates. ^{113m}In is preferred to ^{131}I because the former allows the administration of large quantities of radioactive material, comparable to that of ^{99m}Tc radioisotope. For positron imaging, ^{68}Ga-labeled microspheres have been developed by Hnatowich.[190] The safety of particulate administration and the accompanying radiation dose have been described by Grames et al.[191] There is a wide margin of safety with respect to the number and size of particles used clinically.

The administration of contrast material during coronary angiography causes a hyperemic response;[192,193] there is a greater increase of flow in normal tissue than in ischemic tissue. Hence, the labeled particles should be administered at least 2 min after the last injection of contrast material, so that the resting blood flow pattern can be established. If labeled particles are injected during contrast-induced hyperemia, a "steal" effect from the ischemic tissue to the normal tissue can be observed, and the perfusion pattern may suggest nonviable myocardium. The color-coded composite images of the perfused myocardium by different arteries provide superior pattern recognition.

VI. LABELED CELLS AND SERUM PROTEINS: IMAGING OF
MYOCARDIAL CONTRACTILITY AND VENTRICULAR FUNCTION

The many indexes of left ventricular function, left ventricular ejection fraction, and regional myocardial wall motion are clinically the most useful indexes of ventricular pump performance. Assessment of these indexes is usually done by means of cardiac catheterization and contrast angiography. The gated cardiac blood pool scan with ^{99m}Tc-labeled human serum albumin[194] or red blood cells,[195,196] a gamma camera, a physiologic synchronizer, and a dedicated computer could partly eliminate the invasive and expensive procedure of contrast angiography. Presently, two types of radiotracers are available as blood pool markers: ^{99m}Tc-labeled red blood cells and ^{99m}Tc-HSA.

^{99m}Tc-labeled human serum albumin clears slightly faster than does ^{131}I-labeled human serum albumin; approximately 60% of the administered radioactivity of ^{99m}Tc is retained at 2 hr after injection. The agent is stable 5 to 6 hr after preparation. Because of a lower regional hematocrit level in the myocardium, ^{99m}Tc-labeled red blood cells provide a better delineation of the chambers of the heart. The red blood cells could be labeled in vitro as well as in vivo. The in vitro method involves the use of a heparinized Sn(II)-citrate preparation.[195,196] The preparation involves careful handling so that sterility and pyrogenicity are maintained during labeling. The labeling efficiency is about 95%.

The in vivo labeling procedure[197,198] involves two consecutive injections. First, Sn(II)-pyrophosphate (10 to 30 μg/kg) in saline solution is injected. Since the cation permeability is a million times slower than anion permeability, the stannous ion is given a headstart. In in vivo labeling, the skeleton and kidneys work as sink to remove excess stannous ion so that pertechnetate reacts with stannous ion only inside red blood cells. After an equilibrium time of 20 to 30 min, 15 to 20 mCi of ^{99m}TcO$_4^-$ ion is administered intravenously. The stannous ion inside the red cell reduces TcO$_4^-$. Most of the ^{99m}Tc ion that is bound to hemoglobin remains inside the red cell. The labeling efficiency is 60 to 80%. No expert technical assistance is necessary for in vivo labeling. Sn(II)-complexes are used for preparation of ^{99m}Tc-complexes and in vivo labeling. Dewanjee et al. demonstrated that only a small amount of stannous ion (<1%) binds with red blood cells and 40% of it is taken up in skeleton and the rest is excreted in urine. Dewanjee et al.[173,174] also determined the affinity constant of Tc-pyrophosphate and stannous ion with human serum albumin and hemoglobin. The affinity of stannous ion for hemoglobin is higher than that of albumin. Because the stannous ion localizes into ischemic and infarcted myocardium, which might further aggravate the hypoxic tissue by oxygen depletion for blood pool imaging, caution should be used to minimize the amount of stannous ion given to patients with myocardial infarct.

VII. LABELED CELLS IN MYOCARDIAL INFARCTION,
CARDIOMYOPATHY, INFLAMMATION, TRANSPLANTATION
REJECTION, AND ARTERIAL THROMBOSIS

McAfee and Thakur[200,201] assessed several soluble radioactive compounds and particles for incorporation into leukocytes by either pinocytosis or phagocytosis. They observed that ^{111}In-labeled oxine is an ideal lipid soluble marker for nonspecific cell labeling. Recently, ^{111}In-labeled polymorphonuclear leukocytes have been used in the diagnosis of inflammatory disease, e.g., abscess and myocardial infarction, ^{111}In-labeled platelets for detection of thrombosis in damaged vessel wall, intracardiac thrombi, bacterial endocarditis, carotid artery lesions, thrombogenicity of biomaterials, imaging prosthetic devices, along with studies of platelet consumption by platelet survival technique, and ^{111}In-labeled lymphocytes for renal and cardiac transplantation, and evaluation of lymphocyte-mediated cell killing. The mechanical separation of cells, incubation of cells with [111]indium-oxine in alcohol and acid citrate-dextrose-isotonic-saline and subsequent washing did not significantly alter cell viability, cell migration, and other vital characteristics as determined by 10-min cell recovery, cell survival, and in vivo cellular migration and pooling. Only lymphocytes at higher ^{111}In-oxine levels have been affected. It has been proposed that lipid-soluble ^{111}In-oxine passively diffuses through cell membrane and ^{111}In-oxine undergoes transchelation with cytoplasmic soluble proteins and oxine component is released back into extracellular space.

Using a simple red blood cell model,[229] Dewanjee and Rao have studied the nature of metal-oxine chelates, lipid solubility by partition coefficient and cell membrane permeability and kinetics of metal-oxine hemoglobin transchelation. Although we have evaluated the oxine complexes [111]indium, [67]gallium, [51]chromium, [57]cobalt, [59]iron, and [99m]Tc (both by stannous chloride and dithionite reduction) as cell marker, ^{111}In-oxine is undoubtedly the best available cell marker. We found ^{111}In-oxine in ACD/saline or other aqueous solution is in colloidal form, but these colloids are solubilized by soluble proteins present in extracellular and extracellular space. Using electrophoresis and gel filtration with Sephadex G-50 and Sephacryl® G-200, we found that these ^{111}In-protein complexes involve only ^{111}In and protein; no oxine molecule is involved as an intermediatry bridge between ^{111}In and protein. Cell labeling in the presence of plasma protein is self-defeating and not essential even for platelet labeling; a thin layer of plasma proteins is retained by platelets even after ACD/saline washing. The protein in the incubating media solubilized ^{111}In-oxine colloid, making it inaccessible to cell labeling. The free indium released after cell lysis is not reutilized by other viable cells.

Due to the higher incidence of platelet thrombus formation in biomaterials used in vascular grafts and prosthetic valves, ^{111}In platelets are ideal for in vivo imaging of thrombus formation and its inhibition with platelet inhibitors, e.g.,aspirin, Motrin®, Persantine®, and sulfinpyrazone.[230-234] Figures 25—27 demonstrate platelet deposition in Teflon® sewing ring of Björk-Shiley mitral prosthesis in the canine model. Platelet thrombus in perivalvular damaged cardiac tissue is also demonstrated, although the stainless steel housing and the pyrolytic carbon disc retained few platelet thrombi. We have also quantitated microembolism in brain, kidney, and lung tissue generated by prosthetic devices.

Healthy vessel wall contains a single layer of cells called endothelial cells. Hypertension, hyperlipidemia, cigarette smoking, and rheological parameters damage these endothelial cells. Atherosclerosis develops by the proliferation of underlying smooth muscle cells.[234,239] Ross et al.[238] suggested that platelet-derived growth factor is essential for this uncontrolled growth of smooth muscle cells. Ultimately, this build-up of smooth muscle cells on vessel wall occludes or narrows the vessel lumen. Occasionally, there is ulceration in atherosclerotic plaque leading to the depositon of platelets, formation of thrombus, and subsequent embolism leading to hypoxia, hypertension, and organ failure.

In a normal vessel, the endothelial cells make prostacyclin a potent vasodilator and platelet antiaggregating, short-lived substance.[234] On the other hand, platelets upon activation generate thromboxane A_2,[235] a potent vasoconstrictor and a potent platelet-aggregating substance. This fine balance between prostacyclin and thromboxane is maintained in health and disturbed in different disease processes leading to thrombosis. The understanding of the interaction between cellular components of blood and constituents of vessel walls of artery, vein, and capillary network by tracer and electron microscopic technique will lead to the understanding of the cardiovascular disease process and design modalities for treatment of this prime killer of mankind.

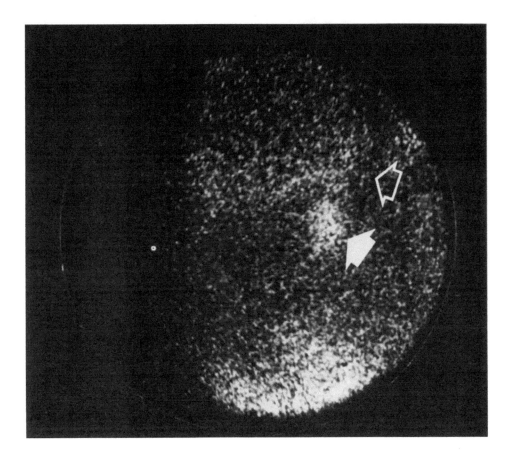

FIGURE 25. Scintiphoto of the thorax of intact dog in left lateral position 25 hr after surgery and 24 hr after intravenous administration of ¹¹¹In-labeled platelets. Uptake shown by solid arrow corresponds to Teflon suture ring and surgically damaged peripheral tissue, open arrow demonstrates the site of hemothorax. Localization of ¹¹¹In-labeled platelets in the upper part of the liver is also evident (lower part of photograph). 25,000 counts were obtained in this view.

VIII. CONCLUSIONS

Recent developments of tracers and instrumentation have made cardiovascular nuclear medicine a very useful noninvasive diagnostic tool. Tracers are available for determination of viable cardiac muscle mass; the exercise test with ²⁰¹Tl provides valuable information about ischemia. In selected cases, perfusion imaging with ^{99m}Tc-labeled microsphere, ¹³¹I-labeled macroaggregates, or ¹³³Xe provides information about regional perfusion and the development of collateral circulation.

For the imaging of blood pool and for providing hemodynamic information about cardiac output, ejection fraction, ejection rate, global and regional wall motion, ^{99m}Tc-labeled human serum albumin and ^{99m}Tc-labeled red blood can be prepared from commercial kits and can be used immediately. For the imaging of dead muscle mass in myocardium and other glycoprotein abnormalities, ^{99m}Tc-pyrophosphate is the obvious choice among the bone-scanning agents. Recent studies of protein binding of ^{99m}Tc-labeled tracers have produced data on the mechanism of localization in necrotic cells. Three-dimensional reconstruction with a single gamma emitter — ^{99m}Tc-pyrophosphate, ²⁰¹Tl ion — and a positron emitter — ¹¹C-labeled palmitic acid — provide quantitative information about the extent of muscle damage. The combined use of these

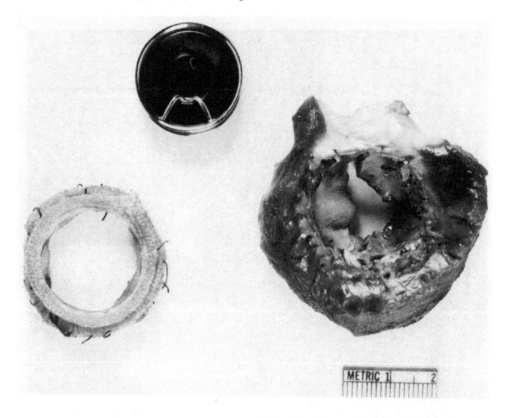

FIGURE 26.   Photographs of pyrolitic carbon-coated disc in stainless steel housing. Teflon sewing ring and surgically damaged tissue were photographed in the same orientation of imaging.

three types of scintigraphic studies presently provides useful information in evaluating patients with suspected coronary artery disease, myocardial infarction and other diseases of myocardium.

ACKNOWLEDGMENTS

The author gratefully acknowledges the constant encouragement of Professor Heinz W. Wahner, the assistance of Dr. Shyam A. Rao for bibliography, Mrs. Urmila Dewanjee for proofreading, and Mrs. Noreen Johnson for excellent typing of the manuscript.

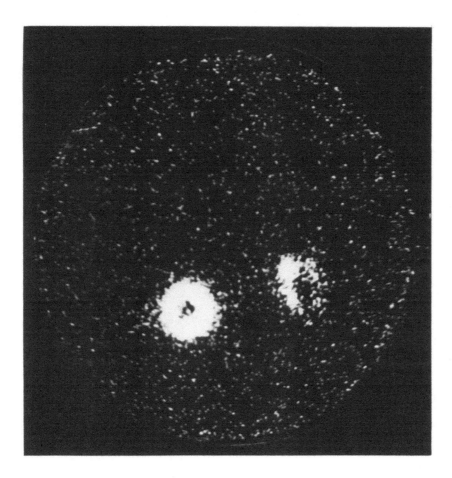

FIGURE 27. Scintiphoto of Teflon sewing ring (left), pyrolitic carbon disc and stainless steel strut, and peripheral surgically damaged tissue. Due to lack of platelet adhesion, the Björk-Shiley housing could not be imaged with a gamma camera. 5,000 counts were obtained in this view.

REFERENCES

1. Blumgart, H. L. and Weiss, S., Studies on the velocity of blood flow. VI. The method of collecting the active deposit of radium and its preparation for intravenous injection, *J. Clin. Invest.*, 4, 389, 1927.
2. Sapirstein, I. A., Fractionation of the cardiac output of rats with isotopic potassium, *Circ. Res.*, 4, 689, 1956.
3. Kety, S. S., Theory of blood-tissue exchange and its application to measurement of blood flow, *Methods Med. Res.*, 8, 223, 1960.
4. Zierler, K. L., Equations for measuring blood flow by external monitoring of radioisotopes, *Circ. Res.*, 16, 309, 1965.
5. Bassingthwaighte, J. B., Strandell, T., and Donald, D. E., Estimation of coronary blood flow by washout of diffusible indicators, *Circ. Res.*, 23, 259, 1968.
6. Griggs, D. M., Jr. and Nakamura, Y., Effect of coronary constriction on myocardial distribution of iodoantipyrine-^{131}I, *Am. J. Physiol.*, 215, 1082, 1968.
7. Conn, H. L., Jr., Equilibrium distribution of radioxenon in tissue; Xenonhemoglobin association curve, *J. Appl. Physiol.*, 16, 1065, 1961.
8. Adelstein, S. J. and Maseri, A., Radioindicators for the study of the heart: principles and applications, *Progr. Cardiovasc. Dis.*, 20, 3, 1977.

9. Chervu, R., Radiopharmaceuticals in radiocardiography, in *Quantitative Nuclear Cardiography*, Pierson, R. N., Jr., Kriss, J. P., Jones, R. H., and MacIntyre, W. J., Eds., John Wiley & Sons, New York, 1975, 254.

10. Cooper, M., Myocardial imaging — an overview, in *Cardiovascular Nuclear Medicine*, Strauss, H. W., Pitt, B., and James, A. E., Jr., Eds., C. V. Mosby, St. Louis, 1974, 149.

11. Dewanjee, M. K., Function of muscle cells and myocardial imaging, in *Principles of Radiopharmacology*, Vol. 3, Colombetti, L. G., Ed., CRC Press, Boca Raton, 1979, 61.

12. Strauss, H. W. and Pitt, B., Thallium-201 as a myocardial imaging agent, *Semin. Nucl. Med.*, 7, 49, 1977.

13. Ritchie, J. D. and Hamilton, G. W., Biologic properties of thallium, in *Thallium-201 Myocardial Imaging*, Ritchie, J. L., Hamilton, G. W., and Wackers, F. J. Th., Eds., Raven Press, New York, 1978.

14. Lenaers, A., [201]Thallium myocardial perfusion scintigraphy during rest and exercise, *Cardiovasc. Radiol.*, 2, 203, 1979.

15. Wackers, F. J. Th., Thallium-201 myocardial scintigraphy in acute myocardial infarction and ischemia, *Semin. Nucl. Med.*, 10, 127, 1980.

16. Leppo, J. A., Scheuer, J., Pohost, G. M., Freeman, L. M., and Strauss, H. W., The evaluation of ischemic heart disease thallium-201 with comments on radionuclide angiography, *Semin. Nucl. Med.*, 10, 115, 1980.

17. Zaret, B. L., Myocardial imaging with radioactive potassium and its analogs, in *Principles of Cardiovascular Nuclear Medicine*, Holman, B. L., Sonnenblick, E. H., and Lesch, M., Eds., Grune & Stratton, New York, 1978, 53.

18. Walsh, W. F., Fill, H. R., and Harper, P. V., Nitrogen-13-labeled ammonia for myocardial imaging, *Semin. Nucl. Med.*, 7, 59, 1977.

19. Holman, B. L., Davis, M. A., and Hanson, R. N., Myocardial infarct imaging with technetium-labeled complexes, *Semin. Nucl. Med.*, 7, 29, 1977.

20. Holman, B. L. and Wynne, J., Myocardial scintigraphy with infarct acid tracers, *Cardiovasc. Radiol.*, 2, 175, 1979.

21. Budinger, T. F. and Rollo, F. D., Physics and instrumentation, *Progr. Cardiovasc. Dis.*, 20, 19, 1977.

22. Budinger, T. F., Physiology and physics of nuclear cardiology, in *Nuclear Cardiology*, Willerson, J. T., Ed., F. A. Davis, Philadelphia, 1979, 9.

23. Perez, I. A., Clinical experience: technetium 99m labeled phosphates in myocardial imaging, *Clin. Nucl. Med.*, 1, 2, 1976.

24. Marcus, M. L. and Kerber, R. E., Present status of the [99m]technetium pyrophosphate infarct scintigram (editorial), *Circulation*, 56, 335, 1977.

25. Levenson, N. I., Adolph, R. J., Romhilt, D. W., Gabel, M., Sodd, V. J., and August, L. S., Effects of myocardial hypoxia and ischemia on myocardial scintigraphy, *Am. J. Cardiol.*, 35, 251, 1975.

26. Kawana, M., Krizek, H., Porter, J., Lathrop, K. A., Charleston, D., and Harper, P. V., Use of [199]Tl as a potassium analog in scanning, *J. Nucl. Med.*, (Abstr.), 333, 1970.

27. Lebowitz, E., Greene, M. W., Fairchild, R., Bradley-Moore, P. R., Atkins, H. L., Ansari, A. N., Richards, P., and Belgrave, E., Thallium-201 for medical use, *J. Nucl. Med.*, 16, 151, 1975.

28. Bradley-Moore, P. R., Lebowitz, E., Greene, M. W., Atkins, H. L., and Ansari, A. N., Thallium-201 for medical use. II. Biological behavior, *J. Nucl. Med.*, 16, 156, 1975.

29. Atkins, H. L., Budinger, T. E., Lebowitz, E., Ansari, A. N., Greene, M. W., Fairchild, R. G., and Ellis, K. J., Thallium-201 for medical use. Human distribution and physical imaging properties, III, *J. Nucl. Med.*, 18, 133, 1977.

30. Strauss, H. W., Harrison, K., Langan, J. K., Lebowitz, E., and Pitt, B., Thallium-201 for myocardial imaging: relation of thallium-201 to regional myocardial perfusion, *Circulation*, 51, 641, 1975.

31. Sheehan, R. M. and Renkin, E. M., Capillary, interstitial, and cell membrane barriers to blood-tissue transport of potassium and rubidium in mammalian skeletal muscle, *Circ. Res.*, 30, 588, 1972.

32. Tancredi, R. G., Yipintsoi, T., and Bassingthwaighte, J. B., Capillary and cell wall permeability to potassium in isolated dog hearts, *Am. J. Physiol.*, 229, 537, 1975.

33. Poe, N. D., Comparative myocardial uptake and clearance characteristics of potassium and cesium, *J. Nucl. Med.*, 13, 557, 1972.

34. Nishiyama, H., Sodd, V. J., Adolph, R. J., Saenger, E. L., Lewis, J. T., and Gabel, M., Intercomparison of myocardial imaging agents: [201]Tl, [129]Cs, [43]K, and [81]Rb, *J. Nucl. Med.*, 17, 880, 1976.

35. Ling, G. N., Thallium and cesium in muscle cells compete for the adsorption sites normally occupied by K^+, *Physiol. Chem. Physics*, 9, 217, 1977.

36. Britten, J. S. and Blank, M., Thallium activation of the (Na^+-K^+)-activated ATPase of rabbit kidney, *Biochim. Biophys. Acta*, 159, 160, 1968.

37. Chandler, H. D., Woolf, C. J., and Hepburn, H. R., Gliding edge dislocations in proteins as a mechanism for active ion transport, *Biochem. J.*, 169, 559, 1978.

38. Robinson, J. D. and Flashner, M. S., The (Na⁺-K⁺) activated ATPase; enzymatic and transport properties, *Biochm. Biophys. Acta,* 549, 145, 1979.

39. Vanleugenhaghe, C., Schwalee, K., and Pourbaix, M., Thallium, in *Atlas of Electrochemical Equilibria in Aqueous Solutions,* Pourbaix, M., Ed., Pergamon Press, London, 1973, 443.

40. Schwartz, J. S., Ponto, R., Carlyle, P., Forstom, L., and Cohn, J. N., Early redistribution of thallium-201 after temporary ischemia, *Circulation,* 57, 332, 1978.

41. Rose, C. P. and Goresky, C. A., Constraints on the uptake of labeled palmitate by the heart; the barriers at the capillary and sarcolemmal surfaces and the control of intracellular sequestration, *Circ. Res.,* 41, 534, 1977.

42. Rose, C. P., Goresky, C. A., and Bach, G. G., The capillary and sarcolemmal barriers in the heart; an exploration of labeled water permeability, *Circ. Res.,* 41, 515, 1977.

43. Deutch, E., Glavan, K. A., Ferguson, D. L., Lukes, S. J., Nishiyama, H., and Sodd, V. J., Development of a Tc-99m myocardial imaging agent to replace ²⁰¹Tl, *J. Nucl. Med.,* 21, 56, 1980.

44. Deutsch, E., Inorganic radiopharmaceuticals, in *Radiopharmaceuticals II,* Proc. 2nd Int. Symp. Radiopharmaceuticals, Society of Nuclear Medicine, New York, 1979, 129.

45. Jones, A. G., Orvis, C., Trop, H. S., Davison, A., and Davis, M. A., A survey of reducing agents for the synthesis of tetraphenyl oxotechnetiumbis (ethanedithiolate) from ⁹⁹ᵐTc-pertechnetate in aqueous solution, *J. Nucl. Med.,* 21, 279, 1980.

46. Eckelman, W. C., Reba, R. C., Gibson, R. E., Rzeszotarski, W. J., Vieras, F., Mazaitis, J. K., and Francis, B., Receptor binding radiotracers: a class of potential radiopharmaceuticals, *J. Nucl. Med.,* 20, 350, 1979.

47. Hunter, W. W., Jr. and Monahan, W. G., ¹³N-ammonia: A new physiologic radiotracer for molecular medicine, *J. Nucl. Med.,* 12, (Abstr.) 368, 1971.

48. Monahan, W. G., Tilbury, R. S., and Laughlin, J. S., Uptake of ¹³N-labeled ammonia, *J. Nucl. Med.,* 13, 274, 1972.

49. Harper, R. V., Lathrop, K. A., Krizek, H., Lembares, N., Stark, V., and Hoffer, R. B., Clinical feasibility of myocardial imaging with ¹³NH₃, *J. Nucl. Med.,* 13, 278, 1972.

50. Lathrop, K. A., Harper, R. V., Rich, B. H., Dinwoodie, R., Krizek, H., Lembares, N., and Gloria. I. Rapid incorporation of short-lived cyclotron-produced radionuclides into radiopharmaceuticals, I. Proc. Ser., STL/PUB/344, 1, 471, 1973.

51. Hoop, B., Jr., Smith, T. W., Burnham, C. A., Correll, J. E., Brownell, G. L., and Sanders, G. A., Myocardial imaging with ¹³NH₄⁺ and a multicrystal positron camera, *J. Nucl. Med.,* 14, 181, 1973.

52. Davidson, S. and Sonnenblick, E. H., Glutamine production by the isolated perfused rat heart during ammonium chloride perfusion, *Cardiovasc. Res.,* 9, 295, 1975.

53. Post, R. L. and Jolly, P. C., The linkage of sodium, potassium, and ammonium active transport across the human erythrocyte membrane, *Biochim. Biophys. Acta,* 25, 118, 1957.

54. Chazov, E., Smirnov, V. N., Mazaev, A. V., Asafov, G. B., Gukowski, D. U., and Krikov, V. I., Myocardial ammonia metabolism in patients with heart disease as revealed by coronary sinus catheterization study, *Circulation,* 47, 1327, 1973.

55. Gould, K. L., Noninvasive assessment of coronary stenosis by myocardial perfusion imaging during pharmacologic vasodilation, I. Physiologic basis and experimental verification, *Am. J. Cardiol.,* 41, 267, 1978.

56. Gould, K. L., Schelbert, H. R., Phelps, M. E., and Hoffman, E. J., Noninvasive assessment of coronary stenoses with myocardial perfusion imaging during pharmacologic coronary vasodilation. V. Detection of 47 percent diameter coronary stenosis with intravenous nitrogen-13 ammonia and emission-computed tomography in intact dogs, *Am. J. Cardiol.,* 43, 200, 1979.

57. Bergmann, S. R., Hack, S., Tewson, T., Welch, M. J., and Sobel, B. E., The dependence of accumulation of ¹³NH₃ by myocardium on metabolic factors and its implications for quantitative assessment of perfusion, *Circulation,* 61, 36, 1980.

58. Huang, G. C., Friedman, A. M., Rayudu, G. V. S., and Clark, P., Radiohalogenated quaternary ammoniums as potential myocardial imaging agents, in 3rd Int. Symp. Radiopharmaceutical Chemistry, St. Louis, June 20-24, 1980, 265.

59. Maynard, L. S. and Cotzias, G. C., The partition of manganese among organs and intracellular organelles of the rat, *J. Biol. Chem.,* 214, 489, 1955.

60. Chauncey, D. M., Schelbert, H. R., and Halpern, S. E., et al., Tissue distribution studies with radioactive manganese: a potential agent for myocardial imaging, *J. Nucl. Med.,* 18, 933, 1977.

61. Ku, T. H., Richards, P., and Stang, L. G., et al., Generator production of manganese-52m for positron tomography, *J. Nucl. Med.,* 20, 682, 1979.

62. Atcher, R. W., Friedman, A. M., Huizenga, J. R., Rayudu, G. V. S., Silverstein, E. A., and Turner, D. A., Manganese-52m, a new short-lived generator-produced radionuclide: a potential tracer for positron tomography, *J. Nucl. Med.,* 21, 565, 1980.

63. Hui, J. C. K., Atkins, H. L., Som, P., Ju, T. H., Fairchild, R. G., Giwa, L. O., and Richards, P., Manganese-52m positron emission transaxial tomography for debating myocardial infarction, *J. Nucl. Med.*, 20 (Abstr.), 648, 1979.

64. Cannon, P. J., Dell, R. B., and Dwyer, E. M., Jr., Measurement of regional myocardial perfusion in man with ^{133}Xe and a scintillation camera, *J. Clin. Invest.*, 51, 964, 1972.

65. Lichtlen, R. R. and Engel, H. J., Assessment of regional myocardial blood flow using the inert gas washout technique, *Cardiovasc. Radiol.*, 2, 203, 1979.

66. Cannon, P. J., Weiss, M. B., and Sciacca, R. R., Myocardial blood flow in coronary artery disease. Studies at rest and during stress with inert gas washout technique, in *Principles of Cardiovascular Nuclear Medicine*, Holman, B. L., Sonnenblick, E. H., and Lesch, M., Eds., Grune & Stratton, New York, 1978, 109.

67. Maseri, A. and L'Abbate, A., Regional myocardial blood flow in the human, in *Cardiovascular Nuclear Medicine*, Strauss, H. W. and Pitt, B., Eds., C. V. Mosby, St. Louis, 1979, 188.

68. Shepherd, J. T. and Vanhoutte, P. M., *The Human Cardiovascular System: Facts and Concepts*, Raven Press, New York, 1979, 46.

69. Evans, J. R., Gunton, R. W., Baker, R. G., Beanlands, D. S., and Spears, J. C., Use of radioiodinated fatty acid for photoscans of the heart, *Circ. Res.*, 16, 1, 1965.

70. Robinson, G. D., Jr. and Lee, A. W., Radioiodinated fatty acids for heart imaging: iodine monochloride addition compared with iodide monochloride addition compared with iodide replacement labeling, *J. Nucl. Med.*, 16, 17, 1975.

71. Poe, N. D., Robinson, G. D., Jr, Graham, L. S., and MacDonald, N. S., Experimental basis for myocardial imaging with ^{123}I-labeled hexadecenoic acid, *J. Nucl. Med.*, 17, 1077, 1976.

72. Zacharias, H. M., Thakur, M. L., and Gottschalk, A., Radioiodination of arachidonic acid: preparation and evaluation, 3rd Int. Symp. Radiopharmaceutical Chemistry, St. Louis, June 20-24, 1980, 75.

73. Otto, C. A., Wieland, D. M., Brown, L. E., and Beierwaltes, W. H., Structure-distribution study of I-125-ω-iodofatty acids, *J. Nucl. Med.*, 21, 14, 1980.

74. Machulla, H. J., Marsmann, M., Dutschka, K., and Van Beuningen, D., Radiohalogenated phenyl fatty acids for metabolic studies of the myocardium, *J. Nucl. Med.*, 21, 57, 1980.

75. Höck, A. Freundlich, I., Vyska, K., Lösse, B., Erbel, R., and Feinendegen, L. E., The metabolism of ω-1-123-heptadecanoic acid in patients with heart disease, *J. Nucl. Med.*, 21, 90, 1980.

76. Korolkovas, A. and Burckholter, J. H., in *Essentials of Medicinal Chemistry*, John Wiley & Sons, New York, 1976.

77. Zanati, G., Gaare, G., and Wolff, M. E., Heterocyclic steroids. V. Sulfur, selenium and tellurium 5α-androstone derivatives and their 7α-methylated congeners, *J. Med. Chem.*, 17, 561, 1974.

78. Knapp, F. F., Ambrose, K. R., Callahan, A. P., Grigsby, R. A., and Irgolic, K. J., Tellurium-123m-labeled isosteres of palmitoleic and oleic acids show high myocardial uptake, in *Radiopharmaceuticals II*, Proc. 2nd Int. Symp. Radiopharmaceuticals, Society of Nuclear Medicine, New York, 1979, 101.

79. Yasuda, T., Knapp, F. F., Elmaleh, D., Kopiwoda, S., McKusick, K. A., and Strauss, H. W., Biodistribution of ^{123m}Tc-9-telluraheptadecanoic acid: large difference of uptake in normal and infarcted myocardium, *J. Nucl. Med.*, 21, 58, 1980.

80. Klein, M. S. and Sobel, B. E., Fatty acid uptake and metabolic imaging of the heart, in *Nuclear Cardiology*, Willerson, J. T., Ed., F. A. Davis, Philadelphia, 1979, 168.

81. Gousios, A., Felts, J. M., and Havel, R. J., The metabolism of serum triglycerides and free fatty acids by the myocardium, *Metabolism*, 12, 75, 1963.

82. Opie, L. H., Metabolism of the heart in health and disease. I. *Am. Heart J.*, 76, 685, 1968.

83. Wakil, S. J., Fatty acid metabolism, in *Lipid Metabolism*, Wakil, S. J., Ed., Academic Press, New York, 1970, 1.

84. Klein, M. S., Goldstein, R. A., and Welch, M. J., External quantification of myocardial metabolism with ^{11}C-labeled fatty acids, *Am. J. Cardiol.*, 41, 378, 1978.

85. Kopec, B. and Fritz, L. B., Properties of a purified carnitine palmitol-transferase and evidence for the existence of other carnitine acyl-transferases, *Can. J. Biochem.*, 49, 941, 1971.

86. Brosnan, J. T. and Fritz, L. B., The permeability of mitochondria to carnitine and acetyl-carnitine, *Biochem. J.*, 125, 94 P, 1971.

87. Wieland, D. M., Brown, I. E., Swanson, D. P., and Beierwaltes, W. H., Synthesis of radio-iodinated ω-(p-amino-phenyl)-substituted fatty acids, *J. Labeled Comp.*, 16, 171, 1979.

88. Eckelman, W. C., Karesh, S. M., and Reba, R. C., New compounds: fatty acid and long chain hydrocarbon derivatives containing a strong chelating group, *J. Pharm. Sci.*, 64, 704, 1975.

89. Livni, E., Davis, M. A., and Warner, V. D., Synthesis and biologic distribution of Tc-99m labeled palmitic acid derivatives, in *Radiopharmaceuticals II*, Proc. 2nd Int. Symp. Radiopharmaceuticals, Society of Nuclear Medicine, New York, 1979, 487.

90. Machulla, H. J., Marsmann, M., Dutschka, K., and Beuningen, D., Radiohalogenated phenyl-fatty acids for metabolic studies of the myocardium, *J. Nucl. Med.*, 21, 57, 1980.

91. Machulla, H. J., Höck, A., Vyska, K., and Feinendegen, L. E., Comparative evaluation of fatty acids labeled with C-11, Cl-34m, Br-77, and I-123 for metabolic studies of the myocardium: concise communication, *J. Nucl. Med.*, 19, 298, 1978.
92. Knust, E. J., Kupfernagel, G. H., and Stöcklin, G., Long-chain F-18 fatty acids for the study of regional metabolism in heart and liver: odd-even effects of metabolism in mice, *J. Nucl. Med.*, 20, 1170, 1979.
93. Ter-Pagossian, M. M., Klein, M. S., Markham, J., Roberts, R., and Sobel, B. E., Regional assessment of myocardial metabolic integrity in vivo by positron-emission tomography with ¹¹C-labeled palmitate, *Circulation*, 61, 242, 1980.
94. Weiss, E. S., Hoffman, E. I., Phelps, M. E., Welch, M. J., Henry, P. D., Ter-Pogossian, M. M., and Sobel, B. E., External detection and visualization of myocardial ischemia with ¹¹C-substrates in vitro and in vivo, *Circ. Res.*, 39, 24, 1976.
95. Hoffman, E. J., Phelps, M. E., Weiss, E. S., Welch, M. J., Coleman, R. E., Sobel, B. E., and Ter-Pogossian, M. M., Transaxial tomographic imaging of canine myocardium with ¹¹C-palmitic acid, *J. Nucl. Med.*, 18, 57, 1977.
96. Gallagher, B. M., Ansari, A. Atkins, H. Casella, V., Christman, D. R., Fowler, J. S., Ido, T., MacGregor, R. R., Som, P., Wan, C. N., Wolf, A. P., Kuhl, D. E., and Reivich, M., Radiopharmaceuticals XXVII. ¹⁸F-labeled 2-deoxy-2-fluoro-D-glucose as a radiopharmaceutical for measuring regional myocardial glucose metabolism in vivo: tissue distribution and imaging studies in animals, *J. Nucl. Med.*, 18, 990, 1977.
97. Schelbert, H., Phelps, M., Hoffman, E., Robinson, G., Hansen, H., Selin, C., and Kuhl, D., Metabolic changes in acute regional myocardial ischemia determined by F-18 deoxyglucose (FDG) and C-11 palmitate (CPA) and positron emission tomography, *J. Nucl. Med.*, 21, 90, 1980.
98. Elmaleh, D. R., Kearfott, K., Goodman, M. M., Varnum, D., Lade, R., Ackerman, R., Strauss, H. W., and Brownell, G. L., A comparison of biodistribution of 2-18-FDG and 3-18-FDG in mice, rats, and dogs, *J. Nucl. Med.*, 21, 13, 1980.
99. Vyska, K., Hock, A., Freundlich, C., Feinendegen, L. E., Kloster, G., and Stöcklin, G., 3-(¹¹C)-methyl glucose, a promising agent for in vivo assessment of function of myocardial cell membrane, *J. Nucl. Med.*, 21, 56, 1980.
100. Pike, V. W., Eakins, M. N., Allen, R. W., and Selwyn, A. P., Preparation of ¹¹C-labeled acetate for the study of myocardial metabolism by emission-computerized axial tomography, in 3rd Int. Symp. Radiopharmaceutical Chemistry, St. Louis, Missouri, 1980, 249.
101. Kloster, G. and Laufer, P., Enzymatic synthesis of L-3- ¹¹C -lactic acid, in 3rd Int. Symp. Radiopharmaceutical Chemistry, St. Louis, June 16-20, 1980, 205.
102. Weinreich, R., Qaim, S. M., and Stöcklin, G., Comparative studies on the production of the positron emitters bromine-75 and phosphorus-30, in 3rd Int. Symp. Radiopharmaceutical Chemistry, St. Louis, June 16-20, 1980, 127.
103. Straatmann, M. G. and Welch, M. J., Enzymatic synthesis of nitrogen-13 labeled amino acids, *Radiat. Res.*, 56, 48, 1973.
104. Cohen, M. B., Spolter, L., MacDonald, N., Masuoka, D. T., Laws, S., Neely, H. H., and Takahasi, J., Production of ¹³N-labelled amino-acids by enzymatic synthesis, Vol. 1, I.A.E.A. Proc. Ser., STI/PUB/344, 1, 483, 1973.
105. Gelbard, A. S., Clarke, L. P., McDonald, J. M., Monahan, W. G., Tilbury, R. S., Kuo, T. Y. T., and Laughlin, J. S., Enzymatic synthesis and organ distribution studies with ¹³N-labeled L-glutamine and L-glutaminic acid, *Radiology*, 116, 127, 1975.
106. Gelbard, A. S., Benua, R. S., McDonald, J. M., Reiman, R. E., Vomero, J. J., and Laughlin, J. S., Organ imaging with N-13-L-glutamate, *J. Nucl. Med.*, 20(Abstr.), 663, 1979.
107. Harper, P. V., Wu, J., Lathrop, K. A., Wickland, T., and Moosa, A., On the mechanism of alanine localization in the heart and pancreas, *J. Nucl. Med.*, 21, 77, 1980.
108. Odessey, R., Khairallah, E. A., and Goldberg, A. L., Origin and possible significance of alanine production by skeletal muscle, *J. Biol. Chem.*, 249, 7623, 1974.
109. Buse, B. G., Biggers, J. F., and Friderici, K. H., Oxidation of branched-chain amino acids by isolated hearts and diaphragms of the rat, *J. Biol. Chem.*, 247, 8085, 1972.
110. Von Euler, U. S., Adrenergic neurohormones, in *Comparative Endocrinology*, Von Eulen, U. S. and Heller, H., Eds., Academic Press, New York, 1963.
111. Raab, W. and Gigee, W., Specific activity of heart muscle to absorb and store epinephrine and norepinephrine, *Circ. Res.*, 3, 353, 1955.
112. Whitby, L. G., Axelrod, J., and Weil-Malherbe, H., The fate of ³H-norepinephrine in animals, *J. Pharmacol. Exp. Ther.*, 132, 193, 1961.
113. Fowler, J. S., MacGregor, R. R., Ansari, A. N., Atkins, H. L., and Wolf, A. P., Radiopharmeceutical: a new rapid synthesis of carbon-11 labeled norepinephrine hydrochloride, *J. Med. Chem.*, 17, 246, 1974.

114. Ansari, A. N., Myocardial imaging with ¹¹C-norepinephrine, in *Cardiovascular Nuclear Medicine,* Strauss, H. W., Pitt, B., and James, A. E., Eds., C. V. Mosby, St. Louis, 1974, 234.

115. Wieland, D. M., Brown, L. E., Worthington, K. C., Rogers, W. L., Swanson, D. P., Beierwaltes, W. H., and Marsh, D. D., Heart imaging with a ¹²³I-noradrenaline storage analog, *J. Nucl. Med.,* 21, 90, 1980.

116. Wieland, D. M., Wu, J., Brown, L. E., Mangner, T. J., Swanson, D. P., and Beierwaltes, W. H., Radiolabeled adrenergic neuron blocking agents: adrenomedullary imaging with ¹³¹I iodobenzyl-guanidine, *J. Nucl. Med.,* 21, 349, 1980.

117. Lefer, A. M., Physiologic mechanisms in acute myocardial infarction, in *Innovations in the Diagnosis and Management of Acute Myocardial Infarction,* Brest, A. Ed., F. A. Davis, Philadelphia, 1975, 25.

118. Bloor, C. M., Functional significance of coronary collateral circulation, *Am. J. Path.,* 76, 563, 1974.

119. Zweifach, B. W., Integrity of vascular endothelium, in *Vascular Endothelium and Basement Membrane,* Allura, B. M., Ed., S. Karger, Basel, 1980, 202.

120. Swanson, D. P., Brady, T. J., Brown, L. E., and Keyes, J. W., Localization of ⁹⁹ᵐTc-colloids in acute myocardial infarcts, *J. Nucl. Med.,* 21, 58, 1980.

121. Kent, S. P., Intracellular plasma protein, a manifestation of cell injury in myocardial ischemia, *Nature (London),* 5042, 1279, 1966.

122. Webb, J., Kirk, K. A., Niedermeier, W., Griggs, J. H., Turner, M. E., and Thomas, N. J., Distribution of 13 trace metals in pig heart tissue, *Bioinorg. Chem.,* 3, 61, 1973.

123. Dewanjee, M. K., Technetium-⁹⁹ᵐpyrophosphate uptake, *Circulation,* 60, 1508, 1979.

124. Kula, R. W., Engel, W. K., and Line, B. R. A., Scanning for soft-tissue amyloid, *Lancet,* 92, 1977.

125. Cochavi, S., Pohost, G. M., Elmaleh, D. R., and Strauss, H. W., Transverse-sectional imaging with Na¹⁸F in myocardial infarction, *J. Nucl. Med.,* 20, 1013, 1979.

126. Buja, L. M., Parkey, R. W., Bonte, F. J., and Willerson, J. T., Pathophysiology of 'cold spot' and hot spot myocardial imaging agents used to detect ischemia and infarction, in *Nuclear Cardiology,* Willerson, J. T., Ed., F. A. Davis, Philadelphia, 105.

127. Carr, E. A., Jr., Beierwaltes, W. H., Patno, M. E., Bartlett, J. D., Jr., and Wegst, A. V., The detection of experimental myocardial infarcts by photoscanning: a preliminary report, *Am. Heart J.,* 64, 650, 1962.

128. Málek, P., Vavrejn, B., Ratuský, J., Kronrád, L., and Kolc, J. A., Detection of myocardial infarction by in vivo scanning, *Cardiologia,* 51, 22, 1967.

129. Hubner, P. J. B., Radioisotopic detection of experimental myocardial infarction using mercury derivatives of fluorescein, *Cardiovasc. Res.,* 4, 509, 1970.

130. Málek, P., Kolc, J., Zástava, V., Žák, F., and Peleška, B., Fluorescence of tetracycline analogues fixed in myocardial infarction, *Cardiologia,* 42, 303, 1963.

131. Eskelson, C. D., Dunn, A. L., Ogobrn, R. E., and McLeay, J. F., Distribution of some radioiodinated tetracyclines in animals, *J. Nucl. Med.,* 4, 382, 1963.

132. Chauncey, D. M., Halpern, S. E., and Alazraki, N. P., Synthesis of radioiodinated tetracyclines: evaluation as tumor scanning agents, *J. Nucl. Med.,* 15, (Abstr.) 483, 1974.

133. Dewanjee, M. K., Fliegel, C., Treves, S., and Davis, M. A., ⁹⁹ᵐTc-tetracyclines: preparation and biological evaluation, *J. Nucl. Med.,* 15, 176, 1974.

134. Holman, B. L., Dewanjee, M. K., Idoine, J., Fliegel, C. P., Davis, M. A., Treves, S., and Eldh, P., Detection and localization of experimental myocardial infarction with ⁹⁹ᵐTc-tetracycline, *J. Nucl. Med.,* 14, 595, 1973.

135. Holman, B. L., Lesch, M., Zweiman, F. G., Temte, J., Lown, B., and Gorlin, R., Detection and sizing of acute myocardial infarcts with ⁹⁹ᵐTc (Sn) tetracycline, *N. Engl. J. Med.,* 291, 159, 1974.

136. Bonte, F. J., Parkey, R. W., Graham, K. D., Moore, J., and Stokely, E. M., A new method for radionuclide imaging of myocardial infarcts, *Radiology,* 110, 473, 1974.

137. Subramanian, G. and McAfee, J. G., A new complex of ⁹⁹ᵐTc for skeletal imaging, *Radiology,* 99, 192, 1971.

138. Perez, R., Cohen, Y., Henry, R., and Panneciere, C., A new radiopharmaceutical for ⁹⁹ᵐTc bone scanning, *J. Nucl. Med.,* 13 (Abstr.), 788, 1972.

139. Williams, C. C., Nishiyama, H., Adolph, R. J., Romhilt, D. W., Sodd, V. J., Saenger, E. L., and Gabel, M., A comparison of technetium etidronate and pyrophosphate for acute myocardial infarct imaging, *J. Nucl. Med.,* 18, 905, 1977.

140. Parkey, R. W., Bonte, F. J., Buja, L. M., Stokely, E. M., and Willerson, J. T., Myocardial infarct imaging with technetium-99m phosphates, *Semin. Nucl. Med.,* 7, 15, 1977.

141. Willerson, J. T., Parkey, R. W., Bonte, F. J., Meyer, S. L., Atkins, J. M., and Stokely, E. M., Technetium stannous pyrophosphate myocardial scintigrams in patients with chest pain of varying etiology, *Circulation,* 51, 1046, 1975.

142. Perez, L. A., Hayt, D. E., and Freeman, L. M., Localization of myocardial disorders other than infarction with ⁹⁹ᵐTc-labeled phosphate agents, *J. Nucl. Med.,* 17, 241, 1976.

143. Chacko, A. K., Gordon, D. H., Bennett, J. M., O'Mara, R. E., and Wilson, G. A., Myocardial imaging with Tc-99m pyrophosphate in patients on Adriamycin treatment for neoplasia, *J. Nucl. Med.*, 18, 680, 1977.

144. Kula, R. W., Engel, W. K., and Line, B. R., Scanning for soft-tissue amyloid (letter to the editor), *Lancet*, 1, 92, 1977.

145. Lancaster, J. L., Tipton, M. D., Parkey, R. W., and Bonte, F. J., Myocardial dosimetry of ^{99m}Tc-pyrophosphate (letter to the editor), *J. Nucl. Med.*, 18, 188, 1977.

146. Davis, M. A. and Jones, A. C., Comparison of ^{99m}Tc-labeled phosphate and phosphonate agents for skeletal imaging, *Semin. Nucl. Med.*, 6, 19, 1976.

147. Wakat, M. A., Chilton, H. M., Hackshaw, B. T., Cowan, R. J., Ball, J. D., and Watson, N. E., Comparison of ^{99m}Tc-pyrophosphate and ^{99m}Tc-hydroxymethylene-diphosphonate in acute mycardial infarction: concise communication, *J. Nucl. Med.*, 21, 203, 1980.

148. Kung, H. F., Blau, M., and Ackerhalt, R., Uptake of radiopharmaceuticals in developing myocardial lesions, in *Radiopharmaceuticals II*, Proc. 2nd Int. Symp. Radiopharmaceuticals, Society of Nuclear Medicine, New York, 1979, 475.

149. Kulkarni, P. V., Parkey, R. W., Wilson, J. F., Lewis, S. E., Buja, L. M., Bonte, F. J., and Willerson, J. T., Modified ^{99m}technetium-heparin for the imaging of acute experimental myocardial infarcts, *J. Nucl. Med.*, 21, 117, 1980.

150. Zweiman, F. G., Holman, B. L., O'Keefe, A., and Idoine, J., Selective uptake of ^{99m}Tc complexes and ^{67}Ga in acutely infarcted myocardium, *J. Nucl. Med.*, 16, 975, 1975.

151. Gustafson, D. E., Berggren, M. J., Singh, M., and Dewanjee, M. K., Computerized transaxial imaging using single gamma emitters, *Radiology*, 129, 187, 1978.

152. Dewanjee, M. K. and Prince, E. W., Cellular necrosis model in tissue culture: uptake of ^{99m}Tc-tetracycline and the pertechnetate ion, *J. Nucl. Med.*, 15, 577, 1974.

153. Dewanjee, M. K. and Kahn, P. C., Mechanism of localization of ^{99m}Tc-labeled pyrophosphonate and tetracycline in infarcted myoscardium, *J. Nucl. Med.*, 17, 639, 1976.

154. Dewanjee, M. K., Localization of skeletal-imaging ^{99m}Tc chelates in dead cells in tissue culture: concise communication, *J. Nucl. Med.*, 17, 993, 1976.

155. Schelbert, H. R., Ingwall, J. S., Sybers, H. D., and Ashburn, W. L., Uptake of infarct-imaging agents in reversibly and irreversibly injured myocardium in cultured fetal mouse heart, *Circ. Res.*, 39, 860, 1976.

156. Buja, L. M., Parkey, R. W., Dees, J. H., Stokely, E. M., Harris, R. A., Jr., Bonte, F. J., and Willerson, J. T., Morphologic correlates of technetium-99m stannous pyrophosphate imaging of acute myocardial infarcts in dogs, *Circulation*, 52, 596, 1975.

157. Reimer, K. A., Martonffy, K., Schumacher, B. L., Henkin, R. E., Quinn, J. L., III, and Jennings, R. B., Localization of ^{99m}Tc-labeled pyrophosphate and calcium in myocardial infarcts after temporary coronary occlusion in dogs, *Proc. Soc. Exp. Biol. Med.*, 156, 272, 1977.

158. Dewanjee, M. K. and Kahn, P. C., Protein binding and the localization of ^{99m}Tc-pyrophosphate and other agents in infarcted myocardium, *Circulation Suppl.*, 2 (Abstr.), 220, 1976.

159. Sobel, B. E., Biochemical and morphologic changes in infarcting myocardium, in *The Myocardium: Failure and Infarction*, Braunwald, W., Ed., H. P. Publishing, New York, 1974, 247.

160. Lie, J. T., Pairolero, P. C., Holley, K. E., McCall, J. T., Thompson, H. K., Jr, and Titus, J. L., Time course and zonal variations of ischemia-produced myocardial cationic electrolyte derangements, *Circulation*, 51, 860, 1975.

161. Shen, A. C. and Jennings, R. B., Kinetics of calcium accumulation in acute myocardial ischemic injury, *Am. J. Pathol.*, 67, 441, 1972.

162. Jennings, R. B. and Ganote, C. E., Mitochondrial structure and function in acute myocardial ischemic injury, *Circ. Res. Suppl.*, 1, 80, 1976.

163. Klotz, I. M., Interaction of organic molecules with proteins, in *Molecular Structure and Biological Specificity*, Pauling, L. C. and Itano, H. A., Eds., American Institute of Biological Sciences, Washington, D.C., 1957, 91.

164. Dewanjee, M. K., and Brueggemann, P. M., Dissociation constants of Tc-99m chelates with serum protein, *J. Nucl. Med.*, 18 (Abstr.), 625, 1977.

165. Kelly, P. J. and Bassingthwaighte, J. B., Studies on bone ion exchanges using multiple-tracer indicator-dilution techniques, *Fed. Proc. Fed. Am. Soc. Exp. Biol.*, 26, 2634, 1977.

166. Fink/Bennett, D., Dworkin, J. H., and Lee, Y. H., Myocardial imaging of the acute infarct, *Radiology*, 113, 449, 1974.

167. Rossman, D. J., Strauss, H. W., Siegel, M. E., and Pitt, B., Accumulation of ^{99m}Tc-glucoheptonate in acutely infarcted myocardium, *J. Nucl. Med.*, 16, 875, 1975.

168. Kramer, R. J., Goldstein, R. E., Hirshfeld, J. W., Jr., Roberts, W. C., Johnston, G. S., and Epstein, S. E., Accumulation of gallium-67 in regions of acute myocardial infarction, *Am. J. Cardiol.*, 33, 861, 1974.

169. Dewanjee, M. K. and Kahn, P. C., Myocardial mapping techniques and the evaluation of new ^{113m}In-labeled polymethylenephosphonates for imaging myocardial infarct, *Radiology*, 117, 723, 1975.

170. Dewanjee, M. K., Beh, R., and Hnatowich, D. J., New Ga-68 labeled skeletal imaging agents for positron scintigraphy, *J. Nucl. Med.* 17, 1003, 1976.

171. Dewanjee, M. K., Brueggemann, P., and Wahner, H. W., Affinity constants of technetium-99m-pertechnetate and Tc-chelates with human serum albumin. in *Radiopharmaceuticals II*, Proc, 2nd Int. Symp. Radiopharmaceuticals, Society of Nuclear Medicine, New York, 1979, 435.

172. Dewanjee, M. K., Localization of ^{113m}In- and ^{67}Ga-labeled polymethylenephosphonates in myocardial infarct, *J. Nucl. Med.*, 16, 151, 1977.

173. Dewanjee, M. K. and Wahner, H. W., Pharmacodynamics of stannous chelates administered with ^{99m}Tc-labeled chelates, *Radiology*, 132, 711, 1979.

174. Dewanjee, M. K., Anderson, G. S., and Wahner, H. W., Pharmacodynamics of stannous chelates administered with ^{99m}Tc-chelates for radionuclide imaging, in *Radiopharmaceuticals II*, Proc. 2nd Int. Symp. Radiopharmaceuticals, Society of Nuclear Medicine, New York, 1979, 421.

175. Sordahl, J. A., Johnson, C. J., Blailock, Z. R., and Schwartz, A., The mitrochondrion, in *Methods in Pharmacology*, Vol. 1, Schwartz, A., Ed., Appleton-Century-Crofts, New York, 1971, 247.

176. Chance, B. and Williams, G. R., The respiratory chain and oxidative phosphorylation, *Adv. Enzymol.*, 17, 65, 1956.

177. Kappas, A. and Maines, M. D., Tin: a potent inducer of heme oxygenase in kidney, *Science*, 192, 60, 1976.

178. Jennings, R. B. and Reimer, K. A., *The Fate of Ischemic Myocardial Cell in Myocardial Infarction*, Corday, E. and Swan, J. H. C., Eds., Williams & Wilkins, Baltimore, 1973, 13.

179. Khaw, B. A., Beller, G. A., Haber, E., and Smith, T. W., Localization of cardiac myosinspecific antibody in myocardial infarction, *J. Clin. Invest.*, 58, 439, 1976.

180. Khaw, B. A., Beller, G. A., and Haber, E., Experimental myocardial infarct imaging following intravenous administration of iodine-131 labeled antibody (Fab′)₂ fragments specific for cardiac myosin, *Circulation*, 57, 743, 1978.

181. Khaw, B. A., Gold, H. K., Leinbach, R. C., Fallon, J. T., Strauss, W., Pohost, G. M., and Haber, E., Early imaging of experimental myocardial infarction by intracoronary administration of ^{131}I-labeled anticardiac myosin (Fab′)₂ fragments, *Circulation*, 58, 1137, 1978.

182. Khaw, B. A., Fallon, J. T., Beller, G. A., and Haber, E., Specificity of localization of myosin-specific antibody fragments in experimental myocardial infarction, *Circulation*, 60, 1527, 1979.

183. Parkey, R. W., Buja, L. M., Kulkarni, P., Stone, M. J., and Willerson, J. T., Localization of a specific I-131 antibody to myoglobin in myocardial tissue and factors which influence myoglobin release from cardiac cells, *J. Nucl. Med.*, 18 (Abstr.), 611, 1977.

184. Kulkarni, P. V., Parkey, R. W., Buja, L. M., Lewis, S. E., Eigenbrodt, E., Stone, M. J., Bonte, F. J., and Willerson, J. T., Localization of antimitochondrial antibody in experimental canine myocardial infarcts, *J. Nucl. Med.*, 21, 90, 1980.

185. Davis, M. A., Particulate radiopharmaceuticals for pulmonary studies, in *Radiopharmaceuticals*, Subramanian, G., Rhodes, B. A., Cooper, J. F., and Sodd, V. J., Eds., Society of Nuclear Medicine, New York, 1975, 267.

186. Utley, J., Carlson, E. L., Hoffman, J. I. E., Martinez, H. M., and Buckberg, G. D., Total and regional myocardial blood flow measurements with 25μ, 15μ, 9μ, and filtered 1-10μ diameter microspheres and antipyrine in dogs and sheep, *Circ. Res.*, 34, 391, 1974.

187. Ashburn, W. L., Braunwald, E., Simon, A. L., Peterson, K. L., and Gault, J. H., Myocardial perfusion imaging with radioactive-labeled particles injected directly into the coronary circulation of patients with coronary artery disease, *Circulation*, 44, 851, 1971.

188. Becker, L. C., Fortuin, N. J., and Pitt, B., Effect of ischemia and antianginal drugs on the distribution of radioactive microspheres in the canine left ventricle, *Circ. Res.*, 28, 263, 1971.

189. Ritchie, J. L., Hamilton, G. W., Gould, K. L., Allen, D., Kennedy, J. W., and Hammermeister, K. E., Myocardial imaging with indium-113m- and technetium-99m-macroaggregated albumin: new procedure for identification of stress-induced regional ischemia, *Am. J. Cardiol.*, 35, 380, 1975.

190. Hnatowich, D. J., Labeling of tin-soaked albumin microspheres with ^{68}Ga, *J. Nucl. Med.*, 17, 57, 1976.

191. Grames, G. M., Jansen, C., Gander, M. R., Wieland, H. C., and Judkins, M. P., Safety of the direct coronary injection of radiolabeled particles, *J. Nucl. Med.*, 15, 2, 1974.

192. Kirk, G. A., Adams, R., Jansen, C., and Judkins, M. P., Particulate myocardial perfusion scintigraphy: its clinical usefulness in evaluation of coronary artery disease, *Semin. Nucl. Med.*, 7, 67, 1977.

193. Heymann, M. A., Payne, B. D., Hoffman, J. I. E., and Rudolph, A. M., Blood flow measurements with radionuclide-labeled particles, in *Principles of Cardiovascular Nuclear Medicine*, Holman, B. L., Sonnenblick, E. H., and Lesch, M., Eds., Grune & Stratton, New York, 1978, 135.

194. Stern, H. S., Zolle, I., and McAfee, J. G., Preparation of technetium (Tc-^{99m})-labeled serum albumin (human), *Int. J. Appl. Radiat. Isot.*, 16, 283, 1965.

195. Dewanjee, M. K., Binding of ^{99m}Tc ion to hemoglobin, *J. Nucl. Med.,* 15, 703, 1974.

196. Smith, T. D. and Richards, P., A simple kit for the preparation of ^{99m}Tc-labeled red blood cells, *J. Nucl. Med.,* 17, 126, 1976.

197. Stokely, E. M., Parkey, R. W., Bonte, F. J., Graham, K. D., Stone, M. J., and Willerson, J. T., Gated blood pool imaging following ^{99m}Tc stannous pyrophosphate imagein, *Radiology,* 120, 433, 1976.

198. Pavel, D. G., Zimmer, A. M., and Patterson, V. N., In vivo labeling of red blood cells with ^{99m}Tc: a new approach to blood pool visualization, *J. Nucl. Med.,* 18, 305, 1977.

199. Dewanjee, M. K. and Wahner, H. W., Localization of stannous ion from ^{99m}Tc-labeled radiopharmaceuticals into myocardial infarct (abstract), presented at Proc. Radiol. Soc. North Am., Chicago, November 26-December 2, 1977, 77.

200. McAfee, J. G. and Thakur, M. L., Survey of radioactive agents for in vitro labeling of phagocytic leukocytes. I. Soluble agents, *J. Nucl. Med.,* 17, 480, 1976.

201. McAfee, J. G. and Thakur, M. L., Survey of radioactive agents for in vitro labeling of phagocytic leukocytes. II. Particles, *J. Nucl. Med.,* 17, 488, 1976.

202. Thakur, M. L., Coleman, R. E., and Welch, M. J., Indium-111 labeled leukocytes for the localization of abscesses. Preparation analysis, tissue distribution and comparison with gallium 67 citrate in dogs, *J. Lab. Clin. Med.,* 89, 217, 1977.

203. Thakur, M. L., Welch, M. I., Joist, J. H., and Coleman, R. E., Indium-111 labeled platelets. Studies on peparation and evaluation of in vitro and in vivo functions, *Thromb. Res.,* 9, 345, 1976.

204. Scheffel, U., McIntyre, P. A., Evatt, B., Dvornicky, J. A., Natarajan, T. K., Bolling, D. R., and Murphy, E. A., Evaluations of In-111 as a new high proton yield gamma-emitting "physiological" platelet label, *The Johns Hopkins Med. J.,* 140, 285, 1977.

205. Joist, H. J., Baker, R. K., Thakur, M. L., and Welch, M. I., Indium-111 labeled human platelets; uptake and loss of label and in vitro functions of labeled platelets, *J. Lab. Clin. Med.,* 92, 829, 1978.

206. Segal, A. W., Deteix, P., Garcia, R., Tooth, P., Zanelli, G. D., and Allison, A. C., Indium-111 labeling of leukocytes: a detrimental effect on neutrophil and lymphocytic functions and an improved method of cell labeling, *J. Nucl. Med.,* 19, 1238, 1978.

207. Wisto, B. A., Grossman, Z. D., McAfee, J. C., Subramanian, G., Henderson, R. W., and Roskopf, M. L., Labeling platelets with oxine complexes of Tc-99m and In-111. In vitro studies and survival in the rabbits, *J. Nucl. Med.,* 19, 483, 1978.

208. Thakur, M. L., Lavender, J. R., Arnot, R. N., Silvester, D. J., and Segal, A. W., In-111 labeled leukocytes in man, *J. Nucl. Med.,* 18, 1014, 1977.

209. Goodwin, D. A., Bushberg, J. T., Doherty, P. W., Lipton, M. I., Conley, F. K., Diamonti, C. I., and Meares, C. F., Indium-111 labeled autologous platelets for localization of vascular thrombi in humans, *J. Nucl. Med.,* 19, 626, 1978.

210. Lavender, J. P., Goldman, J. M., and Arnot, R. N., Kinetics of indium-111 labeled lymphocytes in normal subjects and patients with Hodgkins disease, *Br. J. Med.,* 2, 797, 1977.

211. Thakur, M. L., Segal, A. W., Louis, L., Welch, M. J., Hopkins, J., and Peters, T. J., Indium-111 labeled cellular blood components. Mechanism of labeling and intracellular localization in human neutrophils, *J. Nucl. Med.* 18, 1022, 1977.

212. Phillips, J. P., The reaction of 8-quinolinol, *Chem. Res.,* 56, 271, 1956.

213. Riba, A. L., Thakur, M. L., Gottschalk, A., Andriole, V. T., and Zaret, B. L., Imaging experimental infective endocarditis with indium-111 labeled blood cellular components, *Circulation,* 59, 336, 1979.

214. Riba, A. L., Thakur, M. L., Gottschalk, A., and Zaret, B. L., Imaging acute coronary thrombosis with indium-111 labeled platelets, *Circulation,* 60, 767, 1979.

215. Davis, H. H., Heaton, A. W., Segal, B. A., Mathias, C. J., Joist, J. H., Sherman, I. A., and Welch, M. J., Scintigraphic detection of atherosclerotic lesions and venous thrombi in man by indium-111 labeled autologous platelets, *Lancet,* 1185, 1187, 1978.

216. McIlmoyle, G., Davis, H. H., Welch, M. J., Primeau, J. L., Sherman, L. A., and Siegel, B. A., Scintigraphic diagnosis of experimental pulmonary embolism with In-111 labeled platelets, *J. Nucl. Med.,* 18, 910, 1977.

217. Dewanjee, M. K., Fuster, V., Kaye, M. P., and Josa, M., Imaging platelet deposition with In-111 labeled platelets in coronary artery bypass graft in dogs, *Mayo Clin. Proc.,* 53, 327, 1978.

218. Frost, P., Smith, J., and Frost, H., The radiolabeling of lymphocytes and tumor cells with indium-111, *Proc. Soc. Exp. Biol. Med.,* 157, 61, 1978.

219. Thakur, M. L., Gottschalk, A., and Zaret, B. L., Imaging experimental myocardial infarction with In-111-labeled autologous leukocytes: effects of infarct age and residual regional myocardial blood flow, *Circulation,* 60, 297, 1979.

220. Zakhireh, B., Thakur, M. L., Malech, H. L., Cohen, M. S., Gottschalk, A., and Root, R. K., Viability, random migration chemotaxis, bactericidal capacity and ultrastructure of indium-111 labeled human polymorphonuclear leukocytes, *J. Nucl. Med.,* 20, 741, 1979.

221. Sinn, H. and Silvester, D. J., Simplified cell labeling with ¹¹¹In-acetyl acetone, *Br. J. Radiol.* 52, 758, 1979.

222. Segal, A. W., Thakur, M. L., Arnot, R. N., and Lavender, J. P., Indium-111 labeled leukocytes for localization of abscesses, *Lancet,* 1056, 1976.

223. Robertson, J. S., Dewanjee, M. K., Brown, M. L., Fuster, V., and Chesebro, J. H. A., Distribution and dosimetry of indium-111-labeled platelets, Proc. 65th Annual Meeting Radiol. Soc. North. Am., Atlanta, November 25-30, 1979, 221.

224. Ritchie, J. L., Stratton, J. R., Thiele, B., Harker, L. A., and Hamilton, G. W., Indium-111 platelet imaging for the in-vivo detection of thrombi in abdominal aneurysms and pseudo-aneurysms and for evaluation of platelet active drugs, *J. Nucl. Med.,* 21 (Abstr.), P35, 1980.

225. Davies, R. A., Thakur, M. L., Bergen, H. J., Wackers, F., Gottschalk, A., and Zaret, B., ¹¹¹In-labeled autologous leukocytes for imaging inflammatory response to acute myocardial infarction, *J. Nucl. Med.,* 21, 89, 1980.

226. Bergmann, S. R., Lerch, R. A., Saffitz, J. E., Lee, S., and Sobel, B. E., Noninvasive monitoring of cardiac transplants, *Am. J. Cardiol.,* 45, 410, 1980.

227. Fuster, V., Dewanjee, M. K., Kaye, M. P., Josa, M., Metke, M. P., and Chesebro, J. H., Noninvasive radioisotopic technique for detection of platelet deposition in coronary artery bypass grafts in dogs and its reduction with platelet inhibitors, *Circulation,* 60, 1508, 1979.

228. Dewanjee, M. K., Fuster, V., Kaye, M. P., and Josa, M., Noninvasive radioisotopic technique for detection of platelet deposition in coronary artery bypass grafts in dogs and its reduction with platelet inhibitors, in *Radiopharmaceuticals II,* Proc. 2nd Int. Symp. Radiopharmaceuticals, Society of Nuclear Medicine, New York, 1979, 361.

229. Dewanjee, M. K. and Rao, S. A., Red cell membrane permeability and metal oxine-hemoglobin transchelation: implications in cell labeling, in *2nd Int. Symp. Radiopharmaceuticals,* St. Louis, June 16-20, 1980, 273.

230. Dewanjee, M. K., Kaye, M. P., Fuster, V., and Rao, S. A., Noninvasive radioisotopic technique for detection of platelet deposition in mitral valve prosthesis and renal embolism in dogs, in *Proc. Annual Meeting Am. Soc. Artif. Int. Org.,* New Orleans, April 17-19, 1980.

231. Peterson, K. A., Dewanjee, M. K., Kaye, M. P., and Lim, M. F., Fate of ¹¹¹Indium-labeled platelets during cardiopulmonary bypass with PGI₂, in press.

232. Didisheim, P., Dewanjee, M. K., Kaye, M. P., and Fusten, V., Comparison of in vitro and in vivo blood surface interactions, in press.

233. Fuster, V., Dewanjee, M. K., Kaye, M. P., Fass, D. N., and Bowie, E. J. W., Evaluation of platelet deposition following selective endothelial injury of the carotid artery in normal and Von Willebrand pigs, in press.

234. Moncada, S., Herman, A. G., Higgs, E. A., and Vane, J. R., Differential formation of prostacyclin (PGX or PGI₂) by layers of arterial wall. An explanation for the anti-thrombotic properties of vascular endothelium, *Thromb. Res.,* 11, 323, 1977.

235. Needleman, P., Moncada, S., Bunting, S., Vane, J. R., Hamberg, M., and Samuelsson, B., Identification of an enzyme in platelet microsomes which generate thromboxane A₂ from prostaglandin endoperoxides, *Nature (London),* 261, 558, 1976.

236. Marcus, A. J., Weksler, B. B., and Jaffe, E. A., Synthesis of prostacyclin by cultured human endothelial cells, in *Prostacyclin,* Vane, J. R. and Bergstrom, S., Eds., Raven Press, New York, 1979, 65.

237. Ross, R. and Glomset, J., Atherosclerosis and the arterial smooth muscle cell, *Science,* 180, 1332, 1973.

238. Ross, R., Glomset, J., Kariya, B., and Harker, L., A platelet dependent serum factor that stimulates the proliferation of arterial smooth muscle cells in vitro, *Proc. Natl. Acad. Sci. U.S.A.,* 71, 1207, 1974.

239. Chamley-Campbell, J., Campbell, G. R., and Ross, R., The smooth muscle cell in culture, *Physiol. Rev.,* 59, 1, 1979.

Chapter 3

THE EVALUATION OF HEPATOCYTE FUNCTION WITH RADIOTRACERS

Alan R. Fritzberg

TABLE OF CONTENTS

I. GENERAL ASPECTS OF LIVER PHYSIOLOGY

The liver plays a pivotal role in the handling of a wide variety of naturally occurring and xenobiotic or unnatural compounds. Some compounds such as proteins, triglycerides, phospholipids, glucose, and cholesterol are partly or wholly synthesized in the liver and supplied to the blood. Other relatively nonpolar compounds are converted to metabolites that are conjugated to sugars or amino acids. The resulting derivatives of increased polarity are then more efficiently excreted by the kidneys or the liver. In addition to these synthetic and metabolic activities, the liver also acts as an excretory organ. In this capacity, naturally occurring compounds such as bile acids and bilirubin are removed from the blood and excreted into bile. Similar handling of xenobiotics such as easily measured phthalein and fluorescein dyes has provided a classical means for nonradioactive evaluation of hepatocyte function.

In general, the use of radionuclides to evaluate hepatocyte function has been limited to the labeling of compounds which are removed from the blood by the liver and excreted into bile with or without metabolic transformations. Thus in describing these processes, the vascular and cellular translocation of materials will have to be considered.

II. SPECIFIC ASPECTS OF HEPATOCYTE FUNCTION EVALUATION

A. Hepatocyte Morphology

Materials in the blood are carried into smaller and smaller blood vessels of the liver until the smallest thin-walled vessel, called the sinusoid is encountered (Figure 1).[1] The endothelial cells lining the sinusoid are not tightly adhered to each other so that pores or fenestra result. These pores allow passage of plasma contents, excluding red blood cells, into a perivascular space known as the Space of Disse. It is in this space that the materials to be handled by the hepatocyte come into contact with the surfaces of the hepatocyte. After crossing the hepatocyte membrane they may interact with cytoplasmic proteins or be metabolized by hepatocyte organelles. Those compounds that are excreted into bile must then cross cell membranes that are common with other hepatocytes. A portion of these interhepatocyte membranes is specialized as canilicular or bile secreting, and is referred to as the bile caniliculus. The canilicular space between hepatocytes leads to preductules, ductules, and ducts as the biliary drainage network joins into larger bile-collecting vessels. This ductal network finally ends up in the hepatic ducts which lead to the common bile duct, gall bladder, and other portions of the biliary tract.

B. Biochemical Aspects of Hepatocyte Transport

Hepatocyte uptake across the sinusoidal membrane appears to be selective for a variety of classes of compounds including anions such as bilirubin, the dyes sulfobromophthalein (BSP), indocyanine green (ICG), and rose bengal (RB), bile acids, porphyrins, and free fatty acids, positively charged quarternary amines and neutral steroid sugar conjugates.[2] Specificity of hepatocyte uptake of these classes of compounds is due to membrane proteins, and is indicated by the membrane surface in the Space of Disse shown in Figure 1. In some cases such as bilirubin and BSP, rapid bidirectional flux has been shown to occur between the plasma and hepatocyte[3,4] and suggests that possible carriers on the membrane surface are not unidirectional. Uptake by the sinusoidal membrane has been demonstrated to be saturable for BSP[5] and taurocholate[6,7] in support of a carrier-mediated process. However, uptake of bilirubin or other anions in vitro by rat liver slices has not been shown to be energy dependent,[2] also a property necessary for the demonstration of active transport. Another aspect which is consistent

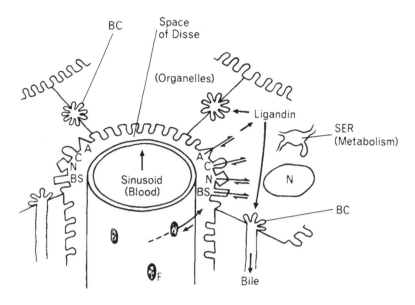

FIGURE 1.   Drawing of hepatocyte showing relationship of cellular parts to blood supply and biliary excretion. Materials in the blood diffuse through pores or fenestra (F) in endothelial lining of sinusoid into the Space of Disse. There they have access to sinusoidal membrane accepters for anions (A), cations (C), neutral compounds (N) or bile salts (BS). After transfer across the membrane they may be bound to intracellular proteins such as ligandin or metabolized at other sites such as the smooth endoplasmic reticulum (SER). Finally, transfer across the canilicular membrane results in passage into the bile via the bile caniliculus (BC).

with active transport across the hepatocyte membrane is the binding of compounds taken up in the liver by plasma proteins. Comparative studies of a series of Tc-99m complexes of derivatives of naphthalene and similar ring systems indicated that the fraction bound to plasma proteins was not related to the rate or extent of liver uptake.[8] However, the fraction bound was inversely related to the fraction of the injected dose excreted in the urine. Dye anions such as BSP[9] and rose bengal[10] which are thought to be actively transported are found predominantly in a protein-bound form in the plasma. Where mediated transport is not involved, protein binding is known to reduce membrane transfer.[9]

Involvement of hepatic cytosolic binding proteins has been shown for bilirubin and anion dyes.[11] These proteins, ligandin (Y protein,[12] which has been shown to be identical to azocarcinogen-binding protein II[13] and glutathione transferase B[14]) and Z protein, form approximately 5% of the soluble protein of rat liver.[15] Ligandin, which is present in larger amounts, binds bilirubin and other dye anions such as BSP, ICG, and rose bengal. It has been suggested,[16] that ligandin influences hepatocyte uptake by controlling the efflux of these compounds into plasma. Since ligandin also has glutathione transferase activity, it is also able to form the glutathione conjugate of BSP, the form in which BSP is excreted into bile. For other types of metabolism before excretion, such as the glucuronidation of bilirubin, a further transfer to the smooth endoplasmic reticulum (SER) must take place.

The final excretory step involves crossing the canilicular membrane. Although active transport at this stage has been supported by the measurement of transport maxima (Tm) and high bile to liver ratios attained by many compounds,[17] experimental attempts to demonstrate energy dependency have been unsuccessful. Thus, the metabolic inhibitor ouabain did not affect the transport of organic acids or bases and is itself transported into bile.[18] 2,4-Dinitrophenol, another metabolic inhibitor, stimulates bile production[19] and does not depress the biliary excretion of bile acids.[20] In any case it has been assumed that at least three active transport systems exist. These systems are selective for organic acids which are anions under physiological conditions and include bilirubin, BSP, and similar aromatic sulfonic and carboxylic acids,[15] cationic quarternary ammonium compounds and neutral compounds such as ouabain, digoxin, and digitoxin.[18] That bile acids are transported by a system separate from dye anions is supported by the mutant corriedale sheep which excrete bile acids normally but cannot excrete other organic acids into bile.[21]

Another aspect of hepatocyte physiology which is important to the understanding of the kinetics of radiolabeled compounds handled by the liver involves the capacity of the liver to excrete materials into the bile. The results of several studies indicate that the maximum hepatic uptake velocity (V_{max}) and excretory transport maximum (Tm) vary considerably. For BSP a V_{max} value of 13 μmole/min·kg was determined[5] while the corresponding BSP Tm was 0.22 μmole/min·kg,[22] a difference factor of 58. This difference has been determined to be twenty-fold for ICG,[23] and fourfold for taurocholate.[24] The capacity of these transport systems also varies from one class to another. The maximal uptake velocity of taurocholate was found to be about five times that of bilirubin and ten times that of ICG.[25] Since the difference between maximal uptake and excretion velocities is greater for ICG than for taurocholate, the Tm capacities for these compounds differ by about fifty-fold.

C. Chemical Property Requirements for Biliary Excretion

Molecular requirements have been elucidated for the efficient handling of compounds by the liver. There are three main factors affecting liver transport: (1) molecular size, (2) polarity, and (3) molecular structure.[26]

A minimum molecular weight appears to be required for biliary excretion. In the rat a threshold level of 325 ± 50 was suggested as a result of studies on a series of sulfonamides.[27,28] Threshold values vary somewhat with species and values of 400 ± 50 in guinea pigs and 475 ± 50 in rabbits were also reported. An upper limit is likely since molecules such as insulin with a molecular weight of 5000 and proteins are only found in trace amounts in bile.[29]

Compounds efficiently excreted in bile generally contain a polar portion and a nonpolar portion. The polar group can be a carboxylic or sulfonic acid group, quarternary amine group, or sugar derivatives. All of these confer water solubility, provide for binding sites with transport proteins, and restrict intestinal reabsorption. Bile acids which participate in enterohepatic cycling are actively transported from the ileum of the intestines and reenter the liver mainly via the hepatic portal circulation.[30] The lipophilic portion of the compounds, which usually consists of polycyclic ring systems, may serve to strengthen protein binding in the plasma, thereby reducing renal excretion via glomerular filtration.

Significant differences resulting from fairly small structural changes have also been observed. Changing the aromatic ring of phthalylsulfathiazole to the partially saturated to the hexahydro, or fully saturated ring, resulted in an increase from 19 to 66% of the dose excreted in the bile.[31] In another comparison, the introduction of another hydroxyl in 4-hydroxybiphenylglucuronide to give 4,4'-dihydroxybiphenylglucuronide increased the biliary excretion from 60 to 90%.[32]

D. The Development of Radiolabeled Compounds for Hepatocyte Evaluation

1. Radioiodinated Compounds

The use of polyhalogenated dyes for the evaluation of hepatobiliary function is based on the report of the biliary excretion of phenolphthalein and its halogenated derivatives in 1909.[33] Subsequently, halogenated phthaleins such as tetraiodophthalein were developed for liver function evaluation, and visualization of the gall bladder and biliary tract by X-rays.[34] Rose bengal (1), a highly halogenated dye, was used for liver function tests as early as 1923.[35]

In 1955, replacement labeling with ^{131}I resulted in ^{131}I-rose bengal and the ability to follow its course by external monitoring techniques.[36] Labeled rose bengal[37] is highly protein bound in plasma, taken up by the hepatocytes at a rate related to their functional status, and excreted into the bile. It is not reabsorbed from the intestines and appears to have negligible renal excretion. A recent study[38] in which the urinary excretion of rose bengal over the first three hours was measured resulted in values of 3 to 7% of the dose in the urine for one manufacturer and less than 1% for another. These values were not related to bilirubin levels indicating that activity in the urine was possibly due to free ^{131}I-iodide or labeled contaminates.

The nonradioactive standard for liver function evaluation, BSP, was labeled with ^{131}I and compared with ^{131}I-rose bengal. Labeling by the iodine monochloride method gave the diiodo compound (2)[39] while preparation with the iodide-iodate method gave a monoiodo product (3).[40] The introduction of iodine into BSP altered

the chemical and biological properties as indicated by comparisons with ^{35}S-BSP. Thus, ^{35}S-BSP has a plasma clearance t½ of 3.5 min, ^{131}I- BSP 6.0 min, and the diiodo compound 10 min.[41] The iodinated BSP compounds were not conjugated with glutathione prior to biliary excretion as is the case with nonlabeled BSP. In comparison, ^{131}I-rose bengal was cleared more slowly from the plasma than either ^{35}S-BSP or monoiodo ^{131}I-BSP.

Although the gamma radiation of ^{131}I allows external detection and imaging, the 364 keV energy of the main gamma radiation is too high for quality images on current gamma cameras, and the associated beta radiation limits the dose that can be administered. These disadvantages are not present in ^{123}I. It decays by electron capture and has a favorable photon energy of 159 keV. In order to exploit these properties rose bengal,[42] BSP,[43] and ICG[44] have been labeled with ^{123}I. In these cases no biological differences are to be expected from the isotopic iodine exchanges, and at least one of these compounds might have been expected to be in routine use if ^{123}I was not so expensive and relatively inconvenient with its 13 hr half-life; and more importantly, the success of Tc-99m-labeled hepatobiliary agents had not come to pass.

2. Technetium-99m-Labeled Compounds

With its ideal radiation properties for imaging and in vivo use in humans and convenient supply, researchers have attempted to replace ^{131}I-labeled radiopharmaceuticals with Tc-99m whenever possible. Since 1972 a variety of Tc-99m complexes have been proposed as alternatives to ^{131}I-RB. Chelating agents have included penicillamine,[45] tetracycline,[46] pyridoxylideneglutamate,[47,48] dihydrothioctic acid,[49] mercaptoisobutyric acid,[50] 8-hydroxyquinoline-7-carboxylic acid,[51] N-(2,6-dimethylacetanilide)iminodiacetic acid,[52] and other iminodiacetic acid derivatives.[8,53,54] Modifications of the biological behavior of these complexes as they have been improved have been both in the specificity for biliary excretion and also in the rates of liver uptake and biliary excretion. Since the renal excretion of these agents generally increases when the hepatocyte function decreases, the increased specificity increases the likelihood of demonstrating biliary activity, and thus ruling out bile duct obstruction. The increased rates of uptake and excretion result in a greater concentration of activity in the bile ducts and better visualization of the intrahepatic duct system. The rapid uptake and rate of clearance of ^{99m}Tc-N-(2,6-diethylacetanilide)iminodiacetate (Tc-diethyl-IDA) can be appreciated by the images of a patient with normal function (Figure 2).[55] As early as 5 min post injection, negligible activity is seen in the blood as indicated by the cardiac blood pool area above the liver. Intrahepatic bile ducts are visible at 5 min and clearly seen at 10 min. Also, by 10 min the gall bladder is clearly visualized. The radioactivity has been nearly completely excreted into the intestines by 30 min.

The first Tc-99m-labeled radiopharmaceutical to demonstrate the clinical utility of these complexes was ^{99m}Tc-pyridoxylideneglutamate (Tc-PG).[56-59] Although it was taken up by the liver and excreted rapidly enough in normal patients, its normal renal excretion of 30% increased in those patients with decreased hepatocyte function. As a result bile activity was not seen with Tc-PG when it was seen with ^{131}I-RB (Figure 3).[60] It was concluded that the superior radiation properties of Tc-PG gave it an advantage in the detection and definition of small bowel activity at 1 hr while the high hepatic uptake and low renal excretion of ^{131}I-RB made it preferable in the detection of colonic activity at 24 hr.

Soon after the clinical evaluation of Tc-PG was underway, an iminodiacetate analog of lidocaine was reported,[52] which exhibited rapid hepatocyte uptake and excretion. Subsequently, a variety of substitution patterns on the phenyl ring were evaluated (Figure 4).[52,53] These different complexes were evaluated in the baboon in which blood, urine, and bile samples were taken. The results shown in Table 1 indicate that the hepatocyte handling of these compounds is sensitive to these structural changes. While all are cleared from the blood relatively rapidly, the amount remaining at 15 min varies over a several fold range with Tc-diethyl-IDA and Tc-diisopropyl-IDA (see Figure 4 for names and abbreviations) being the lowest. Likewise the fraction of the dose excreted in the urine was variable and ranged from 2% for Tc-p-butyl-IDA to 34% for Tc-p-ethoxy-IDA. Interestingly, a relatively small structural change from the p-ethoxy to the p-butoxy group reduced the renal excretion to 7%. Cumulative biliary activity varied from 26% for Tc-p-ethoxy-IDA to 89% for Tc-diethyl-IDA. While the final values for Tc-diethyl-IDA and Tc-diisopropyl-IDA were similar, more of Tc-diisopropyl-IDA was found in bile at 30 and 60 min.

3. Clinical Studies and Hepatocyte Handling

Animal studies are valuable in determining which compounds are the best candidates

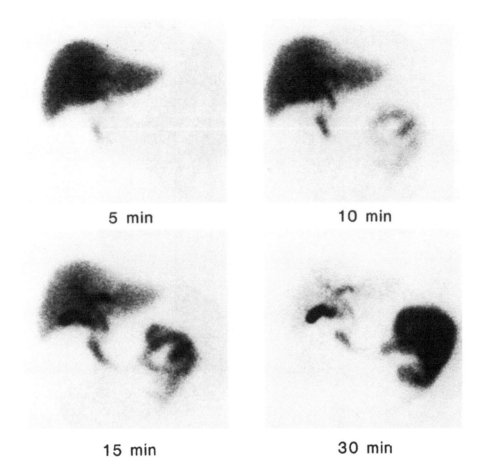

5 min   10 min

15 min   30 min

FIGURE 2.   Images of Tc-diethyl-IDA in a patient with normal function. By five minutes collection into intrahepatic ducts and some extrahepatic duct activity is seen. The ten minute image shows the intrahepatic duct clearly, the gall bladder and intestinal activity. By thirty minutes the liver is almost clear of activity with only biliary tract concentration seen.

for optimal performance in the clinical setting. After reviewing the data, several of them have been developed for clinical evaluation. Several questions need to be answered, however. How relevant are the animal data as indicators of behavior in man? How will they be affected by hepatobiliary pathophysiology? What is the relative importance of rates of liver uptake and biliary excretion, and specificity for hepatobiliary excretion? For example, Tc-*p*-butyl-IDA has the lowest renal excretion, but also a longer hepatocyte transit time resulting in later appearance in the bile.

We have been carrying out clinical evaluations with the object of answering these questions. Paired studies have been performed in order to minimize patient variability. The studies were done with a minimum time separation in order to also minimize changes in hepatocyte function between studies and usually resulted in a time interval of one to three days.

The first iminodiacetate derivative evaluated was Tc-diethyl-IDA. As expected from the relatively high specificity and faster kinetics in animals, Tc-diethyl-IDA was found to be superior to Tc-PG in aspects of image quality and diagnostic information including liver to background, liver to renal bladder, liver to kidneys, hepatic ducts to liver, small bowel to background, time of appearance of activity in small bowel and colon to background at 24 hr post injection.[60] The differences were observed in patients with

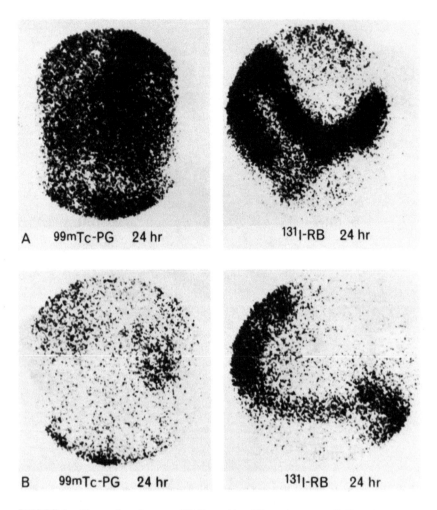

FIGURE 3.   Comparison images of ^{99m}Tc-pyridoxylidene glutamate (^{99m}Tc-PG) and ^{131}I-rose bengal (^{131}I-RB) at 24 hr in patients with decreased hepatocyte function. The colon activity of rose bengal indicates its biological handling superiority in clearly demonstrating bile activity under such conditions. (From Klingensmith, W. C., III, Fritzberg, A. R., Koep, L. J., and Ronai, P. M., A clinical comparison of ^{99m}Tc-diethyl-iminodiacetic acid, ^{99m}Tc Pyridoxylideneglutamate and ^{131}I-rose bengal in liver transplant studies, *Radiology*, 130, 435, 1979. With permission.)

both normal and decreased liver function. Comparison with ^{131}I-rose bengal was as expected with respect to quality of image and kinetics. The renal excretion of Tc-diethyl-IDA, which averaged 7.5% of the dose in normal patients in 3 hr compared to 4.9% for ^{131}I-RB, ranged to 53% in patients with elevated bilirubin levels (Figure 5). In the same patients with total serum bilirubin levels from 0.7 to 22.4 mg/dℓ, the renal excretion of ^{131}I-RB was independent of the bilirubin levels. The effect of these differences was visualization of the kidneys with Tc-diethyl-IDA in patients with reduced hepatocyte function, and increased likelihood of nonvisualization of the colon on delayed images.

More interesting perhaps were comparisons between similar superior Tc-99m-iminodiacetates. The initial comparison was done with the *p*-isopropyl and diethyl derivatives.[38] The baboon data show only slightly lower fractions of the dose in the bile at 30 and 60 min for *p*-isopropyl-IDA indicating little difference in rates of biliary excretion. However, in patients, a consistently slower blood clearance and later biliary ap-

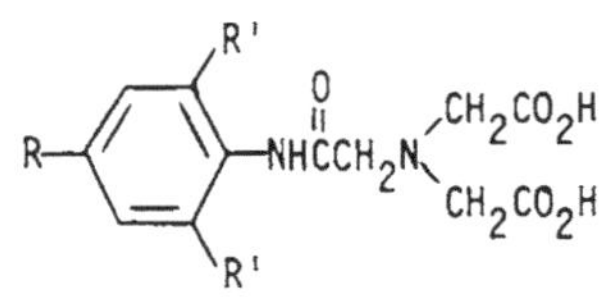

R = H, R' = CH$_3$:   N-(2,6-Dimethylacetanilide)iminodiacetic Acid
                        (HIDA)

R = H, R' = CH$_2$CH$_3$:   N-(2,6-Diethylacetanilide)iminodiacetic Acid
                              (Diethyl-IDA, EHIDA)

R = H, R' = CH(CH$_3$)$_2$:   N-(2,6-Diisopropylacetanilide)iminodiacetic Acid
                                (DIPIDA, diisopropyl-IDA, disofenin)

R = CH(CH$_3$)$_2$, R' = H: N-(p-Isopropylacetanilide)iminodiacetic Acid
                                (PIPIDA, p-isopropyl-IDA)

R = (CH$_2$)$_3$CH$_3$, R' = H: N-(p-Butylacetanilide)iminodiacetic Acid
                                    (BIDA, p-butyl-IDA)

R = OCH$_2$CH$_3$, R' = H: N-(p-Ethoxyacetanilide)iminodiacetic Acid
                               (p-ethoxy-IDA)

R = O(CH$_2$)$_3$CH$_3$, R = H: N-(p-Butoxyacetanilide)iminodiacetic Acid
                                    (p-butoxy-IDA)

R = I, R' = H: N-(p-Iodoacetanilide)iminodiacetic Acid
                   (p-iodo-IDA)

FIGURE 4.   Structures, chemical names, and abbreviations of iminodiacetate chelating agents evaluated in baboons and patients.

Table 1

COMPARISON OF BIODISTRIBUTION PARAMETERS OF ^{99m}Tc-IMINODIACTATE COMPLEXES IN THE BABOON[52,53] [a]

^{99m}Tc-Complex	Blood activity % at 15 min	Urine activity 3 hr total	Bile activity		
			30 min	60 min	90 min
HIDA	7	20	12	51	64
diethyl-IDA	3	5	38	65	89
diisopropyl-IDA	—[b]	4.5	44	76	87
p-Isopropyl-IDA	—	—	35	63	74
p-butyl-IDA	—	2	28	58	70
p-ethoxy-IDA	10	34	5	20	26
p-butoxy-IDA	—	7	41	66	82
p-iodo-IDA	10	15	10	36	48
rose bengal	6	—	13	40	58

[a]   Data are expressed as percent injected dose.
[b]   Data not included

pearance was observed for the Tc-p-isopropyl-IDA complex (Figure 6). While this difference was not large enough to impair interpretations in most cases, the increased hepatocyte transit time decreased the concentration of radiotracer in the hepatic ducts, thereby reducing the quality of images of the intrahepatic ducts.

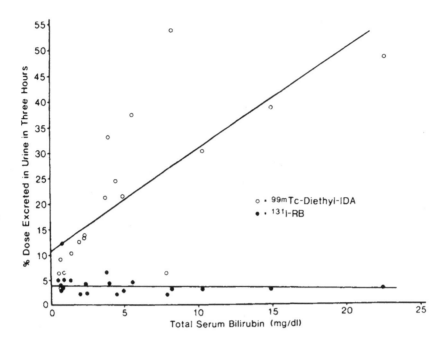

FIGURE 5.   Renal excretion in percent dose excreted in 3 hr of ⁹⁹ᵐTc-diethyl-IDA and
¹³¹I-rose bengal (¹³¹I-RB) plotted versus total serum bilirubin. The renal excretion of the
technetium complex increases with bilirubin levels while rose bengal is invariant over this
collection period. (From Klingensmith, W. C., III, Fritzberg, A. R., and Koep, L. J.,
Comparison of Tc-99m diethyl-iminodiacetic acid and I-131 rose bengal for hepatobiliary
studies in liver transplant patients, *J. Nucl. Med.*, 20, 314, 1979. With permission.)

The low renal excretion of Tc-*p*-butyl-IDA has been the basis for its recommended
use in patients with high bilirubin levels and expected reduced hepatocyte function. In
the study presently underway, distinct differences between Tc-*p*-butyl-IDA and Tc-
diethyl-IDA have been observed.[55] While the hepatic uptake of Tc-*p*-butyl-IDA is
slower than Tc-diethyl-IDA, it usually reaches a higher liver-to-background ratio. A
significantly longer time is required for the Tc-*p*-butyl-IDA to reach the bile, indicating
an increased hepatocyte transit time relative to Tc-diethyl-IDA (Figure 7). In patients
with impaired hepatocyte function, the decreased rate of hepatic uptake was also accompanied by increased cellular transit times. Thus, in comparisons in patients with
bilirubin levels of 0.2 to 17.0 mg/dℓ, activity reached the bile much sooner in inter-
pretable levels with Tc-diethyl-IDA. Thus, despite the higher liver to background levels
reached with Tc-*p*-butyl-IDA, the question of biliary patency vs. hepatocyte function
was answered much more rapidly with Tc-diethyl-IDA. One patient with a serum bilirubin level of 50 mg/dℓ showed bile activity at 24 hr with Tc-*p*-butyl-IDA only. Since
the greater hepatic uptake and retention of Tc-*p*-butyl-IDA may indicate it to be superior in patients with worse hepatocyte function than those already studied, the comparative evaluation is still underway.

The final comparison currently underway involves Tc-diethyl-IDA and Tc-diisopropyl-IDA.[55] As indicated by the baboon data, these agents are similar in specificity and
differ only by an apparent greater biliary excretion of Tc-diisopropyl-IDA at the 30
and 60-min periods. The initial experience in 14 paired studies indicate similar behavior. In the patients who demonstrated biliary excretion, the agents were judged equal
in intestinal arrival time in seven cases, Tc-diisopropyl-IDA faster in three cases, and
Tc-diethyl-IDA faster in two cases (Figure 8). Renal excretion has been consistently
lower with Tc-diisopropyl-IDA.

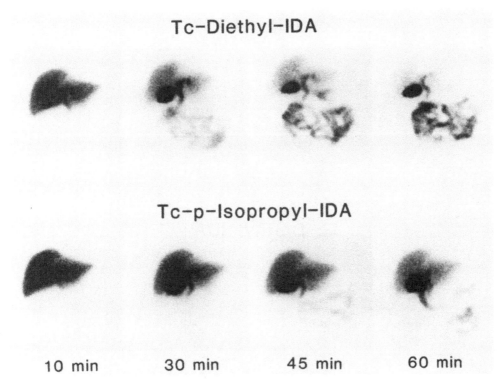

Tc–Diethyl–IDA

Tc–p–Isopropyl–IDA

| 10 min | 30 min | 45 min | 60 min |

FIGURE 6. Anterior projection images of Tc-diethyl-IDA and Tc-*p*-isopropyl-IDA at 10, 30, 45, and 60 min after projection in a patient with good hepatic function (total serum bilirubin = 0.8 mg/d*l*). Tc-diethyl-IDA demonstrates higher liver to background, hepatic ducts to liver, and small bowel to background activity ratios. The arrival time of Tc-diethyl-IDA in the small bowel is also earlier indicating a shorter hepatocyte transit time. (From Klingensmith, W. C., III, Fritzberg, A. R., Spitzer, V. M., and Koep, L. J., Clinical comparison of ⁹⁹ᵐTc diethyl-IDA and ⁹⁹ᵐTc PIPIDA for evaluation of the hepatobiliary system, *Radiology*, 134, 195, 1980. With permission.)

E. Mechanism of Hepatobiliary Transport of Tc-99m-Iminodiacetate Radiopharmaceuticals

Clinical studies of several of these Tc-99m iminodiacetates have established the value of these agents for evaluation of hepatocyte function. Physiological interpretations of the behavior of these radiopharmaceuticals in patients, however, require an understanding of their mode of transport by the hepatobiliary system. Since they are transition metal complexes, it is not obvious whether they should be expected to be handled by one of the organic compound systems mentioned previously, or by a system unique to metal complexes.

Structural studies of Tc-HIDA[62,63] showed it to have a 2:1 ligand-to-metal ratio and a net −1 charge (Figure 9). The net negative charge gives it similarity to the organic dye anions such as BSP and bile acids. However, transition metals such as technetium have a three-dimensional arrangement in the structures of their complexes that suggests interactions with transport proteins may be quite different from the interactions with sodium salts of sulfonic (BSP, ICG) and carboxylic (rose bengal) acids. Interaction of the imino nitrogens with technetium by unshared electron pair donation may suggest structural resemblance to the quaternary amine cation, procaineamide ethobromide (4).

$$\left[H_2N - \bigcirc - \overset{\overset{O}{\parallel}}{C}NHCH_2CH_2\overset{+}{N} \overset{CH_2CH_3}{\underset{CH_2CH_3}{\overset{\diagup}{\underset{\diagdown}{-}}CH_2CH_3}} \right] Br^- $$

procaine amide ethobromide

Tc-p-Butyl-IDA

Tc-Diethyl-IDA

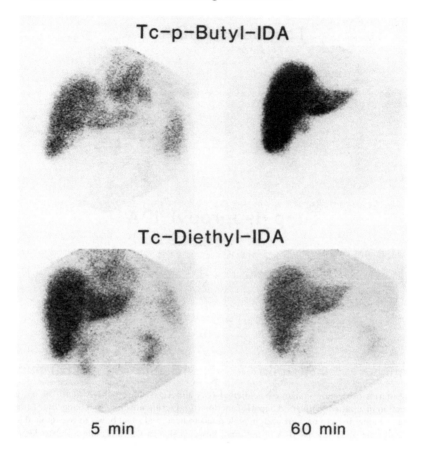

5 min   60 min

FIGURE 7.   Comparison anterior projection images of Tc-diethyl-IDA and Tc-*p*-bu-
tyl-IDA at 5 and 60 min in a patient with reduced hepatocyte function (total bilirubin
10 mg/dℓ). At 5 min the liver to blood (cardiac area and background) is slightly higher
for diethyl-IDA. At 60 min the ratio is markedly higher for Tc-*p*-butyl-IDA. Biliary
activity in a T-tube is evident in the Tc-diethyl-IDA 60 min image but not in the other.

1. In Vivo Studies

Studies have been performed in both dogs[62] and rats[64] to elucidate the transport
pathway involved. These were carried out by administering saturating levels of members of each pathway, and observing the effect on the hepatobiliary excretion of the
Tc-99m-labeled agents. In dogs, BSP reduced the 1-hour cumulative hepatobiliary excretion of ^{131}I-rose bengal from 60 to 20%, and Tc-HIDA from 55.2 to 1.5%. The
organic cation, oxyphenonium, on the contrary, had no effect on Tc-HIDA. In rats,
experiments were carried out under both continuous infusion and bolus injection conditions. Saturating level of BSP reduced biliary excretion of Tc-diethyl-IDA from 80%
of the 10-min infused dose in bile collected in 10 min to 45% and ^{131}I-RB from 52%
to 12%. Bolus injection of the radiopharmaceuticals after infusion of BSP resulted in
25% of Tc-diethyl-IDA in the bile compared to control value of 60% at 10 min. At
18 min, the corresponding doses in bile were 53% of the treated vs. 90% for control.
At 18 min ^{131}I-RB was reduced from 16% to 2%. These data also indicate the rapidity
of biliary excretion of Tc-diethyl-IDA compared to ^{131}I-RB, and also that the complex
competes more effectively under competitive conditions. The bile salt, sodium taurocholate, only transiently increased bile output of both agents, indicating transport independent of bile acids. Saturating doses of procaineamide ethobromide, as a member
of the organic cation pathway, also did not depress biliary output under either experimental condition.

Tc-Diisopropyl-IDA

Tc-Diethyl-IDA

5 min 10 min 15 min

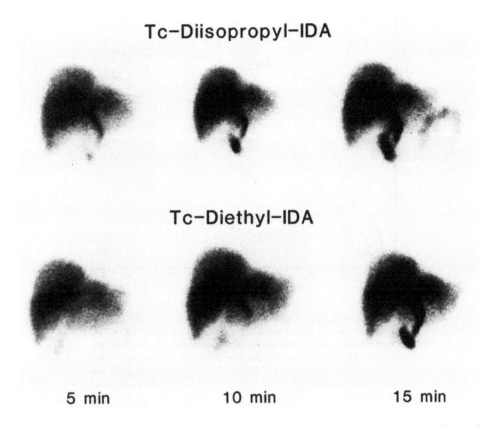

FIGURE 8. Comparison anterior projection images of Tc-diisopropyl-IDA and Tc-diethyl-IDA at 5, 10, and 15 min in a patient with normal hepatocyte function. In this example Tc-diisopropyl-IDA shows slightly greater biliary concentration in early images (see text for numerical results of patients studied).

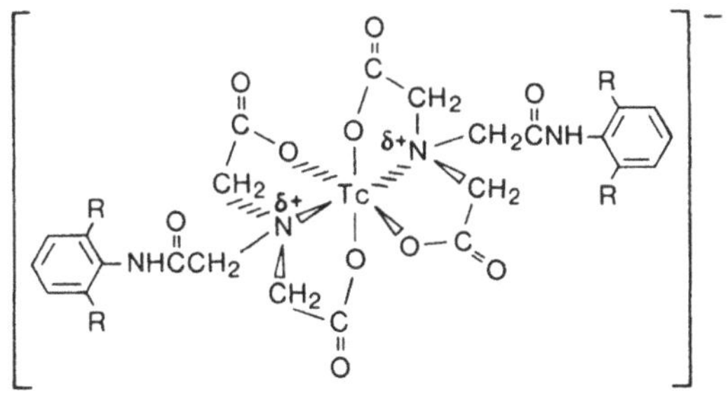

FIGURE 9. Structure of Tc-N-(acetanilide)imino-diacetates indicating 2:1 ligand to metal ratio and net -1 overall charge on complex.

Since the decrease in biliary excretion observed after treatment with BSP could also have been due to general hepatotoxicity, the effects of another anion dye, indocyanine green (ICG), was also studied. A decrease in biliary excretion is observed at both 16 and 24 mg/kg doses (Figure 10). These dose levels were reported to be saturating without reducing bile flow in the rat.[66] The results with ICG further support the use of the dye anion transport pathway by the Tc-99m iminodiacetates.

The effects of saturating doses of BSP and ICG on blood disappearance rates have also been evaluated. In dogs 2.7% of a dose of Tc-HIDA remained in the blood at 30 min before BSP treatment. After BSP infusion 8.6% was found.[62] In rats the blood

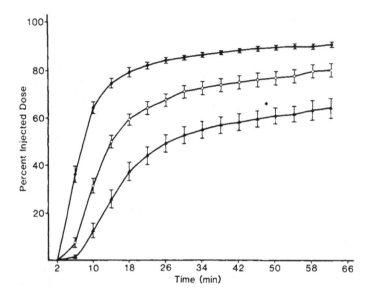

FIGURE 10.   Comparative rates of biliary appearance of Tc-diethyl-IDA under conditions post injection of 16 mg/kg (o—o) and 24 mg/kg (▲—▲) of indocyanine green and control (•—•) in rats. Bile was collected from the common bile duct and data are expressed as percent ± S.E.M. dose in bile versus time. The appearance of Tc-diethyl-IDA is later, appears at slower rates, and total accumulation in bile is reduced when compared to control values.

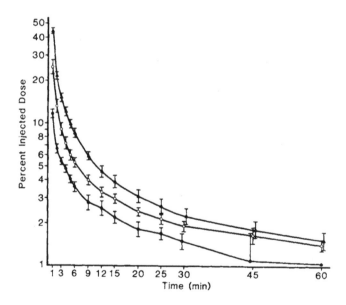

FIGURE 11.   Comparative blood disappearance rates of Tc-diethyl-IDA after injection of 16 mg/kg (o—o) and 24 mg/kg (▲—▲) doses of indocyanine green and controls (•—•) in rats. Data are presented as percent injected dose ± S.E.M. versus time post injection of radiopharmaceutical. Disappearance rates are reduced by indocyanine green indicating competitive hepatocyte uptake.

was sampled from 1 to 60 min post injection, giving the disappearance curves of Tc-diethyl-IDA shown in Figure 11. An increasing reduction on blood disappearance rates is seen as the dosage levels of ICG is increased. Thus, early values at 1 to 5 min are about twofold higher for the 16 mg/kg dose, and about fourfold higher for the 24 mg/kg dose. Later values show smaller differences since estimates based on the ICG V_{max} values indicate that the 16 mg/kg dose would be expected to be saturating for only 5 min and the 24 mg/kg dose for 8 min.[67]

Biodistribution studies of Tc-HIDA using Sn-113 and C-14 radiolabels for the stannous iminodiacetate complex and carrier compounds, respectively, have shown that only the Tc-99m complex exhibits biliary excretion; the others are excre excreted predominantly by the kidneys.[68,69] Interestingly, HIDA complexed to ^{51}Cr-Cr(III) showed similar behavior to the Tc-99m complex demonstrating that the net charge on the complex and spatial arrangement of bonds can be satisfied by other metals.[70] Similarly, evaluation of ruthenium complexes of HIDA and p-isopropyl-IDA for utilization of the longer half-life of Ru-97 (t½ 2.9 days, Eγ 216 keV of 86% abundance) resulted in substantial but somewhat less biliary excretion than the Tc-99m complexes.[70] Thus, [103]Ru-p-isopropyl-IDA has 28% in the bile at 15 min, 33% at 30 min, 44% at 60 min, and 47% at 120 min, while ^{99m}Tc-p-isopropyl-I had 44%, 52%, and 60% at the same time periods in rats. Other organs were similar although the percent excreted in the urine was not reported.

2. Implications for Hepatocellular Pathology Evaluation

Since only the Tc-99m complexed material of these iminodiacetates is excreted into the bile, nanogram level amounts are administered in these studies. Thus our in vivo observations and clinical use are based on radiotracer levels. Unless one wants to work with milligram amounts of Tc-99 (t½ = 200,000 years), further transport characterization of V_{max} and T_m values is not practical by classical dose level means. This tracer level limitation also is important in the diagnosis of mild hepatocellular disease. Traditional hepatic function evaluation procedures with BSP or ICG have involved various dose levels and sensitivity for detecting liver disease has been found to be increased when the dose of ICG was increased.[72] Thus, while 100% of patients with cirrhosis, 75% with viral hepatitis, and 42% with fatty liver had disappearance rate constants more than two standard deviations below the normal mean at 0.64 μmol·ICG/per kilogram body weight, all patients in each class were more than two standard deviations below at 6 μmol·ICG/per kilogram dose levels. Implications of these results for Tc-99m iminodiacetates may mean insensitivity for mild liver disease unless a means for stressing the liver such as coinjection of a competitive dye such as ICG is employed.

3. In Vitro Studies

Only a few studies of the handling of Tc-99m iminodiacetates by hepatocytes using in vitro models have been reported. An isolated hepatocyte system has been used to compare the relative uptake of different labeled compounds.[63] Results indicated no uptake for ^3H_2O, or ^{99m}Tc-DTPA (diethylenetriamine pentaacetic acid) as expected. Uptake of hepatobiliary agents was 22% for Tc-HIDA, 25% for ^{131}I-rose bengal, 35% for Tc-p-isopropyl-IDA, and 45% for Tc-diethyl-IDA at 30 min. Labeled taurocholic acid (C-14), on the contrary, approached 90% by 15 min. These results are consistent with comparative blood disappearance rates of these compounds. Hepatic extraction efficiencies have been reported as 55%[73] for Tc-diethyl-IDA, and 90% for taurocholic acid;[72] also in line with the differences observed. The isolated hepatocyte uptake was also reported for a few compounds in the presence of bilirubin. Uptake of rose bengal and Tc-HIDA was markedly reduced by 4 mg/dℓ or more while that of Tc-p-isopropyl-IDA required 8 mg/dℓ or more for significant reduction of uptake.[63]

An important consideration for Tc-iminodiacetate hepatobiliary transport characterization is intracellular interactions. As mentioned earlier, bilirubin, BSP, ICG, rose bengal, and other compounds transported by the dye anion system bind to the intracellular protein, ligandin, or glutathione transferase-B. Correspondingly, binding of these complexes would also be expected. One reported attempt to show binding of Tc-HIDA to intercellular proteins was not successful.[75] Comparative studies have indicated a longer hepatocyte transit time for the para substituted complexes (p-isopropyl-IDA, p-butyl-IDA) than for the 2,6-substituted complexes (diethyl-IDA, diisopropyl-IDA, dimethyl -IDA). It seems possible that the complexes with the longer hepatocyte transit times may be binding more strongly to ligandin. Should this be the case, the use of Tc-dimethyl-IDA may have made it more difficult to observe intracellular binding. We are currently isolating and purifying ligandin in order to answer these questions. Using the purified protein, we plan to measure the binding of the series of complexes alone and also with competitive levels of BSP in order to evaluate the relative ligandin binding of the Tc-iminodiacetates.

F. Localization of Radiolabeled Materials in the Liver via Membrane Receptors

A new area of interest in the hepatic localization of radiopharmaceuticals involves the binding of modified proteins to hepatic cell membrane receptors. These proteins normally have a terminal sequence consisting of sialic acid in the end position, and then galactose and N-acetylglucosamine, respectively. When the sialic acid portion is removed by treatment with the enzyme neuraminidase, the resulting asialo protein binds to the hepatic membrane binding protein. When ceruloplasmin which has ten such sequences was treated, the resulting asialoceruloplasmin was rapidly cleared from the circulation of rats.[76,77] The binding is irreversible, and occurs prior to endocytosis and lysosomal catabolism.[78] The utility of this approach in achieving liver images was demonstrated with radioiodine and Tc-99m-labeled asialoceruloplasmin, asialofetiun, and asialoorosmucoid.[78] Albumin has been modified by coupling with galactose to give neoglycoalbumin, which also shows the hepatic binding behavior with a Tc-99m label. The clinical utilization of this receptor binding remains to be seen at this time.

III. FUTURE DEVELOPMENTS

It would appear that the best of the current Tc-99m-labeled iminodiacetate hepatobiliary agents can be improved on very slightly by similar agents. However, the combination of rapid hepatocyte transit as seen with the 2,6-disubstituted iminodiacetate complexes with low renal excretion as seen in the case of ^{99m}Tc-p-butyl-IDA in patients with reduced levels of hepatocyte function has not been reported yet. Since the hepatic extraction efficiency of ^{99m}Tc-diethyl-IDA of 55% is considerably lower than that of taurocholate at 90%, structural modifications may increase the efficiency of hepatic uptake.

Current imaging hepatobiliary agents utilize the dye anion transport pathway, and thus will be forced to compete with bilirubin for hepatic uptake. Since patients with reduced hepatocyte function have increased bilirubin levels, this causes further reduction in the efficiency of hepatocyte handling. A way to circumvent the common transport with bilirubin would be to design new labeled materials that are handled differently. As mentioned earlier, the capacity of hepatocytes for uptake and excretion of bile acids exceeds that of dye anions such as BSP, ICG, and bilirubin. Thus, a twofold gain may be realized by utilizing the bile acid pathway, noncompetition with bilirubin and greater transport capacity. Bile acids, which retain normal hepatic handling characteristics, have been labeled with selenium-75.[79] These have been used to follow absorption after oral administration, but apparently behavior after intravenous administration has not been extensively evaluated. The radiotracer properties of Se-75 are not optimal, and imaging applications would be expected to be limited. Another potential alternative is the cation pathway utilized by quaternary amines. At this time, the hepatocyte transport of these compounds are relatively little known, and the evaluation of labeled analogs may serve to realize the potential of this pathway.

REFERENCES

1. Lane, B. P., Functional analysis of the liver, in *The Liver, Normal and Abnormal Functions*, Part A, Becker, F. F., Ed., Marcel Dekker, New York, 1974, chap. 1.
2. Arias, I. M., Transfer of bilirubin from blood to bile, *Semin. Hematol.*, 9, 55, 1972.
3. Goresky, C., The hepatic uptake and excretion of sulfobromophthalein and bilirubin, *Can. Med. Assoc. J.*, 92, 851, 1965.
4. Goresky, C. and Back, G. C., Membrane transport and hepatic circulation, *Ann. N.Y. Acad. Sci.*, 170, 18, 1970.
5. Goresky, C. A., Initial distribution and rate of uptake of sulfobromophthalein in the liver, *Am. J. Physiol.*, 207, 13, 1964.
6. Glasinovic, J. C., Dumont, M., Duval, M., and Erlinger, S., Hepatic uptake of taurocholate in the dog, *J. Clin. Invest.*, 55, 419, 1975.
7. Reichen, J. and Paumgartner, G., Kinetics of taurocholate uptake by the perfused rat liver, *Gasteroenterology*, 68, 132, 1975.
8. Fritzberg, A. R., Eshima, D., Klingensmith, W. C., and Whitney, W., Comparison of ^{99m}Tc-hepatic biliary agents based on napthalene and similar ring systems, *J. Nucl. Med.*, 19, 694, 1978.
9. Goldstein, A., Aronow, L., and Halman, S. M., *Principles of Drug Action: The Basis of Pharmacology*, 2nd ed., John Wiley & Sons, New York, 1974, 202 and 218.
10. Jansholt, A. L., Krohn, K. A., Stadalnik, R. C., Matolo, N. M., and DeNardo, G. L., Catabolism and protein binding of ^{99m}Tc-pyridoxylidene glutamate, *J. Nucl. Med.*, 19, 1036, 1978.
11. Levi, A. T., Gatmaitan, Z., and Arias, I. M., Two hepatic cytoplasmic protein fractions, Y and Z, and their possible role in the hepatic uptake of bilirubin, sulfobromophthalein, and other anions, *J. Clin. Invest.*, 48, 2156, 1969.
12. Litwack, G., Ketterer, B., and Arias, I. M., Ligandin: An abundant liver protein which binds steroids, bilirubin, carcinogens and a number of exogenous anions, *Nature (London)*, 234, 466, 1971.
13. Morey, K. S. and Litwack, G., Isolation and properties of cortisol metabolite binding proteins of rat liver cytosol, *Biochemistry*, 8, 4813, 1969.
14. Habig, W., Pabst, M., Fleischner, G., Gatmaitan, Z., Arias, I. M., and Jakoby, W., The identity of glutathione transferase B with ligandin, a major binding protein of the liver, *Proc. Natl. Acad. Sci. U.S.A.*, 71, 3879, 1974.
15. Jakoby, W. B., Ketley, J. N., and Habig, W. H., Rat glutathione S-transferases: binding and physical properties, in *Glutathione: Metabolism and Function*, Arias, I. M. and Jakoby, W. B., Eds., Raven Press, New York, 1976, 213.
16. Arias, I. M., Fleischner, G., Kirsch, R., Mishkin, S., and Gatmaitan, Z., On the structure, regulation, and function of ligandin, in *Glutathione: Metabolism and Function*, Arias, I. M. and Jakoby, W. B., Eds., Raven Press, New York, 1976, 175.
17. Schanker, L. S., Secretion of organic compounds into bile, in *Handbook of Physiology, Alimentery Canal*, Section 6, Vol. 5, Code, C. F. and Heidel, W., Eds., American Physiology Society, Washington, D.C., 1968, chap. 114.
18. Kupferberg, H. J. and Schanker, L. S., Biliary secretion of oubain-^{3}H and its uptake by liver slices in the rat, *Am. J. Physiol.*, 214, 1048, 1968.
19. Pugh, P. M. and Stone, S. L., The effect of 2,4-dinitrophenol and related compounds on bile secretion, *J. Physiol.*, 198, 39, 1968.
20. Boyer, J. L., Canilicular bile formation in the isolated perfused rat liver, *Am. J. Physiol.*, 221, 1156, 1971.
21. Alpert, S., Mosher, M., Schanske, A., and Arias, I. M., Multiplicity of hepatic excretory mechanisms for organic anions, *J. Gen. Physiol.*, 53, 288, 1969.
22. Wheeler, H. O., Meltzer, J. I., and Bradley, S. E., Biliary transport and hepatic storage of sulfobromophthalein sodium in the unanesthetized dog, in normal man, and patients with hepatic disease, *J. Clin. Invest.*, 39, 1131, 1960.
23. Paumgartner, G., The handling of indocyanine green by the liver, *Schweiz. Med. Wochenschr.*, 105, 337, 1967.
24. Paumgartner, G., Herz, R., Sauter, K., and Schwarz, H. P., Taurocholate excretion and bile formation in the isolated perfused rat liver, *Naunyn-Schmiedebergs Arch. Pharmacol.*, 285, 165, 1974.

25. Paumgartner, G. and Reichen, J., Kinetics of hepatic uptake and excretion of organic anions, in *The Hepatobiliary System*, Taylor, W., Ed., Plenum Press, New York, 1976, 287.
26. Nielsen, P. and Rasmussen, F., Relationships between molecular structure and excretion of drugs, *Life Sci.*, 17, 1495, 1975.
27. Milburn, P., Smith, R. L., and Williams, R. T., Biliary excretion of foreign compounds; sulfonamide drugs in the rat, *Biochem. J.*, 105, 1283, 1967.
28. Aziz, F. T. A., Hirom, P. C., Millburn, P., Smith, R. L., and Williams, R. L., The biliary excretion of anions of molecular weight 300—800 in the rat, guinea pig, and rabbit, *Biochem. J.*, 125, 25P, 1971.
29. Dalgaard-Mikkelsen, S., On the renal excretion of salicylate, *Acta Pharmacol. Toxicol.*, 7, 243, 1951.
30. Weiner, I. M. and Lack, L., Bile salt absorption; enterohepatic circulation, in *Handbook of Physiology, Alimentary Canal*, Section 6, Vol. 5, Code, C. F. and Heidel, W., Eds., American Physiology Society, Washington, D.C., 1968, chap. 73.
31. Hirom, P. C., Millburn, P., Smith, R. L., and Williams, R. T., Molecular weight and chemical structure as factors in the biliary excretion of sulfonamides in the rat, *Xenobiotica*, 2, 205, 1972.
32. Millburn, P., Smith, R. L., and Williams, R. T., Biliary excretions of foreign compounds; Biphenyl, stilboestrol, and phenolphthalein in the rat, molecular weight, polarity and metabolism as factors in biliary excretion, *Biochem. J.*, 105, 1275, 1967.
33. Abel, J. J. and Rountree, L. G., On the pharmacological action of some phthaleins and their derivatives, with special reference to their behavior as purgatives: I, *J. Pharmacol. Exp. Ther.*, 1, 231, 1909.
34. Graham, E. A. and Cole, W. H., Roentgenologic examination of the gallbladder: preliminary report of a new method utilizing the intravenous injection of tetrabromophthalein, *J. Am. Med. Soc.*, 82, 613, 1924.
35. Delprat, G. D., Jr., Studies on liver function: rose bengal elimination from blood as influenced by liver injury, *Arch. Intern. Med.*, 32, 401, 1923.
36. Taplin, G. V., Meredith, O. M., Jr., and Kade, H., Radioactive (^{131}I-tagged) rose bengal uptake-excretion test for liver function using external gamma-ray scintillation counting techniques, *J. Lab. Clin. Med.*, 45, 665, 1955.
37. Nordyke, R. A., Metabolic and physiologic aspects of ^{131}I-rose bengal in studying liver function, *Semin. Nucl. Med.*, 2, 157, 1972.
38. Klingensmith, W. C., III, Fritzberg, A. R., Spitzer, V. M., and Koep L. J., Clinical comparison of ^{99m}Tc-diethyl-IDA and ^{99m}Tc-PIPIDA for evaluation of the hepatobiliary system, *Radiology*, 134, 195, 1980.
39. Jirsa, M. and Jones, E. A., Proc. of the 19th Symposium of Colson Research Society, 139, 1967; as cited in Ito, M., Yamada, H., Kitani, K., and Sasaki, Y., in *Nuclear Hepatology*, Igaku Shoin, Tokyo, 1973, 50.
40. Tubis, M., Nordyke, R. A., Psonick, E., and Blahd, W. H., The preparation and use of ^{131}I-labelled sulfobromophthalein in liver function testing, *J. Nucl. Med.*, 2, 282, 1961.
41. Ito, M., Yamada, H., Kitani, K., and Sasaki, Y., Labeled dye metabolism and liver function study, in *Nuclear Hepatology*, Igaku Shoin, Tokyo, 1973, chap. 4.
42. Christy, B., King, G., and Smook, W. M., Preparation of iodine-123 labeled rose bengal and its distribution in animals, *J. Nucl. Med.*, 15, 484, 1974.
43. Goris, M. D., ^{123}I-Iodobromosulphalein as a liver and biliary scanning agent, *J. Nucl. Med.*, 14, 820, 1973.
44. Ansari, A. N., Atkins, H. L., and Lambrecht, R. M., ^{123}I-indocyanine green (^{123}I-ICG) as an agent for dynamic studies of the hepatobiliary system, in *Dynamic Studies with Radioisotopes in Medicine*, Vol. 1, International Atomic Energy Agency, Vienna, 1975, 111.
45. Krishnamurthy, G. T., Tubis, M., and Endow, J. S., ^{99m}Tc-penicililamine — a new radiopharmaceutical for cholescintigraphy, *J. Nucl. Med.*, 13, 447, 1972.
46. Fliegel, C. P., Dewanjee, M. K., and Holman, L. B., ^{99m}Tc-tetracyclin as a kidney and gallbladder imaging agent, *Radiology*, 110, 407, 1974.
47. Baker, R. J., Bellen, J. C., and Ronai, P. M., ^{99m}Tc-pyridoxylideneglutamate: a new rapid cholescintigraphic agent, *J. Nucl. Med.*, 15, 476, 1974.
48. Baker, R. J., Bellen, J. C., and Ronai, P. M., Technetium-99m-pyridoxylideneglutamate: a new hepatobiliary radiopharmaceutical, 1. Experimental aspects, *J. Nucl. Med.*, 16, 720, 1975.
49. Tonkin, A. K. and Deland, F. H., Dihydrothioctic acid: a new polygonal cell imaging agent, *J. Nucl. Med.*, 15, 539, 1974.
50. Lin, T. H., Khentigan, A., and Winchell, H. S., A ^{99m}Tc-labeled replacement for ^{131}I-rose bengal in liver and biliary tract studies, *J. Nucl. Med.*, 15, 613, 1974.
51. Fritzberg, A. R., Lyster, D. M., and Dolphin, D. H., ^{99m}Tc-bioquin-7CA, a potential new hepatobiliary scanning agent, *J. Nucl. Med.*, 17, 907, 1976.

52. Harvey, E., Loberg, M., and Cooper, M., Tc-99m-HIDA: a new radiopharmaceutical for hepatobiliary imaging, *J. Nucl. Med.*, 16, 533, 1975.

53. Wistow, B. W., Subramanian, G., Van Heertum, R. L., Henderson, R. W., Gagne, G. M., Hall, R. C., and McAfee, J. E., An evaluation of ^{99m}Tc-labeled hepatobiliary agents, *J. Nucl. Med.*, 18, 455, 1977.

54. Wistow, B. W., Subramanian, G., Gagne, G. M., Henderson, R. W., McAfee, J. G., Hall, R. C., and Grossman, Z. D., Experimental and clinical trials of new ^{99m}Tc-labeled hepatobiliary agents, *Radiology*, 128, 793, 1978.

55. Klingensmith, W. C., III, Fritzberg, A. R., Spitzer, V. M., and Kuni, C. C., unpublished results.

56. Ronai, P. M., Baker, R. J., Bellen, J. C., Collins, P. J., Anderson, P. J., and Lander, H., Technetium-99m-pyridoxylideneglutamate: a new hepatobiliary radiopharmaceutical. II. Clinical aspects, *J. Nucl. Med.*, 16, 728, 1975.

57. Matolo, N. M., Stadalnik, R. C., Krohn, K. A., et al., Biliary tract scanning with ^{99m}Tc-pyrodoxylideneglutamate: a new gallbladder scanning agent, *Surgery*, 80, 317, 1976.

58. Poulose, K. P., Eckelman, W. C., and Reba, R. C., et al., Evaluation of ^{99m}Tc-pyridoxylideneglutamate for the differential diagnosis of jaundice, *Clin. Nucl. Med.*, 1, 70, 1974.

59. Stadalnik, R. C., Matolo, N. M., and Jansholt, A. L., Technetium-99m-pyridoxylideneglutamate cholescintigraphy, *Radiology*, 121, 657, 1976.

60. Klingensmith, W. C., III, Fritzberg, A. R., Koep, L. J., and Ronai, P. M., A clinical comparison of ^{99m}Tc-diethyl-iminodiacetic acid, ^{99m}Tc-pyridoxylideneglutamate, and ^{131}I-rose bengal in liver transplant studies, *Radiology*, 130, 435, 1979.

61. Klingensmith, W. C., III, Fritzberg, A. R., and Koep, L. J., Comparison of Tc-99m diethyl-iminodiacetic acid and I-131 rose bengal for hepatobiliary studies in liver transplant patients, *J. Nucl. Med.*, 20, 314, 1979.

62. Loberg, M. D. and Fields, A. T., Chemical structure of technetium-99m-labeled N-(2,6-dimethylphenylcarbamoyl)-iminodiacetic acid (Tc-HIDA), *Int. J. Appl. Rad. Isot.*, 29, 167, 1978.

63. Loberg, M. D., Porter, D. W., and Ryan, J. W., Review and current status of hepatobiliary imaging agents, in *Radiopharmaceuticals II*, Proc. 2nd Int. Symp. Radiopharmaceuticals, Society of Nuclear Medicine, New York, 1979, 519.

64. Harvey, E., Loberg, M., Ryan, J., Sirkorski, S., Faith, W., and Cooper, M., Hepatic clearance mechanism of Tc-99m-HIDA and its effect on quantitation of hepatobiliary functions, *J. Nucl. Med.*, 20, 310, 1979.

65. Fritzberg, A. R., Whitney, W. P., and Klingensmith, W. C., III, Hepatobiliary transport mechanism of Tc-99m-N, α-(2,6-diethylacetanilide)-iminodicarboxylic acid (Tc-99m-diethyl-IDA), in *Radiopharmaceuticals II*, Proc. 2nd Int. Symp. Radiopharmaceuticals, Society of Nuclear Medicine, New York, 1979.

66. Klaasen, C. D. and Plaa, G. L., Plasma disappearance and biliary excretion of indocyanine green in rats, rabbits, and dogs, *Toxicol. Appl. Pharmacol.*, 15, 374, 1969.

67. Paumgartner, G., Probst, P., Kraines, R., and Levy, C. M., Kinetics of indocyanine green removal from the blood, *Ann. N.Y. Acad. Sci.*, 170, 134, 1970.

68. Callery, P. S., Faith, W. C., Loberg, M. D., Fields, A. T., Harvey, E. B., and Cooper, M. D., Tissue distribution of technetium-99m and carbon-14 labeled N-(2,6-dimethylphenylcarbamoylmethyl)iminodiacetic acid, *J. Med. Chem.*, 19, 962, 1976.

69. Ryan, J., Cooper, M., and Loberg, M. D., Technetium-99m-labeled N-(2,6-dimethylcarbamoylmethyl)iminodiacetic acid (Tc-99m HIDA): a new radiopharmaceutical for hepatobiliary imaging studies, *J. Nucl. Med.*, 18, 997, 1977.

70. Burns, H. D., Worley, P., Wagner, H. N., Jr., Marzilli, L., and Risch, V., Design of technetium radiopharmaceuticals, in *The Chemistry of Radiopharmaceuticals*, Heindel, N. D., Burns, H. D., Honda, T., and Brady, L. W., Eds., Masson, New York, 1978, 269.

71. Schachner, E. R., Gil, C., Atkins, H. L., Som, P., Srivastava, S. C., Sacker, D. F., and Richards, P., Evaluation of Ru-HIDA and Ru-PIPIDA as potential agents for delayed studies in the biliary tract, *J. Nucl. Med.*, 20, 682, 1979.

72. Paumgartner, G., The handling of indocyanine green by the liver, *Schweiz. Med. Wochenschr.*, 105 (Suppl.), 3, 1975.

73. Henriksen, J. H. and Winkler, K., Pharmacokinetics of ^{99m}Tc-diethyl-IDA in man, in *Proc. Symp. Hepatobiliary Scintography by Means of IDA Derivatives*, Biersach, H. J. and Mahlstedt, J., Eds., Gitr. Verlag, E. Ebler, Darmstadt, West Germany, 1978.

74. O'Maille, E. R. L., Richards, T. G., and Short, A. H., The influence of conjugation of cholic acid on its uptake and secretion: hepatic extraction of taurocholate and cholate in the dog, *J. Physiol.*, 189, 337, 1967.

75. Chervu, L. R., Robbins, E. B., Hug, S. S., and Blaufox, M. D., In vivo and in vitro studies of Tc-99m HIDA uptake, *J. Nucl. Med.*, 20, 655, 1979.

76. Gregoriadis, G., Morell, A., Sternlieb, I., and Scheinberg, I., Catabolism of desialylated ceruloplasmin in the liver, *J. Biol. Chem.*, 245, 5833, 1970.

77. Morell, A., Gregoriadis, G., Scheinberg, I., Hickman, J., and Ashwell, G., The role of sialic acid in determining the survival of glycoproteins in the circulation, *J. Biol. Chem.*, 246, 1461, 1967.

78. Vera, D. R., Krohn, K. A., and Stadalnik, R. C., Radioligands that bind to cell-specific receptors: hepatic binding protein ligands for hepatic scintigraphy, in *Radiopharmaceuticals II*, Proc. 2nd Int. Symp. Radiopharmaceuticals, Society of Nuclear Medicine, New York, 1979, 565.

79. Boyd, G. S. and Merrick, M. V., New radiopharmaceuticals for assessment of hepatic and G.I. function, *J. Nucl. Med.*, 20, 684, 1979.

Chapter 4

RADIO ION EXCHANGE IN BONE

M. W. Billinghurst

TABLE OF CONTENTS

I. INTRODUCTION

To understand the uptake of radioactive tracers into bone it is first necessary to understand the structure, composition, and metabolism within the bone. Bone is a living matrix consisting of living cells surrounded by an organic matrix which is impregnated with an inorganic crystalline matrix. All three of these may play an important part in the localization of radioactive tracers in bone while in some cases only one may play a significant role.

A. Bone Cells

There are three distinct types of bone cells called osteoblasts, osteoclasts, and osteocytes. The function of each of these will be discussed separately.

1. The osteoblasts are responsible for the formation of new bone, and are found primarily on the external surfaces and in the growing regions of the bone. They are also especially plentiful in the region of a fracture during the healing process.
2. The osteoclasts are responsible for the resorption of bone. These are found primarily on the internal surfaces of the bone, along the bone marrow cavity, and in other areas where significant bone remodeling is going on.
3. The osteocyte or mature bone cell is found throughout the mature bone. These cells are capable of both bone formation and bone resorption, and are responsible for the maintenance of the mature bone through continual resorption and reformation of the bone. Thus, due to the normal function of these cells, mature bone is not a static matrix but is under constant remodeling and the organic matrix and crystalline matrix are constantly being dissolved and then relaid.

B. Organic Matrix

Approximately 30% of the bone consists of the organic matrix which consists mainly of the protein, collagen. It is the organic matrix of the bone which is responsible for the tensile strength of the bone. The fibers of collagen run in all directions, with particular emphasis on the directions in which stress would normally be exerted, forming a fibrous mesh with considerable tensile strength, although very flexible and elastic as demonstrated by the fact that demineralized bone, i.e., bone from which the crystalline substance has been removed (also called decalcified bone), is tough, flexible, and elastic to the extent that it may be stretched and even tied in a knot.

C. Inorganic Crystalline Matrix

Approximately 70% of the bone is composed of an inorganic substance which is a crystalline calcium phosphate compound of a structure known as hydroxyapatite. The hydroxyapatite which has a chemical formula $Ca^{2+}_{10-x}(H_3O^+)_{2x}(PO_4^{3-})_6(OH^-)_2$ forms in flat plate-like crystals which are approximately 400 Å in length, 100 Å wide, and 10 to 30 Å thick. Each of the plate-like crystals is individually bound to a collagen fiber. It is the filling of the organic matrix with the hydroxyapatite crystals which gives the bone its rigidity and compressive strength. Thus, the structure of bone may be compared to that of reinforced concrete where the steel rods, like the collagen, supply the tensile strength; and the concrete, like the hydroxyapatite, supplies the compressive strength. While the hydroxyapatite is made up primarily of calcium and phosphate ions, some substitution of other cations and anions is known to occur. The cations, magnesium, sodium, and potassium occur in fairly constant proportions in normal hydroxyapatite, magnesium 0.44%, sodium 0.73%, and potassium 0.06%, (calcium 27.24%) as percentage of weight, while the anions, carbonate, citrate, chloride, and fluoride are consistently present in low percentages in normal bones. Other ions may also be present on a random basis due to the ionic absorption properties of this type of crystal; in fact, at least nine of the 14 major radioactive products of the hydrogen bomb are known to be absorbed onto the hydroxyapatite of bone.

D. Mechanism of Bone Formation

New bone is laid down by the osteoblast, and to a lesser extent by the osteocytes. These cells secrete the precursor of the organic matrix, an amorphous protein polysaccharide, which undergoes chemical alteration and polymerization to form collagen. The next step in bone formation is the building of hydroxyapatite crystals on these overlapping collagen fibers. The mechanism by which the calcium phosphate precipitation is initiated is not well understood. It is known that the concentration of calcium and phosphate in the blood is above that necessary for crystal formation, i.e., the blood is supersaturated with respect to calcium and phosphate. However, no precipitation occurs presumably due to the large number of ions which must come together to form a molecule of hydroxyapatite, or possibly due to the lack of a suitable center of nucleation. It may be that the formation of the hydroxyapatite crystals on the collagen matrix arises simply from the fact that the collagen fibers provide a suitable center for nucleation for the calcium and phosphate ions which are present in the extracellular fluid in the bones. Of course, the calcium and phosphate ions in the extracellular fluid in the bone are in dynamic equilibrium with the calcium and phosphate ions in the blood passing through the walls of the capillary blood vessels which supply the bone. An alternative theory holds that the osteoblasts secrete a substance that neutralizes an inhibitor of hydroxyapatite crystallization (possibly pyrophosphate), and that once neutralization occurs, the natural affinity of collagen for calcium salts causes precipitation. Both of these approaches are supported by the fact that if purified collagen fibers are added to plasma they cause precipitation of the calcium phosphate. Another theory is based on the fact that osteocytes are known to concentrate large quantities of calcium and phosphate ions in their mitochondria, and even to precipitate calcium phosphate ions there. These calcium phosphate-containing vesicles can then break away and migrate to the walls of the cell, and there extrude the preformed calcium phosphate salts into the surrounding extracellular fluid. This theory then holds that it is these preformed calcium salts which become attached to the collagen fibers. The significant difference between these theories amounts to the extent to which the bone-forming cells, the osteoblasts and osteocytes, are involved with the laying down of the hydroxyapatite.

E. Uptake of Radioactive Tracer

The uptake of any radiotracer which is injected into the bloodstream into a given area of bone is dependent on one or more of the following factors.

1. Blood flow. Clearly if the blood does not perfuse a given area, then it is impossible, at least initially, for the radioactive tracer to reach that area.
2. Capillary membrane permeability. If the radioactive tracer is to be localized in the bone, it must cross the capillary wall of the blood vessel, either directly to the bone or, more likely, first into the extracellular bone fluid and then onto the bone itself. Thus, the permeability of the capillary membrane may influence the local uptake of the tracer. With some radiotracers, e.g., fluorine-18 fluoride, this factor is believed to be relatively insignificant since the fluoride passes very readily through the membrane. However, with other tracers such as the technetium-99m polyphosphates, it is generally believed to be a significant factor in the local tracer uptake.
3. Volume of exchangeable bone. Since most bone uptake involves either an ex-

Table 1

NUCLEAR PROPERTIES OF THE ALKALINE EARTH RADIONUCLIDES

Radionuclide	Half-life	Beta energy (MeV)	% Abundance	Gamma energy	% Abundance
Ca^{47}	4.5 day	1.98	16	0.490	5
		1.48	2	0.810	5
		.67	82	1.29	71
Sr^{85}	64 day	None		0.514	100%
Sr^{87m}	2.83 hr	None		0.388	100%
Ba^{131}	12 day	None		124	
				216	
				373	
				496	
Ba^{135m}	28.7 hr	None		268	100%

change of a radiotracer with a nonradioactive atom or molecule on the bone surface, or possibly the incorporation of the radioactive tracer into newly forming bone, then obviously the interior of the individual hydroxyapatite crystals and areas within the bone which are not perfused by the blood or extracellular fluid are not subject to radioactive tracer uptake. Thus, the local bone uptake of the radioactive tracer will be related to the volume of the bone which is capable of taking up the tracer (called the volume of exchangeable bone), rather than the total bone volume. The term "volume of exchangeable bone" originates in radiotracer studies using truly tracer quantities of the calcium radionuclide calcium-47 and considering calcium tracer dilution studies. This concept holds for all radiotracers which localize in the inorganic matrix of the bone.

4. Bone metabolism. Since bone is a living substance and is in a constant dynamic equilibrium with bone being formed and resorbed, the local bone uptake will be dependent to some extent on the local balance between the formation of new bone and the resorption of old bone tissue. A radioactive tracer atom, which exchanges with a nonradioactive atom on the surface of a hydroxyapatite crystal that is currently undergoing dissolution, will not remain bound to the bone very long since it will undergo dissolution with the rest of that crystal; thus sites that are subject to resorption without equal bone formation will not show any significant radioactive tracer uptake.

II. ALKALINE EARTHS

The earliest radionuclide bone studies were carried out using the alkaline earths. Since it was known that the basic organic component of bone was calcium phosphate, then the possibility of doing radioactive tracer studies by substituting radioactive calcium for natural calcium was immediately apparent, and many such studies are reported in the literature.[1-5] Equally obviously, the substitution of other alkaline earths for the calcium has a very good probability of success, and since calcium 47 with the nuclear properties shown in Table 1 is far from an ideal radionuclide for in vivo diagnostic imaging, a number of the other radionuclides of the alkaline earths were investigated.[5-13] These include the various radioisotopes of barium and strontium whose nuclear properties are also shown in Table 1. In general it may be said that the principal problem with the localization of the radionuclides of the alkaline earths lies in the relatively slow clearance of that fraction that is not absorbed on the bone, and the fact that a good portion of this fraction is eventually excreted in the feces. This problem arises because the alkaline earths readily become bound to various proteins and are not exclusively present either in the ionic state in the blood or precipitated in hydroxyapatite crystals in the bone. The approaches which can be used to minimize the interference due to this are dependent on the radioisotope used. Calcium-47, strontium-85, barium-131, and barium 135m are best imaged 48 hr post-injection when the principal problem is the radioactivity in the colon, rectosigmoid and cecum. This may be cleared by the use of laxatives and enemas. It is also worth noting that, in general, the proportion excreted in the feces increases as the atomic size increases, i.e., the problem is less severe for calcium than for barium, with strontium being in an intermediate position. On the other hand, if strontium-87m is used, the short half-life precludes waiting for good blood clearance and imaging must be performed earlier, normally ½ to 2 hr post-injection, and the relatively high blood background radioactivity must be accepted. It is, however, important to make sure that the patient voids immediately prior to imaging to avoid visualization of the urinary tract. In general, none of these radionuclides are currently used to any significant extent in bone studies since each has its own serious drawbacks. Calcium-47 has gamma rays whose energies are too high to be conveniently used for in vivo imaging with commercially available instruments, and because of the associated beta radiation gives an unacceptably high radiation dose to the subject unless very small amounts of radioactive tracer are used, which then result in unacceptably long imaging times. Strontium-85 has too long a physical half-life, so that to keep the radiation dose within reason, the quantity which may be injected is limited to the order of 100 μCi resulting in poor counting statistics and excessively long imaging times.

Furthermore, the 514 keV gamma, although not as bad as those of calcium-47, is still well above the optimum energy for commercially available imaging equipment. Strontium-87m, available as the product of the yttrium 87/strontium-87m generator, is limited to some extent by the cost of the generator, which is not insignificant, since the yttrium-87 is cyclotron produced, and, by the life of the generator, which is less than two weeks since the parent yttrium-87 has a half-life of only 3.33 days. Furthermore, the clarity of the resultant image is less than ideal due to the high blood background at the early scanning times necessitated by the 2.83-hour half-life of the strontium-87m. While the barium isotopes, barium-131 and barium 135m, both appeared to show some promise of overcoming this difficulty despite the high proportion excreted in the gastrointestinal canal, studies using barium[8-13] have not been pursued due to the introduction of the technetium-99m phosphorous compounds for bone scanning.[14]

Kinetic data on the alkaline earths[5,6,12] suggest that, although possibly not identical one to another, the kinetics of the uptake by the bone follows similar paths with a two-component exponential uptake curve. The first component, responsible for the majority of the uptake, having a half-time from 10 to 30 min, while the second component, accounting for possibly 5 to 15% of the uptake in normals, has a half-time of several hours. While the first component is probably entirely due to isotopic exchange, the second component could possibly relate to a metabolic phenomenon. With all the alkaline earths, approximately 50% of the radionuclide is excreted via the urine and feces, with the ratio of the percent in urine to percent in feces being larger for the lighter members than for the heavier members.

III. RARE EARTHS OR LANTHANIDES

Early studies on the biological fate of the lanthanides[15] indicated that when injected as the citrate in carrier-free conditions, the heavier lanthanides localized primarily in the bone, while the lighter lanthanides localized primarily in the liver. Based on this observation, O'Mara et al.[16] investigated some of the radionuclides of the lanthanides, lutetium-176m, lutetium-177, samarium-153, and erbium-171, in various chemical

Table 2

RADIONUCLIDIC PROPERTIES OF THE LANTHANIDES

		Beta particle		γ Ray	
Radionuclide	T½	Energy MeV	% Abundance	Energy MeV	% Abundance
^{176m}Lu	3.7 hr	1.22	~65	0.088	~10
		1.31	~35		
^{177}Lu	6.7 hr	0.17	6.7	0.113	100
		0.38	7	0.208	171
		0.497	86	0.250	3.3
				0.321	3.4
^{153}Sm	47 hr	0.68	32	0.070	17
		0.70	48	0.103	100
		0.80	20		
^{171}Er	7.5 hr	0.59	3.6	0.112	25
		1.065	91	0.116	3
		1.49	2.3	0.296	28
				0.308	63
^{157}Dy	8.1 hr	None		0.326	100

forms. They found that, as might be expected, if they injected a strong chelate such as DTPA (diethylenetriaminepentaacetic acid), rapid excretion via the kidneys took place, while if injected as a weak chelate or salt then significant localization occurred in the reticuloendothelial system, presumably due to the in vivo formation of a radiocolloid. However, if a chelate of intermediate stability was used, such as hydroxyethylene diamine tetraacetic acid (HEDTA), then the formation of the radiocolloid was apparently inhibited without significantly inhibiting the bone uptake, with the result that approximately 50% was rapidly taken up in the bone while the remainder was cleared from the plasma by the kidneys. A study of the radionuclidic properties of these isotopes, Table 2, suggests that while they were superior to the alkaline earths with respect to their physical half-life, number of usable gamma emissions per disintegration, and radiation dose to the patient, they were still not ideal. Thus, the more expensive cyclotron-produced radioisotope dysprosium-157 was investigated.[17,18] Although initial studies looked promising, they were not pursued due to the introduction of the technetium phosphorous compounds for bone scanning at that time.

IV. GALLIUM - 68

At various times, a small amount of interest has been shown in the possibility of using gallium - 68 as a bone scanning agent. Its primary attractions lie in the facts that gallium - 68 is a positron emitter and is available in the form of a germanium-68/gallium-68 generator. The long life of the germanium-68 parent (275 days) means that the gallium-68 may be readily made available in areas which do not have ready access to a cyclotron, while the short half-life of the gallium - 68 (1.14 hr), results in a low radiation dose to the patient. The fact that the 511 keV gammas arising from the positron have an energy which is above the optimum energy for commercial imaging devices, did not totally suppress the interest in gallium-68 since there were few isotopes for bone scanning with significantly better energies before the advent of the technetium-99m phosphorous compounds; and, more recently the growing interest in positron tomography has renewed interest in gallium-68 as a possible positron emitting bone agent which would not have the distribution problems of fluorine-18. In the early work,[5,19,20] gallium was used as the citrate, and coupled with the addition of an appropriate amount of gallium carrier, to suppress the protein binding of the gallium. The effect of the presence of carrier gallium is to remarkably reduce the general background, presumably by saturating the preferred binding site in the serum proteins, thus enabling the bone uptake to be the predominant feature. One of the problems with the use of the germanium-68/gallium-68 generator to supply gallium 68 for bone scanning is that the usual generator yields the gallium-68 in the form of the EDTA (ethylenediaminetetraacetic acid) complex while it is required in the citrate form for bone scanning. Thus, the EDTA complex must be converted to the citrate form and the EDTA completely removed from the preparation, a procedure that calls for careful chemical manipulation.[20] However, recent advances in the design of gallium-68 generators has resulted in the production of a germanium-68/gallium-68 generator which yields ionic gallium-68.[21]

The need to enhance the bone uptake of gallium by adding significant amounts of gallium carrier led to the exploration of the possibility of using some of the bone seeking phosphorous chelates developed initially for the technetium-99m phosphorous bone agents as bone localizing chelates for gallium-68. The chelates ethylenediaminetetramethylene phosphorate (EDTMP), and diethylenetriaminepentamethylene phosphorate (DTPMP),[22] have been shown to be clinically useful as bone localizing chelates of gallium-68, and will undoubtedly become important if positron bone imaging becomes a widely adopted technique.

V. LEAD - 203

There have been a number of reports[23-25] of the possible skeletal imaging potential of lead-203 in the literature. Lead-203, which decays by electron capture with a 52 hr half-life and the emission of a 99% abundant 280 keV gamma, appears to be a relatively suitable radionuclide for in vivo imaging with a half-life that is reasonably short, but long enough for good blood clearance of the nonabsorbed fraction, i.e., scanning at 72 hr post-injection is practical, while the 280 keV energy of the gamma ray is within a reasonable energy range for currently available commercial imaging equipment. A number of chemical forms have been superficially investigated, including acetate, chloride, chelates such as HEDTA and EDTA, and the phosphorous compounds used in technetium-99m bone scanning. The major nonskeletal localization is observed in the liver, except in the case of the chelates where it is the kidneys. The EDTA complex, as expected, leads to poor skeletal uptake due to the predominance of the kidney excretion; while the salts, e.g., chloride and acetate, result in more liver uptake than is acceptable, so that the best prospects for reasonable bone uptake appear to be the bone localizing phosphorous compounds and the intermediate strength chelate HEDTA.

VI. IRON-59

The iron radionuclides have normally been considered to localize in the bone marrow when injected intravenously in the ionic form. However, there is a report in the literature[26] which shows that iron, when injected as the ferrous salt, is taken up to some extent by the bone, as well as the bone marrow. Autoradiography of the dissected bones showed that the bone uptake was due to uptake by the osteocytes and/or the associated lacunae. The utilization of iron radionuclides for bone scanning is, however, unlikely since the principal uptake is undoubtedly that of the bone marrow which would mask any bone uptake unless the marrow is first completely removed from the bone.

Table 3
NUCLEAR REACTIONS LEADING
TO THE PRODUCTION OF
FLUORINE-18

Reactor Production

^{6}Li   (n, α) ^{3}H ^{16}O   (t, n) ^{18}F

Reaction carried out using ^{6}Li enriched Li_2CO_3

Cyclotron Production

^{16}O	(^{4}He, np)	^{18}F		
^{16}O	(^{4}He, d)	^{18}F		
^{16}O	(^{4}He, 2n)	^{18}Ne	→	^{18}F
^{16}O	(^{3}He, p)	^{18}F		
^{18}O	(p, n)	^{18}F		
^{19}F	(γ n)	^{18}F		
^{19}F	(n, 2n)	^{18}F		
^{20}Ne	(d, α)	^{18}F		

VII. FLUORINE-18

The first anion which was employed for bone localization was the fluoride ion in the form of fluorine-18 fluoride. It is thought that the fluoride ion is incorporated into the hydroxyapatite crystal by ion exchange with the hydroxide ion in the normal hydroxyapatite crystal.

Fluoride, unlike the cations that have been discussed, is not bound to the serum proteins so that it clears from the blood more rapidly, with the result that the bone images obtained show relatively little background compared to studies done with cationic radionuclides at comparable times post-injection. Fluorine-18 is a positron-emitting radionuclide with a half-life of 1.8 hr; and, as a result the imaging must be carried out with equipment suitable for imaging the 511 keV annihilation gamma ray, i.e., either a positron camera or routine imaging equipment fitted with special high energy collimators. The relatively short half-life of 1.8 hr has always been the main problem limiting the wide utilization of fluorine-18, in that it dictates that there must be a producing facility in the immediate vicinity of the user, and even given the fact that fluorine-18 can be produced with either a reactor or cyclotron, (see Table 3 for some of the nuclear reactions that lead to the production of fluorine-18), the majority of nuclear medicine laboratories can not obtain a satisfactory supply of fluorine-18.

The great advantage of fluorine-18 is that it is rapidly cleared from the blood as a result of the rapid uptake by the bone with the bone uptake being close to complete within 1 hr,[27] and by the rapid excretion by the kidneys, 50% of the radioactivity is excreted by the kidney within 6 hr of injection.[27] The fact that fluoride is a highly diffusible anion with little affinity for proteins or red blood cells probably accounts for a great deal of the rapidity of its blood clearance. Studies of the blood clearance curves of fluorine-18[28-30] have shown that it approximates a biexponential curve with greater than 70% of the radioactivity leaving the blood within 2 min. Conceptually, Blau et al.[31] views this as a result of the passage of fluoride from, (1) the plasma through (2) the extracellular fluid space in the bone, into (3) the shell of bound water molecules around each hydroxyapatite crystal, (4) onto the crystal surface, and (5) into the interior of the crystal. The first three of these processes would clearly proceed rapidly and be subject to the establishment of equilibria. The fourth step would be somewhat slower, and therefore could account for the second component of the blood

MODEL OF FLUORIDE - 18 KINETICS

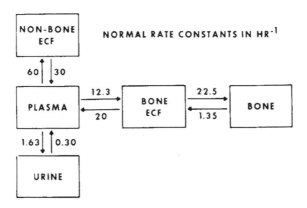

FIGURE 1. A five compartment model of fluoride kinetics (From Charkes, N. D. and Philips, C. M., A new model for [18]F Fluorides kinetics in humans, in *Medical Radionuclide Imaging*, Vol. 2, International Atomic Energy Agency, Vienna, 1977, 138. With permission.)

clearance curve, since as the fluoride concentration in the bound water decreased, due to its being bound on the surface of the actual hydroxyapatite crystal, it would be subject to the establishment of the fluoride equilibrium levels in the extracellular fluid and in the blood. Physiologically, of course, once the fluoride was absorbed by the bound water, it could be considered essentially part of the bone. The final phase, the incorporation into the interior of the hydroxyapatite crystal, is too slow to be observed with fluorine-18 as the radioactive tracer. Such a concept naturally leads to the conclusion that these areas of bone that are perfused to a greater extent by the extracellular fluid would show a greater uptake unless saturation levels of the fluoride ion were approached. It is a well-established fact that the bone uptake of tracer is nonuniform with the greater uptake being in spongy bone, the subperisteal area, and the epiphyseal area,[27] thus with respect to the spongy bone vs. normal compact bone, some support for the importance of the bone extracellular fluid is found. Furthermore, Hosain et al. [13] showed that for a single bone, the radioactive uptake corresponded with the extracellular fluid content.

Charkes and Philips[30] developed a kinetic model, Figure 1, to explain the blood clearance of fluoride.

Superficially, the principal difference between this model and the concept of Blau et al.; is the inclusion of the nonbone extracellular fluid; this they found necessary to allow for an extracellular fluid space outside of the blood large enough to account for the rapid drop in approximately 2 min of the fluoride blood concentration to approximately 27% of injected dose if the drop was to be based on diffusion. In actual fact, this nonbone extracellular fluid plays the primary role in this model since its volume is larger than that of the bone extracellular fluid, and it has been assigned the largest rate constant leaving the blood pool. Some practical support of this is seen in a 15 min post injection scan of a normal subject, shown in Figure 2, in which primarily soft tissue background is observed at a time when the radioactivity in the blood is known to be only about 20% of the injected radioactivity. According to the computer-generated curves based on their kinetic model, Charkes and Philips suggest that at this time the percentage of the dose in the bone is approximately equivalent to that in the nonbone extracellular fluid with all other components being considerably lower (Figure 3). It is also generally observed that the rate of extraction of fluoride from the blood

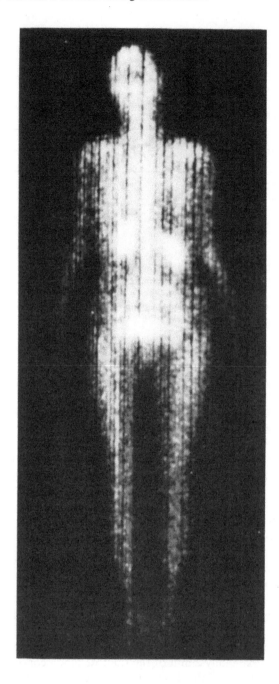

FIGURE 2.   A [18]F Fluoride scan of a normal subject 15 min post-injection. Note that the skeleton has not yet accumulated sufficient radioactivity to be clearly evident. (From Van Dyke, D., Anger, H. O., Parker, H., McRae, J., Dobson, E. L., Yano, Y., Noets, J. P., and Linfoot, J., Markedly increased bone blood flow in myelofibrosis, *J. Nucl. Med.*, 12, 506, 1971. With permission.)

by the bone is very high, leading to the belief that the fluorine-18 bone image primarily represents the blood flow to the bone. In fact some have suggested that the fluoride is totally extracted on the first pass through the bone; however, this was shown not to be the case by Costeas et al.,[32] who demonstrated that the bone uptake occurs rapidly,

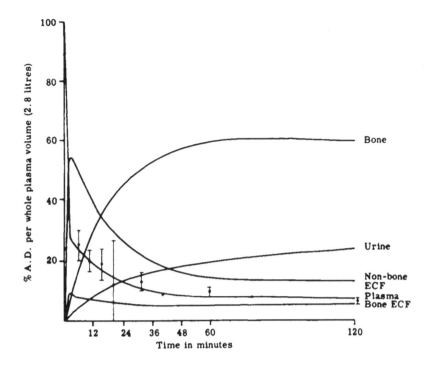

FIGURE 3. A computer generated model of ¹⁸F kinetics in humans based on the five compartment model (Figure 1). (From Charkes, N. D. and Philips, C. M., A new model for ¹⁸F Fluoride kinetics in humans, in *Medical Radionuclide Imaging*, Vol. 2, International Atomic Energy Agency Vienna, 1977, 138. With permission.)

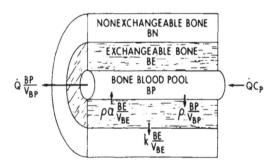

FIGURE 4. A model for the local bone uptake of radionuclides. (From King, M. A., Kilpper L. W., and Weber, D. A., A model for the local accumulation of bone imaging radiopharmaceuticals, *J. Nucl. Med.*, 18, 1106, 1977. With permission.)

reaching an equilibrium between the bone surface and the blood. Such an observation is not inconsistent with the concept of Blau et al., the kinetic model of Charkes and Philips, although in itself it does not necessitate the presence of the extracellular fluid, nor the simpler model proposed by King et al.,[33] Figure 4, which considers the bone uptake rather than the blood clearance of the fluorides. These latter workers noted that the interposition of the extracellular fluid compartment between the blood and the bone had little effect on the computer-generated curves for the bone uptake, and therefore, omitted it for simplicity. While this may seem a drastic departure from the model of Charkes and Philips, it is not such a significant change in reality, since an examination of the rate constants they obtained shows that it is the nonbone extracellular fluid which plays the principal role in the early blood clearance, and since radioactivity from the nonbone extracellular fluid must re-enter the blood before entering the bone extracellular fluid, and King et al. are considering bone uptake rather than blood clearance. It is perhaps significant to note at this point that while the blood radioactivity drops to about 30% in the first 2 min post-injection, the bone uptake does not reach 50% of its maximum level until approximately 20 min post injection, again supporting the significance of the nonbone extracellular fluid in the kinetics of bone uptake of fluoride.

In the literature on the clinical use of fluorine-18, questions have been raised with respect to what the actual species is which is present in the radiopharmaceutical. It has been suggested that the active species is a complex ionic species formed by the interaction of HF and the container used in preparing the radiopharmaceutical. In addition, rat studies show some discrepancies in the biological distribution according to the source of the fluorine-18.[30]

Since it is virtually impossible to identify species in aqueous solution at these very low tracer levels, a definitive resolution of this problem has not been arrived at; although, it is generally believed that the fluorine-18 which was supplied by Hammersmith was not initially in the form of fluoride ion.

VIII. PHOSPHOROUS COMPOUNDS

Unlike those localizing agents so far discussed, where the localization is primarily a property of the radionuclide itself, the phosphorous compounds are a group of compounds which can bind to various radionuclides, and then when injected, carry that radionuclide onto the binding sites in the bone, i.e., the bone localization is a property of the phosphorous compound while the radioactivity is associated with a radionuclide which is bound to the phosphorous compound. It will, of course, be recognized that phosphorous does not have an isotope with suitable radionuclidic properties for in vivo diagnostic work. In studies already discussed, it will be remembered that in the case of certain radionuclides, which although they showed some skeletal localization, did not have acceptable target to non-target ratios when used as the ionic salt, chelates with phosphorous compounds were employed to provide better skeletal localization. Phosphorous compounds were introduced by Subramanian in 1971[14] as a complexing agent which would carry technetium-99m to the bone, and since technetium-99m was a readily available inexpensive radionuclide with close to ideal radionuclidic properties for in vivo diagnostic studies, the interest from the nuclear medicine community was immediate and substantial. While Subramanian's original work dealt with the use of tripolyphosphate a wide range of phosphorous compounds were quickly evaluated over the next few years, including phosphates of various molecular weights from pyrophosphate to long chain polyphosphates with up to 46 phosphate units in the chain, and a large number of organic phosphonates. The structures of some of these phosphorous compounds are shown in Figure 5. Today, the number of phosphorous compounds in regular routine used as bone localizing agents has decreased to three: one inorganic phosphate, pyrophosphate, and two organic phosphonates; methylene diphosphonate (MDP), and hydroxyethylenediphosphonate (HEDP). Despite all the interest which has been shown in this group of compounds, no theory of the mechanism of localization has become generally accepted. In part, this is due to the fact that so many different compounds have been investigated in order to select the best ones to use, with the result that the amount of well-documented data on any single compound is not all that plentiful. Even with a single phosphorous compound, formulation differences such as variation in the ratio of phosphorous compound to the stannous ion, present to act as

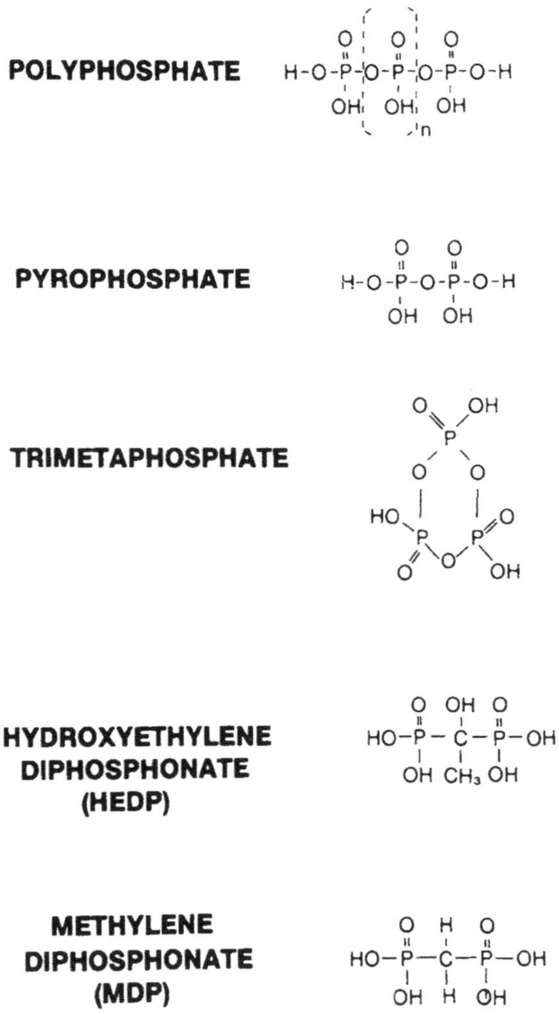

POLYPHOSPHATE

$$
\begin{array}{c}
\quad O \quad\; O \quad\; O \\
\quad \parallel \quad\; \parallel \quad\; \parallel \\
H-O-P-O-P-O-P-O-H \\
\quad | \quad\;\; | \quad\;\; | \\
\quad OH \;\; OH \;\; OH
\end{array}\Bigg]_n
$$

PYROPHOSPHATE

$$
\begin{array}{c}
\quad O \quad\; O \\
\quad \parallel \quad\; \parallel \\
H-O-P-O-P-O-H \\
\quad | \quad\;\; | \\
\quad OH \;\; OH
\end{array}
$$

TRIMETAPHOSPHATE

HYDROXYETHYLENE DIPHOSPHONATE (HEDP)

$$
\begin{array}{c}
\quad O \quad OH \quad O \\
\quad \parallel \quad\; | \quad\;\; \parallel \\
HO-P - C - P-OH \\
\quad | \quad\; | \quad\;\; | \\
\quad OH \;\; CH_3 \;\; OH
\end{array}
$$

METHYLENE DIPHOSPHONATE (MDP)

$$
\begin{array}{c}
\quad O \quad H \quad O \\
\quad \parallel \quad | \quad\; \parallel \\
HO-P-C-P-OH \\
\quad | \quad | \quad\;\; | \\
\quad OH \;\; H \;\; OH
\end{array}
$$

FIGURE 5.   Chemical structures of some of the bone seeking phosphorous compounds.

a reductant for the pertechnetate, and thus enable technetium-99m radionuclide to bind to the phosphorous compound, and in the pH of the final preparation cause apparent differences in the biological localization,[34] and thus further complicate the comparison of data from different studies. Thus, the development, or the substantiation, of a mechanistic model from published data is seriously hampered. Despite the lack of a generally accepted mechanistic theory, certain information has been clearly established; the remainder of this section will review those facts together with some of the alternative concepts of the mechanism of localization of these phosphorous compounds.

A. Blood Clearance of Technetium Phosphorous Compounds

As already mentioned, a large number of bone-localizing phosphorous compounds have been investigated, from the very long chain polyphosphates to smaller molecules like pyrophosphate and methylenediphosphonate with some very different chemical properties, such as the readily hydrolyzable P—O—P bonds of the phosphate compounds to the nonhydrolyzable P—C—P bonds. Yet, surprisingly, there seems to be a reasonably uniform blood clearance pattern for all the technetium phosphorous compounds. The blood clearance has been shown to be a triexponential function.[35-37] A number of investigators have reported a biexponential blood clearance;[28,38-40] however, a retrospective examination of the experimental data on which they based this conclusion in conjunction with the half-time of the first exponential reported by other workers[35-37] shows that they would have missed the first exponential clearance by allowing too much time to elapse after injection of the radioactivity before they started their clearance study. This first exponential has a half-time of 3 to 6 min and accounts for 70 to 80% of the injected dose, cf. Fluorine-18 with a first exponential clearance of greater than 70% and a half-time of about 2 min. Similar to the rapid blood clearance of fluoride, the rapid blood clearance of the phosphorous compounds is believed to be due to distribution throughout the extracellular fluid space as well as the blood, the somewhat slower half-time being a result of the fact that the labeled phosphorous compound would clearly diffuse a lot less rapidly than the highly diffusable fluoride ion. The second exponential of the blood clearance for the phosphorous compounds has a half time of from 15 to 30 min and accounts for 15 to 20% of the injected radioactivity. This phase is believed to represent the uptake by the bone, presumably via the bone extracellular fluid leading to the reestablishment of the blood extracellular fluid distribution equilibrium, and a resultant drop in both the blood and extracellular fluid radioactivity. The third exponential of the blood clearance accounting for only 1 to 7% of the injected radioactivity has a half-life of from 150 to 500 min. This very wide range probably reflects the chemical differences of the phosphorous compounds; for example polyphosphate is thought not to leave the blood stream until it has been degraded into smaller units, presumably pyrophosphate units[41] and, more importantly, the inorganic phosphates lead to significant levels of red blood cell labeling,[28,35,38] with as much as 30% of the circulating radioactivity being bound to the red cells 2 hr post injection, while the organic phosphonates do not lead to significant red blood cell labeling.[35,39] However, part of the wide range in determined half-times undoubtedly reflects the uncertainty inherent in determining a half-time of a component as small as 1 to 7% of the total radioactivity in a complex biological system. It is generally believed that this third exponential represents the renal clearance of the fraction remaining in the blood. Although this seems to be in marked contrast to the fact that 60 to 70% of the radioactivity is excreted in the urine in the first 6 hr following injection, it must be remembered that, (1) the renal clearance is a direct function of the plasma concentration, thus the renal clearance rates would be much higher during the early periods when there is a much higher blood radioactivity, and (2) the blood radioactivity remains in equilibrium with the radioactivity in the extracellular fluid which presumably returns to the blood as the radioactivity is cleared from the blood and is, in turn, excreted via the kidneys.

Alternative explanations of this third exponential could be that it relates to the metabolic formation of new bone or that it relates to adsorption on the collagen as opposed to ion exchange on the surface of the hydroxyapatite crystals which is assumed to account for the majority of the uptake of the phosphorous compounds. Figure 6 shows a generalized blood clearance curve for technetium-99m phosphorous bone-seeking compounds.

B. Bone Uptake Kinetics

There are few published papers which deal specifically with bone uptake kinetics although there are a good number of articles[36,40,42,43] that have reported qualitatively that very little additional bone uptake seems to occur after the first two hours. Most reports agree that despite the lack of additional bone uptake after two hours, the clarity of the scan does improve at later times, i.e., 5 to 6 hr post injection,[37,44,45] but this is

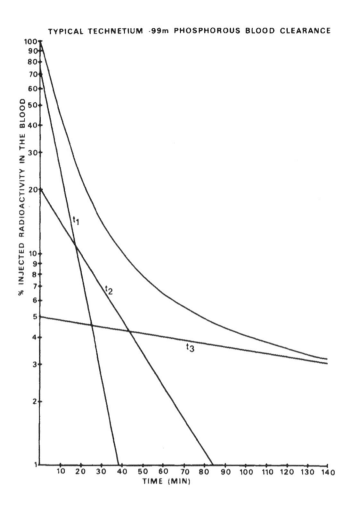

FIGURE 6. A generalized blood clearance curve for technetium-99m when injected in the form of technetium phosphorous compounds.

generally believed to be primarily due to the lower background as a result of the clearance of the residual radioactivity from the blood and nonbone extracellular fluid. This observation, that the primary bone uptake is complete within 2 hr, fits very well with the concept that the second exponential of the blood clearance curve which has a half-time of 15 to 30 min represents bone uptake, since the two hours would represent 4 to 8 half-times, i.e., the bone uptake would have reached 94 to 99.6% of its final value.

Although very few research workers have addressed the question of the kinetics of phosphorous compounds in bone directly, a number of significant facts have emerged.

Garnett et al.[44] have shown that the uptake by normal bone of the technetium phosphorous compounds, unlike fluoride, is not essentially complete in a single pass through the bone but rather only approximately 60% is extracted. Thus, unlike fluorine-18 studies which are thought to represent primarily the blood flow to the bone, there is a possibility that technetium phosphorous bone studies may indicate some other factor since any change within the bone which alters the 60% extraction efficiency on a single pass will affect the local bone uptake, and in fact Garnett and co-workers were able to show higher abnormal to normal bone radioactivity ratios for technetium-99m pyrophosphate than for fluorine-18 fluoride in the same bone lesion. This observation serves to remind us that if we are seeking "hot spots" in radionuclide scans as markers of abnormality, and are not necessarily interested only in the blood supply to the target organ, then total extraction of the radiotracer in the normal organ during the first pass of the blood through that organ is definitely undesirable; rather, we must seek a radiotracer that is more actively accumulated by abnormal tissue than by normal tissue. This point may have major significance in the selection of the most effective bone scanning radiotracer for positron scanning. Currently, the choice is between fluorine-18 fluoride and gallium-68 in the form of a chelate with a phosphorous bone-seeking compound. Although fluorine-18 data are already widely documented, it is possible that the gallium-68 may have the advantage of a better abnormal to normal ratio in "hot" lesions as well as the advantage of a greater ease of distribution.

The fact that the blood flow alone is not responsible for the variation in local uptake of technetium-99m labeled phosphorous compounds is supported by the work of Piccone et al.,[46] who showed that despite an artificially induced fourfold increase in blood flow to a bone only a 70% increase in uptake of technetium-99m labeled phosphorous compound was induced, i.e., the bone uptake of the phosphorous compound, in this case MDP, appears to be saturated. In addition, Lavender et al.[47] showed substantially increased bone uptake of technetium-99m MDP where there was only a modest increase in blood flow.

Citrin et al.[40] studied the uptake of ^{99m}Tc HEDP by normal bone and established that the bone uptake followed a biexponential function with half-times of approximately 12 and 53 min, tending to suggest that there may be two components to the bone uptake. This observation was made in a study of the rate of uptake in the spine of normal subjects, and thus does not distinguish in any way between bone and bone extracellular fluids; therefore, it is possible that it could be interpreted as representing uptake by bone extracellular fluid, T½ 12 min, and uptake by the bone, T½ 53 min. However, such an interpretation implies that there would have to be a substantially different rate of exchange between the blood and the bone extracellular fluid as compared to the exchange between the blood and the nonbone extracellular fluid, since the blood clearance data require the latter to be of the order of 3 to 6 min. Alternatively, it may be considered that the two expontentials represent two different mechanisms of uptake within the actual bone matrix. Citrin and co-workers further noted that the uptake in tumors continued to increase with respect to normal bone throughout the whole of their 4 hr study. In a similar study Castronovo et al.[36] showed that the normal bone uptake rises to 60 to 70% of its 1 hr level in approximately ½ min followed by a slower relatively steady uptake that tends to level off as it approaches 1 hr. The rapid rise in the first ½ to 2 min post-injection is considered to represent a "vascular" phase of the radioactivity in the bone representing the arrival of the tracer in the blood at the bone, and it is presumably largely retained once it arrives by distribution relatively immediately into the bone extracellular fluid. It is, perhaps, more than coincidental that this rapid appearance of radioactivity in the bone should be 60 to 70% while Garnett et al.[44] reported on 60% extraction from the blood in a single pass through bone. The observations of Castronovo et al.[36] on two abnormal cases also provide some very interesting data which any overall mechanistic theory must explain. In a patient with Paget's disease, they noted that the initial rise in radioactivity in the bone at ½ to 2 min post injection surpassed the 1 hr radioactivity, being approximately 130% of the 1 hr level, but dropped to 90% of the 1 hr level at around 2 min post injection followed by a slow, steady accumulation that leveled off as it approached 1 hr post-injection. While the initial very high radioactivity is readily explained by the fact that Paget's disease is known to be characterized by a very marked increase in vascularity, it is clear that this increased vascularity, and therefore greater blood supply to the bone, is not matched by an equal extraction of the radioactivity reaching the bone. Apparently the uptake by the bone is limited by some other factor, an observation very much in agreement with the results of Piccone et al.[46] In the case of bone metastases, Castronovo and co-workers observed similar patterns irrespective of whether the metastases were "hot" or "cold". The initial peak of radioactivity between ½ and 2 min post injection was approximately equal to the 1 hr levels, but dropped right back to the baseline around 2 min post injection, suggesting that there was virtually no extraction of radioactivity from the blood on the first pass. This is then followed by a relatively rapid uptake which continues past the 1 hr time frame of the study, cf. Citrin et al.,[40] who noted that the radioactivity accumulated by bone tumors continued to increase beyond 4 hr post injection. The lack of extraction of the radioactivity from the blood on the "first pass" argues against the suggestion of Garnett et al.[44] that the reason for the less than 100% extraction efficiency of the phosphorous compounds is related to the permeability of the capillary membrane, since, as they point out, it is difficult to imagine a bone lesion in which the permeability of the capillary membrane is decreased, and especially decreased to the point of virtually no initial extraction. A more probable model would involve the bone extracellular fluid in the immediate or "first pass" uptake, and thus the lack of initial uptake would suggest a marked reduction of the extracellular fluid in the area of the bone metastases.

C. Hydroxyapatite Uptake

One belief is that the uptake of technetium phosphorous compounds takes place in a manner similar to that of the alkaline earths, i.e., it occurs on the hydroxyapatite crystals by ionic exchange or adsorption processes. Van Langevelde et al.[48] studied the uptake of technetium-99m labeled HEDP on in vitro hydroxyapatite crystals and showed that the mechanism of uptake involved the phosphorous compound acting as a carrier for the technetium to the location of the hydroxyapatite, where the hydroxyapatite successfully competes with the technetium for the phosphorous compound setting the reduced technetium free to bind independently. Similar studies were conducted by Tofe and Francis[49] who showed that 1 mg of HEDP in conjunction with 0.02 mg of stannous ion bound to 100 mg of hydroxyapatite, but if the level of HEDP was maintained and the stannous level increased, the binding of both stannous ion (as percent of total stannous ion) and technetium to the 100 mg of hydroxyapatite was decreased. This was shown not to be a result of the increased Sn:HEDP ratio, since, if the stannous ion was maintained at 0.02 mg, the HEDP could be reduced to 0.1 mg without loss of binding to the hydroxyapatite, i.e., it is not a result of some of the tin and technetium becoming disassociated with the HEDP. Thus, it would appear that there is competition for the binding sites on the hydroxyapatite, although whether all three compete for the same sites on the hydroxyapatite has not been unequivocally demonstrated. Additional support for the concept that the HEDP acts as a carrier for the technetium, but releases it and becomes bound to hydroxyapatite, is found in the work of McRae et al.,[50] who found that exchange of technetium from technetium HEDP to gluconate could be induced by the presence of calcium ions, indicating that the HEDP preferred to complex with calcium ions rather than technetium, in the work of Byun et al.,[51] who observed the in vivo localization of technetium in the area of a previous intramuscular injection of iron dextran, and in the work of Cox[52] who showed that the role of the phosphorous compound in bone localization was a secondary one, probably preventing the formation of colloidal species by the reduced technetium and the associated reducing agent.

D. Collagen Uptake

As discussed earlier, the other major component of the bone matrix is the collagen, and some evidence has been found that supports the argument for the involvement of the collagen in the final uptake of the technetium-99m in the bone following injection of technetium phosphorous compounds. This approach has been developed by Rosenthall and co-workers,[53-57] who found very little correlation between the early bone uptake, up to 20 min post-injection, and the 5 hr bone to soft tissue ratio, clearly undermining any theory which suggested the primary rate of blood supply in determining the final bone uptake. In in vitro experiments[54] they showed that while radioactive calcium, ^{32}P pyrophosphate and ^{14}C diphosphonate are primarily localized in the inorganic bone, i.e., the hydroxyapatite crystals, the technetium–99m–labeled bone-seeking radiopharmaceuticals lead to technetium-99m being found in both the organic and inorganic phases of the bone. That is to say that there are two components to the uptake of technetium-99m phosphorous compounds by bone, one similar to that of the uptake of the parent phosphorous compound involving the hydroxyapatite, and another, presumably related to the technetium involving the collagen. Localization independent of the phosphorous compound is also suggested by Cox,[52] although it was not noted as to whether it related to hydroxyapatite or the collagen. In fact, on the basis of the work of Van Langevelde et al.,[48] it is suggested that if reduced technetium is in the presence of hydroxyapatite, and is not chelated to another molecule, it will become bound to the hydroxyapatite unless the hydroxyapatite has all its binding sites occupied. Rosenthall et al. also obtained suggestive evidence that the localization of technetium in the organic phase of bone is dependent on the amount of immature collagen present, and in a patient study[56] they were able to correlate the degree of bone uptake of the technetium-99m from the technetium phosphorous compound to the hydroxyproline levels in the serum and urine. Since increased excretion of hydroxyproline is known to reflect collagen breakdown and turnover, this observation supports the contention that the binding of technetium is related to the amount of immature collagen.

These observations were all made primarily on cases with metabolic bone disorders, and thus it is not surprising that they appear at odds with the general observations based on studies in normal subjects; however, they are supported by the studies of Castronovo et al.[36] and Citrin et al.[4] in those observations they made in cases of bone disease. Additional support is also found in the report by Williams et al.[58] that with neonatal subcutaneous fat necrosis the bone scans were positive before there was radiographic evidence of calcification, suggesting that the uptake was possibly occurring on the multiple unbound side chains of the newly forming collagen in the necrotic fatty deposit.

E. Soft Tissue Uptake

There are many reports in the literature of the accumulation of technetium-99m radioactivity in soft tissue following the injection of the technetium-99m labeled phosphorous compounds. In some of these cases,[59-61] abnormal levels of calcium were observed in the area of localization, but in others[62-67] there was no evidence of calcification. Thus, clearly, there must be some mechanism of localization of technetium-99m phosphorous compounds which does not involve calcium.

It has also been established that technetium-99m labeled phosphorous compounds, particularly the pyrophosphate, localize in infarcted myocardial tissue, and since it was known that there are many biochemical changes in the damaged myocardial tissue initially, this uptake was thought to relate to the presence of hydroxyapatite crystals in the area of the infarct. However, some subcellular work carried out by Dewanjee and co-workers[68,69] showed that the majority of the technetium-99m was associated with the soluble protein fraction, irrespective of whether the cells were dead or alive, indicating that the cellular localization is not related to the hydroxyapatite that is found associated with the mitochondria. These studies revealed only relatively minor differences in the subcellular distribution of the technetium-99m radioactivity between the dead and live cells, despite the fact that the dead cells took up 15 to 20 times the radioactivity taken up by the live cells. Thus, either the availability of the binding sites on all the cellular components must be comparably affected by cell death, or the effect relates to the increased permeability of the cell membrane as a result of cell damage. These authors also noted that if the cells were pretreated with the unlabeled phosphorous compound, the uptake by the dead cells was decreased by approximately 50%, although the uptake by the live cells was unaffected. An observation which is consistent with the permeability of the cell wall being the governing factor, since if the unlabeled phosphorous compound, like its labeled counterpart, was unable to readily cross the cell wall to saturate the binding sites on the various components within the cell, then no suppression of uptake would be expected as a result of pretreatment of the cell with the unlabeled compound. However, in the case of the damaged cell, penetration of the cell wall by the unlabeled compound would saturate the binding sites on the cell components, and uptake of the labeled compound would be restricted to exchange of the labeled compound for the unlabeled compound at the various binding sites.

In a study of the myocardial localization of a number of phosphorous compounds labeled with technetium-99m, Jones and Davis[70] showed that the ratio of the uptake in normal myocardium to uptake in damaged myocardium was relatively constant for a large number of the technetium-99m labeled phosphorous compounds, despite the fact that the actual uptake varied widely. In particular, the inorganic phosphate compound, technetium pyrophosphate, resulted in a much greater uptake of the technetium-99m than the labeled organic phosphonates. When this is considered in conjunction with the observation of McRae et al.[50] that the technetium pyrophosphate bond is weaker than the technetium phosphonate bond, it suggests that the actual mechanism of cellular localization of the radionuclide involves the transfer of the technetium from the phosphorous compound to the cellular component, while presumably the increased localization in damaged cells relates to the increased ease of the complex crossing the cell membrane once damage has occurred.

F. Enzymatic Inhibition

Another theory which has been advanced to explain the various localizations of the labeled phosphorous compounds is that of enzymatic inhibition. This theory was suggested by Zimmer et al.[71] when they observed that in normal and pathological breast tissue which showed uptake of technetium-99m diphosphonate, biopsies showed no microscopic calcification or significant stainable calcium while histochemical stains for phosphatase showed high concentrations of acid phosphatase. In support of this, they were able to show that the enzymatic activity of both acid and alkaline phosphatase was inhibited by incubation with stannous phosphonates, although the inhibition of the alkaline phosphatase was completely reversed in the presence of calcium or magnesium ions. Thus, the phosphatase enzymes appear to be able to act as acceptors for the phosphonates, and since the highest concentrations of these enzymes are found in areas where technetium phosphorous compounds are found to localize, it seems possible that the mechanism of localization may involve the receptor function of the enzymes. However, the general applicability of this theory has been challenged by the work of Weigmann et al.[55] and by Krishnamurthy et al.[72] who independently were unable to find any correlation between the alkaline phosphatase levels and the bone uptake of technetium pyrophosphate in cases of parathyroid disease.

IX. SUMMARY

It is generally believed that the uptake of the alkaline earths, rare earths, ionic gallium, ionic lead, and fluoride, by the bone takes place on the hydroxyapatite crystals of the bone by the process of ion exchange. In the case of fluoride, the uptake is very rapid and is believed to reflect the blood supply to bone. Iron-59 is thought to be taken up by the bone due to some interaction of the cells in the bone, principally the osteocytes; however, this radionuclide will be of little practical importance since the predominate uptake of iron is in the bone marrow. The mechanism of uptake of the labeled phosphorous compounds by the bone seems to remain a debated point, at least as far as the mechanism involved in both normal and abnormal states. However, it appears that the hydroxyapatite crystals play a significant role, probably the predominant role in normal bone, while it seems equally probable collagen plays a role in the total uptake, at least in the case of uptake involving bone disease where significantly elevated amounts of immature collagen are present. The evidence for the involvement of enzymes does not appear compelling although it may have significance in some cases, while it seems likely that many of the observed soft tissue localizations with this class of compounds may be explained on the basis of the permeability of the cell wall.

REFERENCES

1. Cohn, S. H., Lippincott, S. W., Gusmano, E. A., et al., Comparative kinetics of Ca[47] and Sr[85] in man, *Radiat. Res.*, 19, 104, 1963.
2. Neer, R., Berman, M., Fisher, L., et al., Multicompartmental analysis of calcium kinetics in normal adult males, *J. Clin. Invest.*, 46, 1364, 1967.
3. Massin, J. P., Vallee, G., and Savoie, J. C., Compartmental analysis of calcium kinetics in man: application of a four compartment model, *Metabolism*, 23, 399, 1974.
4. Rich, C., Ensinck, J., and Fellows, H., The use of continuous infusions of calcium 45 and strontium 85 to study skeletal function, *J. Clin. Endocrinol.*, 21, 611, 1961.
5. Weber, D. A., Greenberg, E. J., Dimich, A., et al., Kinetics of radionuclides used for bone studies, *J. Nucl. Med.*, 10, 8, 1969.
6. Galasko, C. S. B., False positives and negatives with [87m]Sr, *J. Nucl. Med.*, 12, 142, 1971.
7. Charkes, N. D., Some differences between bone scans made with [87m]Sr and [85]Sr, *J. Nucl. Med.*, 10, 491, 1969.
8. Spencer, R. P., Lange, R. C., and Treves, S., [131]Ba: An intermediate - lived radionuclide for bone scanning, *J. Nucl. Med.*, 11, 95, 1970.
9. Lange, R. C., Treves, S., and Spencer, R. P., [135m]Ba and [131]Ba as bone scanning agents, *J. Nucl. Med.*, 11, 340, 1970.
10. Subramanian, G., Barium 135m: preliminary evaluation of a new radionuclide for skeletal imaging, *J. Nucl. Med.*, 11, 650, 1970.
11. Hosain, F., Syed, I. B., and Wagner, H. N., Ionic [135]Ba for bone scanning, *J. Nucl. Med.*, 11, 328, 1970.
12. Spencer, R. P., Lange, R. C., and Treves, S., Use of [135m]Ba and [131]Ba as bone scanning agents, *J. Nucl. Med.*, 12, 216, 1971.
13. Hosain, F., Syed, I., Som, P., et al., Mechanism of the uptake of radioactive substances by bone, *J. Nucl. Med.*, 12, (Abstr.), 367, 1971.
14. Subramanian, G. and McAfee, J. G., A new complex of ^{99m}Tc for skeletal imaging, *Radiology*, 99, 192, 1971.
15. Durbin, P. W., Asing, C. W., Johnston, M. E., et al., The metabolism of the lanthanons in the rat. II. Time studies of the tissue deposition of intravenously administered radioisotopes, *U.S.A.E.C.*, Rep. ORINS, 12, 171, 1956.
16. O'Mara, R. E., McAfee, J. G., and Subramanian, G., Rare earth nuclides as potential agents for skeletal imaging, *J. Nucl. Med.*, 10, 49, 1969.
17. Subramanian, G., McAfee, J. G., Blair, R. J., et al., [157]Dy-HEDTA for skeletal imaging, *J. Nucl. Med.*, 12, 558, 1971.
18. Yano, Y., Van Dyke, D. C., Verdon, T. A., Jr., et al., Cyclotron-produced [157]Dy compared with ^{18}F for bone scanning using the whole-body scanner and scintillation camera, *J. Nucl. Med.*, 12, 815, 1971.
19. Hayes, R. L., Carlton, J. E., and Byrd, B. L., Bone scanning with gallium-68. A carrier effect, *J. Nucl. Med.*, 6, 605, 1965.

20. Hayes, R. L., Radioisotopes of gallium, *U.S.A.E.C.*, 603, 1966.
21. Neirinckx, R. D. and Davis, M. A., Generator for ionic gallium 68, *J. Labelled Compd. Radiopharm.*, 16, 109, 1979.
22. Dewanjee, M. K., Hnatowich, D. J., and Beh, R., New ^{68}Ga labelled skeletal-imaging agents for positron scintigraphy, *J. Nucl. Med.*, 17, 1003, 1976.
23. Rao, D. V. and Goodwin, P. N., [203]Pb: A potential radionuclide for skeletal imaging, *J. Nucl. Med.*, 14, 872, 1973.
24. Hoving, J., Versluis, A., and Woldring, M. G., Lead 203 for skeletal imaging, *J. Nucl. Med.*, 16, 170, 1975.
25. Lathrop, K. A., Gloria, I. V., Harper, P. V., et al., Manipulation of skeletal localization of Pb-203 in the mouse, *J. Nucl. Med.*, 16 (Abstr.), 544, 1975.
26. Thomson, R. A. E., Corriveau, O. J., and Rubin, P., ^{59}Fe labelling in bone, *J. Nucl. Med.*, 15, 161, 1974.
27. Dworkin, H. J. and La Fleur, P. D., Fluorine-18: Production by neutron activation and pharmacology, *U.S.A.E.C.*, 635, 1966.
28. Krishnamurthy, G. T., Thomas, P. B., Tubis, M., et al., Comparison of ^{99m}Tc polyphosphate and [18]F. 1. Kinetics, *J. Nucl. Med.*, 15, 832, 1974.
29. Ackerhalt, R. E., Blau, M., Bakshi, S., et al., A comparative study of three ^{99m}Tc labelled phosphorous compounds and ^{18}F-fluoride for skeletal imaging, *J. Nucl. Med.*, 15, 1153, 1974.
30. Charkes, N. D. and Philips, C. M., A new model of ^{18}F fluoride kinetics in humans, in *Medical Radionuclide Imaging*, Vol. 2, International Atomic Energy Agency, Vienna, 1977, 137.
31. Blau, M., Ganatra, R., and Bender, M. A., [18]F - fluoride for bone imaging, *Semin. Nucl. Med.*, 2, 31, 1972.
32. Costeas, A., Woodward, H. W., and Laughlin, J. S., Depletion of ^{18}F from the blood flowing through bone, *J. Nucl. Med.*, 11, 43, 1970.
33. King, M. A., Kilpper, R. W., and Weber, D. A., A model for the local accumulation of bone imaging radiopharmaceuticals, *J. Nucl. Med.*, 18, 1106, 1977.
34. Eckelman, W. C., Reba, R. C., Kubota, H., et al., ^{99m}Tc pyrophosphate for bone imaging, *J. Nucl. Med.*, 15, 279, 1974.
35. Subramanian, G., McAfee, J. G., Blair, R. J., et al., Technetium-99m methylene diphosphonate - a superior for skeletal imaging: comparison with other technetium complexes, *J. Nucl. Med.*, 16, 744, 1975.
36. Castronovo, F. P., Guiberteau, M. J., Berg, G., et al., Pharmacokinetics of technetium 99m diphosphonate, *J. Nucl. Med.*, 18, 809, 1977.
37. Wellman, H. M., Browne, A., Kavula, M., et al., Optimization of a new kit prepared skeletal-imaging agent ^{99m}Tc-Sn-EHDP, compared to ^{18}F, in *Radiopharmaceuticals and Labelled Compounds*, International Atomic Energy Agency, Vienna, 1973, 63.
38. Krishnamurthy, G. T., Huebotter, R. J., Walsh, C. F., et al., Kinetics of ^{99m}Tc labelled pyrophosphate and polyphosphate in man, *J. Nucl. Med.*, 16, 109, 1975.
39. Krishnamurthy, G. T., Walsh, C. F., Shoop, L. E., et al., Comparison of ^{99m}Tc-polyphosphate and ^{18}F II Imaging, *J. Nucl. Med.*, 15, 837, 1974.
40. Citrin, D. L., Bessent, R. G., McGinley, E., et al., Dynamic studies with ^{99m}Tc HEDP in normal subjects and in patients with bone tumors, *J. Nucl. Med.*, 16, 886, 1975.
41. Bowen, B. M. and Garnett, E. S., Analysis of the relationship between ^{99m}Tc-Sn-polyphosphate and ^{99m}Tc-Sn-pyrophosphate, *J. Nucl. Med.*, 15, 652, 1974.
42. Snow, R. M. and Weber, D. A., Time dependent image quality using ^{99m}Tc pyrophosphate, *J. Nucl. Med.*, 16, 879, 1975.
43. Potsaid, M. S., Guiberteau, M. J., and McKusick, K. A., Quality of bone scans compared to time between dose and scan, *J. Nucl. Med.*, 18, 787, 1977.
44. Garnett, E. S., Bowen, B. M., Coates, G., et al., An analysis of factors which influence the local accumulation of bone seeking radiopharmaceuticals, *Invest. Radiol.*, 10, 564, 1975.
45. Weber, D. A., Keyes, J. W., Benedette, W. J., et al., ^{99m}Tc pyrophosphate for diagnostic bone imaging, *Radiology*, 113, 131, 1974.
46. Piccone, J. M., Charkes, N. D., and Makler, P. T., Jr., Skeletal tracer uptake and bone blood flow in dogs, *J. Nucl. Med.*, 19, 705, 1978.
47. Lavender, J. P., Khan, R. A. A., and Hughes, S. P. F., Blood flow and tracer uptake in normal and abnormal canine bone: comparison with Sn-85 microspheres Kr-81m and Tc99m MDP, *J. Nucl. Med.*, 20, 413, 1979.
48. Van Langevelde, A., Driessen, O. M. J., Pauwels, E. K. J., et al., Aspects of [99m]technetium binding from an ethane-1-hydroxy-11-diphosphonate-^{99m}Tc complex to bone, *Eur. J. Nucl. Med.*, 2, 47, 1977.
49. Tofe, A. J. and Francis, M. D., Optimization of the ratio of stannous tin: ethane-l-hydroxy-1 1-diposphate for bone scanning with ^{99m}Tc pertechnetate, *J. Nucl. Med.*, 15, 69, 1974.

50. McRae, J., Hambright, P., Valk, P., et al., Chemistry of ^{99m}Tc tracers II. In vitro conversion of tagged HEDP and pyrophosphate (bone seekers) into gluconate (renal agent). Effects of Ca and Fe (II) on in vivo distribution, *J. Nucl. Med.*, 17, 208, 1976.

51. Byun, H. H., Rodman, S. G., and Chung, K. E., Soft tissue concentration of ^{99m}Tc phosphates associated with injections of iron dextron complex, *J. Nucl. Med.*, 17, 374, 1976.

52. Cox, P. H., ^{99}Tc^m complexes for skeletal scintigraphy. Physico-chemical factors affecting bone and bone-marrow uptake, *Br. J. Med.*, 16, 40, 1975.

53. Rosenthall, L. and Kaye, M., Technetium-99m pyrophosphate kinetics and imaging in metabolic bone disease, *J. Nucl. Med.*, 16, 33, 1975.

54. Kaye, M., Silverton, S., and Rosenthall, L., Technetium-99m pyrosphophate: Studies in vivo and in vitro, *J. Nucl. Med.*, 16, 40, 1975.

55. Wiegmann, T., Rosenthall, L., and Kaye, M., Technetium 99m pyrophosphate bone scans in hyperparathyrodism, *J. Nucl. Med.*, 18, 231, 1977.

56. Rosenthall, L. and Kaye, M., Observations on the mechanism of ^{99m}Tc labelled phosphate complex uptake in metabolic bone disease, *Semin. Nucl. Med.*, 6, 59, 1976.

57. Wiegmann, T., Kirsh, J., Rosenthall, L., et al., Relationship between bone uptake of ^{99m}Tc pyrophosphate and hydroxyproline in blood and urine, *J. Nucl. Med.*, 17, 711, 1976.

58. Williams, J. L., Capitano, M. A., and Harke, H. T., Jr., Bone scanning in neo-natal subcutaneous fat necroses, *J. Nucl. Med.*, 19, 861, 1978.

59. Gay, W. and Crowe, W. J., Splenic accumulation of ^{99m}Tc diphosphonate in a patient with sickle cell disease: case report, *J. Nucl. Med.*, 17, 108, 1976.

60. Gates, G. J., Ovarian carcinoma imaged by ^{99m}Tc pyrophosphate: case report, *J. Nucl. Med.*, 17, 29, 1976.

61. Lyons, K. P., Kuperas, J., and Green, H. W., Localization of Tc-99m-pyrophosphate in the liver due to massive liver necrosis: case report, *J. Nucl. Med.*, 18, 550, 1977.

62. Shom, R., Sain, A., Silver, L., et al., Localization of ^{99m}Tc phosphate compounds in renal tumors, *J. Nucl. Med.*, 18, 311, 1977.

63. Campeau, R. J., Gottlieb, S., and Kallos, N., Aortic aneurysm detected by ^{99m}Tc pyrophosphate imaging: case report, *J. Nucl. Med.*, 18, 272, 1977.

64. Winter, P. T., Splenic accumulation of ^{99m}Tc Diphosphonate, *J. Nucl. Med.*, 17, 850, 1976.

65. Hardy, J. G., Anderson, G. S., and Newble, G. M., Uptake of ^{99m}Tc pyrophosphate by metastatic extragenital seminoma, *J. Nucl. Med.*, 17, 1105, 1976.

66. Lowenthal, I. S., Tow, D. E., and Chang, Y. C., Accumulation of ^{99m}Tc polyphosphate in two squamous cell carcinomas of the lung: case report, *J. Nucl. Med.*, 16, 1021, 1975.

67. Lentle, B. C., Percy, J. S., Regal, W. M., et al., Localization of Tc-99m pyrophosphate in muscle after exercise, *J. Nucl. Med.*, 19, 223, 1978.

68. Dewanjee, M. K. and Kahn, P. C., Mechanism of localization of ^{99m}Tc labelled pyrophosphate and tetracycline in infarcted myocardium, *J. Nucl. Med.*, 17, 639, 1975.

69. Dewanjee, M. K., Localization of skeletal imaging ^{99m}Tc chelates in dead cells in tissue culture: concise communication, *J. Nucl. Med.*, 17, 993, 1976.

70. Jones, A. G. and Davis, M. A., The affinity of radiolabelled bone seeking compounds for injured myocardium in the rat, *Int. J. Appl. Rad. Isot.*, 28, 203, 1977.

71. Zimmer, A. M., Isitman, A. T., and Holmes, R. A., Enzymatic inhibition by diphosphonate: a proposed mechanism of tissue uptake, *J. Nucl. Med.*, 16, 352, 1975.

72. Krishnamurthy, G. T., Brickman, A. S., and Blahd, W. H., Technetium-99m-Sn-pyrophosphate pharmaco-kinetics and bone image changes in parathyroid disease, *J. Nucl. Med.*, 18, 236, 1977.

Chapter 5

MECHANISMS OF FIXATION OF BONE IMAGING RADIOPHARMACEUTICALS

James S. Arnold

TABLE OF CONTENTS

I. INTRODUCTION

At the present time, our clinical use of bone imaging agents is somewhat empirical, inasmuch as we really do not know just how they concentrate in bone and bone lesions. Perhaps we know the least about the extensively used ^{99m}Tc-labeled phosphate agents. We know these agents concentrate in forming bone, because we see them concentrate in the epiphyseal lines of growing bones and fracture callus where new bone and minerals are being actively deposited. We know that blood flow to the bone is important, because decreased blood flow produced by arterial disease or experimental intervention[1] reduces bone uptake. In addition, increased blood flow to bone produces an increased uptake of agent in bone whether due to interruption of the vasoconstricting sympathetic nerve supply, or to an experimentally produced increase of mechanical perfusion pressure.[2] Logically, we have explained the concentration of the bone imaging agents in bone lesions on the basis of increased blood flow and/or bone formation.

II. CURRENT THEORY

A. Inadequacy of Current Theories
1. Osteomalacia
The use of these ^{99m}Tc-diphosphonates in the study of metabolic bone disease has resulted in some disturbing inconsistencies regarding our currently accepted theories of mechanisms of the increase in uptake of ^{99m}Tc-diphosphonates in forming bone. Fogelman et al.[3] reported very high whole body retention levels of ^{99m}Tc-HEDP in patients suffering from osteomalacia. Osteomalacia is a form of adult rickets where tetracycline histokinetic studies have demonstrated that bone formation at individual bone surfaces is markedly reduced.[4] Here we have a paradoxical situation in which an agent that is known to concentrate in forming bone is concentrating in bone in a disease where the bone formation rate is decreased. There is a very large increased surface area of bone tissue engaged in bone formation in osteomalacia as compared to normal. The problem in osteomalacia is that the deposition of mineral at these increased bone formation sites has all but stopped, so that the total new mineral being deposited in the skeleton is usually reduced.[4] The discrepancy between bone deposition rate and ^{99m}Tc-diphosphonate bone uptake suggests that the uptake of the imaging agents is not associated with the bulk deposition of mineral, but rather involves selective adsorption at surfaces of bone engaged in bone formation, irrespective of the rate at which bone apposition is taking place at these surfaces.

2. Heterotopic Calcification
Our current concept of ^{99m}Tc-phosphate uptake is that it binds to hydroxyapatite (HA) crystal surfaces. High concentrations of these imaging agents occur in numerous situations such as myocardial infarction,[5] tissue injury,[6] tumor calcification,[7, 8] and the lung in pulmonary calcification of chronic renal dialysis patients with secondary hyperparathyroidism.[9] These conditions have in common the fact that most of the mineral present, at least initially, is in the form of amorphous calcium phosphate (ACP)[11-13] rather than HA, as well as the fact that the actual quantity of mineral may be very small and recognized only on microscopic or chemical examination. This finding suggests that ACP may play a previously unrecognized and important role in the concentration of bone imaging agents in soft tissue calcification as well as in bone formation.

3. Blood Flow and Bone Uptake
As Charkes et al. have pointed out,[15] as blood flow is increased the uptake of a

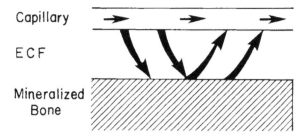

Capillary

ECF

Mineralized
Bone

FIGURE 1. Schematic diagram of the relation of capillary, intravening ECF space and the mineralized surface of bone. The bone imaging agent carried by blood (arrows indicate direction of flow) passes from the capillary by diffusion into the ECF, which bathes the bone surface. The agent that is either not bound to, or released from, the bone surface is returned to the efferent limb of the capillary.

bone imaging agent increases linearly up to some point. Further nonphysiologic increase in blood flow results in a nonlinear increase or plateau in uptake.[14] From the kinetic standpoint we can assume that the concentration of an agent in the extracellular fluid (ECF) bathing bone surfaces, and that after the first few minutes after injection, an agent is very close to being in equilibrium with the concentration of the agent in blood (see Figure 1). An increase in the rate of flow of the blood through the supplying capillary should have little effect on the rate of bone uptake. If equilibrium does not exist between the ECF and the bone surface (as in the case of areas of active bone formation, because of continuous irreversible binding), less activity will be returned to the efferent limb of the capillary than entered at the afferent limb. The controlling factor in retention is logically the extent of irreversible binding or lack of equilibrium between the ECF and bone surface activity and not blood flow. As pointed out by Charkes,[15] it is only when there is reduced blood flow that bone uptake is linearly related to blood flow.

The understanding of how bone imaging agents deposit in bone requires that we assemble all the known facts and attempt to make a reasonable model that is consistent with the available data. The model should then be directly tested and revised if necessary in order to fit the observed behavior of the system. Fortunately, there are available both old and recent data which bear on the problem, and which have as yet not been considered by workers in the field.

B. Proposed Mechanism for Tc-Phosphate Bone Uptake

The model which the author is proposing for the mechanism of uptake of Tc-phosphate bone imaging agents in bone forming lesions is based on the following reasoning. The agent concentrates specifically in lesions forming osteoid-lamellar type or woven bone.[16] Since modest increases in blood flow cannot explain the fivefold increase in uptake seen in healing fractures,[17] it must be concluded that there is an increase in the efficiency of extracting the agent from the perfusing blood. The increased extraction efficiency of agent from blood must, in turn, be due to an increase in avidity of binding of the agent[18] by chemoadsorption.[19] The proposed mechanism suggests that ACP rather than some form of HA may be the material in the "mineralization front" of osteoid and woven bone which avidly binds bone imaging agents. Compartmental analysis of kinetic studies reported in this communication indicate that the rate of release of the firmly bound agent is under biologic control having a half-life of 1.5 to 3 hr in normal and reactive remodeling bone, but is markedly increased (slower loss) in osteomalacia and woven bone lesions. Passive desorption of agent bound to HA would not be expected to be under biologic controls. If (1) ACP were the initial form of deposition of bone mineral at mineralization fronts of osteoid covered surfaces of lamellar bone formation, and (2) the imaging agents were principally bound to the ACP, the observed biologic controls of desorption of bound agent could be explained. ACP has a different chemical composition than HA, and therefore must dissolve in order to change into HA. In vitro this takes place with a half-time of about ½ to 1 hr,[20] which is comparable to the observed rate of loss of bone agents. It is suggested that a bone agent bound to the surface of ACP may be released at the time of dissolution (to form HA) and be returned to the bone ECF and partially lost back to blood. In osteomalacia the reduced rate of mineral apposition[4] would be expected to be accompanied by a reduction in the rate of conversion of ACP to HA, which would explain the reduced rate of loss of the tightly bound agents in this condition. In woven bone formation it is suggested that ACP may be incompletely converted to HA at the time of mineral deposition, thus providing a large amount of ACP available to virtually irreversibly bind bone agents, accounting for the absence of desorption observed in some metastatic bone lesions and callus of healing fractures.

C. Soft Tissue Background

From the standpoint of bone imaging, the soft tissue background is of great importance because it represents the "noise" level above which the bone "signal" must be detected. Correction of ^{99m}Tc-diphosphonate soft tissue images for contained blood activity reveals the true soft tissue background. Kinetic study of the soft tissue activity demonstrates that, like bone, the soft tissue binds the agent, only more loosely, and its loss lags behind that anticipated on the basis of simple volume dilution in the ECF. The soft tissue background curves reported for ^{18}F, on the other hand, drop quite rapidly following injection, lagging only slightly behind the fall in blood levels. Charkes[21] has shown that this compartment for ^{18}F corresponds to the ECF chloride space kinetically.

D. Basic Concepts of Uptake of Bone Imaging Agents

1. Binding Sites Essential

The uptake of any pharmaceutical or radiopharmaceutical in bone (or any organ) depends on the presence of binding sites for the agent. If it were otherwise, the blood-borne agent would simply equilibrate with the ECF and rapidly wash out again as the blood levels drop.

2. Capillary Blood to Bone

As depicted schematically in Figure 1, the agents enter the tissue level of bone in a capillary and the diffusible agent is lost, not only by diffusion, but also with the filtered fluid. Diffusion as well as bulk fluid movement brings the agent in contact with the surface cells of bone. Since the pericapillary ECF has been shown to freely pass between the intracellular gaps between cells (see Figure 2) the ECF fluid freely bathes the mineralized surfaces of bone. The agent then binds to the mineralized surfaces of bone if binding sites are present or encountered. If not bound to a bone element, the agent is swept out of the ECF space by both bulk fluid flow and back diffusion into the efferent limb of the perfusing capillary. Since the distances involved are less than 100 μm, the rates of movement by diffusion alone between the capillary and bone surface would be expected to reach equilibrium in a matter of seconds.

3. Extraction Efficiency

The efficiency of the bone for extracting the agent from the blood is the fraction the bound agent represents of the agent delivered through the afferent capillary. Thus,

FIGURE 2. SEM photograph of osteoblasts attached to a bone forming surface of a rat's tibia. Note the space separating the flat osteoblast, which, if not artifactual, could represent a direct route for the passage of agent in the ECF space to bone surfaces. (From Jones, J. L., Davis, W. L., Jones, R. G., Miller, G. W., and Mathews, J. L., *Calcif. Tissue Res.*, 24, 1, 1977. With permission.)

the extraction efficiency is related to the probability of the agent contacting a binding site as well as to the avidity of binding to the site.

4. Binding Affinity

Binding of an agent may be broadly classified as (1) loose, (2) tight, and (3) irreversible. Isoion exchange has been shown to be an example of loose binding of radioisotopes of calcium and orthophosphate ions in bone, as well as their analogues radium, strontium and barium. When exchange occurs between solution and solid ions of a different species (such as fluoride and hydroxide ions), a complication arises as to the difference in the binding energy of the new and original ion. The highly electron negative fluoride ion would be less dissociable than the less negative OH ion and much less likely to be displaced from the mineral surface by reexchange with another OH ion. On the other hand, the displacement of fluoride ion at a mineral surface would be much more likely by another fluoride ion.

Binding that involves more than one site of atomic attachment is probably more complicated.

The binding of a molecule of pyrophosphate to the surface of HA has been shown to involve displacement of one phosphate group and the incorporation of a single calcium ion from the crystal hydration shell.[22]

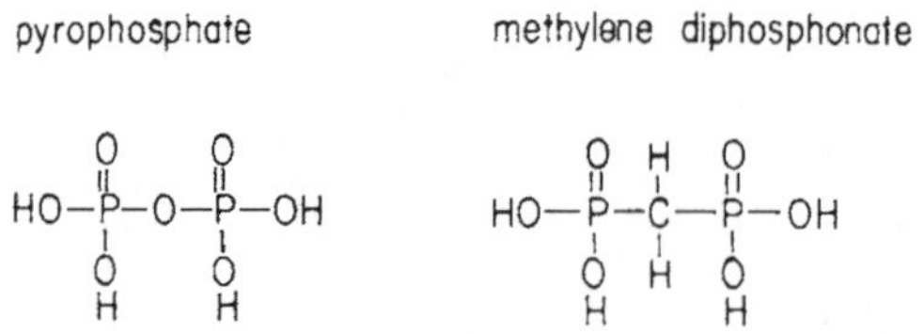

pyrophosphate   methylene diphosphonate

FIGURE 3. The structural formulas of pyrophosphate and methylene diphosphonate show they differ by the two phosphate groups being linked by an oxygen and carbon atom, respectively.

Large charged colloidal particles, when bound, tend to be irreversibly bound, perhaps because they repel smaller ions that would compete for their atomic sites of attachment. Perhaps the tetracyclines and alizarin fall into this category of bone-seeking materials.

5. Chemical Level Binding of Pyrophosphate and Analogues

The effect of pyrophosphate ions in vitro is (1) to prevent ACP from transforming into HA,[23] and (2) to prevent the growth of HA crystals from available calcium and orthophosphate ions in solution.[23]

In all calcifying biologic systems it appears that calcium phosphate mineral is initially precipitated from biologic fluids as ACP. The pyrophosphate being continually produced in the phosphate energy metabolism cycle might well be expected to poison the system at that stage. Fleisch[23] and others have suggested that the principal purpose of the production by the osteoblast of alkaline phosphatase might well be to hydrolyze any pyrophosphate in the areas of bone formation to prevent poisoning of the production of HA.

ACP has a different chemical structure than HA and must dissolve for it to be converted to HA. Since pyrophosphate prevents the dissolution of ACP, it probably does so by adsorbing to its surfaces as well as to the growth surfaces of HA crystals.

Diphosphonates differ from pyrophosphate in having a carbon linking the phosphate atoms instead of an oxygen (see Figure 3). Diphosphonates act in vitro like pyrophosphate to inhibit HA formation.[24] In addition, at high dosage levels they inhibit bone mineral deposition in vivo[24] since they are not affected by phosphatase.

6. Chemical Structure of Tc-Phosphate Agents

Clinical bone imaging is based on the well-known fact that ^{99m}Tc complexes of pyrophosphate and the diphosphonates concentrate in bone when injected. Presumably the Tc-complexed form of pyrophosphate protects it from hydrolysis by pyrophosphatase enzyme. This is suggested by the fact that ^{32}P-labeled pyrophosphate behaves metabolically like ^{32}P-orthophosphate while ^{99m}Tc-diphosphonates behave like ^{14}C-labeled diphosphonates.

The question of just how pyrophosphate and the diphosphonate complex with technetium has not been clearly answered. Hosain et al. [25] have demonstrated that carbamyl phosphate similarly complexes with technetium and concentrates in bone in a similar fashion, although at a somewhat lesser concentration.[25] Spencer suggests[26] that the site of complexing of Tc may be between the double-bonded oxygens of the carboxy and phosphonate group in the case of carbamyl phosphate by a yet undefined fashion (Figure 4).

7. Binding of ^{99m}Tc-Phosphate Imaging Agents to Bone

The detailed nature of the binding of Tc-pyrophosphate and diphosphonate to bone

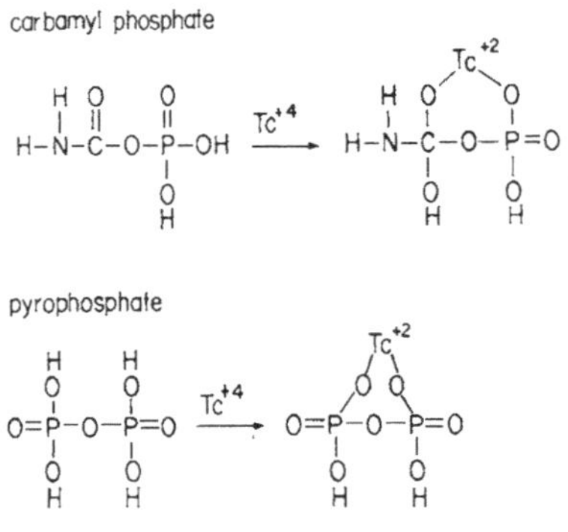

FIGURE 4. The structural formula of carbamyl phosphate and a possible configuration of a technetium complex of this monophosphate[25] like that of phosphonoacetic acid[49] concentrate in bone to a comparable extent as that of pyrophosphate shown below. The technetium apparently bonds through the double-bonded oxygen which is reduced to an alcohol. Apparently two hydroxyl groups, whether from a phosphate or carbon compound, are required for binding to bone mineral.

is, to a great extent, unknown. Most speculations assume that the ^{99m}Tc complex is bound as a unit to mineral surfaces. However, Van Langevelde et al. [27] have also suggested that the technetium atom may be released from the agent at the time of binding and secondarily bound to a phosphate group of the mineral surface. Since no such dissociation has been detected in vitro or in vivo, the latter is thought to be unlikely.

The generally held concept is that a phosphate group of the agent may substitute for a phosphate group of the mineral surface or that an oxygen of a phosphate hydroxyl group may bind to a calcium of the mineral surface. If two pyrophosphate or diphosphonate molecules are linked to a single technetium atom, the possibility exists that all four phosphorous, or any combination of the four groups, might bind to the mineral surface. The steric three-dimensional restrictions imposed by the bond angles within the Tc-complex and the position of the surface calcium atoms of HA crystals would suggest that binding would not be likely through all of the phosphorus groups. The better the fit of the phosphate hydroxyl group to the calcium of the mineral surface, the higher the binding energy. Substitution of various groups for the hydrogens on the linking carbon of methylene diphosphonate (MDP) might well be expected to alter the bonding angles between phosphorus groups, and therefore, the hydroxyl groups which bind to mineral surface calcium. Since the precision of the spacing of the hydroxyl groups probably determine binding energy, substitutions on the linkage carbon of diphosphonates should produce large changes in binding energy of the technetium complexes to bone mineral. As we will demonstrate, differences in binding avidity of ^{99m}Tc hydroxyethane diphosphonate (HEDP), methane diphosphonate (MDP), and hydroxymethane diphosphonate (HMDP) have been demonstrated in compartmental analysis of kinetic studies in humans.

ACP, being noncrystalline, probably contains a wide variety of spacing between the surface calcium and phosphate groups. In addition, these groups probably are mobile. Suitable binding sites for almost any spacing between hydroxyl groups would be expected to be present in this heterogeneous mineral system. The possibility of multiple phosphorous atoms of a complex being bound to the mineral would be potentially greater because of the mobility of the mineral groups. The binding of imaging agents would be expected to be somewhat more nonspecific. The configuration of the bonding angles in the complex would be expected not to be a critical or a limiting factor because of the permissible variability in the solid phase. Further, for a single agent there would be a wide spectrum of different possible binding arrangements, and binding energies of the agent to ACP. By contrast, binding of an agent to HA crystals would be expected to occur at two or three discrete energies corresponding to the fit of the hydroxyl groups of the agent to the calcium spacings at the surfaces corresponding to the three axes of the crystal.

8. Effect of Bone Formation

All bone imaging agents concentrate in forming bone of the epiphyseal growth plates and healing fractures. In the case of the radioactive alkaline earths (^{47}Ca ^{85}Sr, etc.), the concentration in forming bone results from the calcium of the blood being radioactive at a time when calcium of the blood is being utilized to form the mineral of the new bone. The radiotracers of "native" mineral constituents of bone thus concentrate in areas of new bone because they are swept along with the bulk deposition of mineral ions from blood which happens to contain a radioactive isotope.

9. Three Types of Bone Formation

Before discussing the mechanisms of deposition of bone image agents in forming bone, it should be pointed out that there are three different histological types of bone which probably have widely different, but as yet not clearly defined, schemes of mineral deposition. The three types of bone are (1) calcified cartilage, (2) woven bone, and (3) lamellar bone.

The first two are extremely important in (1) embryonal life where the skeleton is first formed of cartilage and then transformed into embryonal woven bone, and (2) in fracture repair at any age where the initial callus is formed as calcified cartilage and woven bone. Woven bone is produced around primary and metastatic bone tumors and in chronic osteomyelitis.[16]

The third type of bone is lamellar bone that composes virtually the entire skeleton after the first few months after birth except for the epiphyseal growth plates.

There are two obvious morphologic differences between woven and lamellar bone that are illustrated in Figure 5. In woven bone the collagen is deposited in random directions (thus its name), while in lamellar bone collagen is deposited in layers oriented in the same direction at slightly oblique angles in alternating layers (lamella). The second difference is that in woven bone the collagenous polysaccharide matrix begins to mineralize immediately after secretion by the osteoblast that forms it (see Figure 5), while in lamellar bone the collagenous matrix remains uncalcified for four to eight days and then rapidly calcifies within a single day. The unmineralized matrix is called osteoid and the zone of rapid mineralization is the "mineralization front".

The third difference between woven and lamellar bone is in their pattern of uptake of markers for bone mineralization. Tetracycline is concentrated intensely at the mineralization front in lamellar bone formation, but in woven bone formation is diffusely deposited throughout a broad area with no concentration in a discrete line or front. ^{45}Ca radioautography and tetracycline ultraviolet fluorescence microscopy studies both show a sharp line of concentration at the "mineralization front" of lamellar bone formation and a broad diffuse concentration in forming woven bone. This diffuse uptake might logically be explained by a slow progressive deposition of mineral in woven bone. This is completely incorrect, since we know from microradiographic studies that woven bone is almost fully mineralized right up to the edge of matrix deposi-

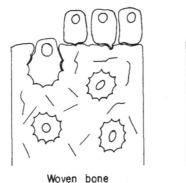 

Woven bone   Lamellar bone

FIGURE 5.   A schematic sketch showing that lamellar bone (on the right) contains an osteoid layer depicted as a hatched area beneath the surface osteoblasts. Woven bone by contrast contains no osteoid seam, the matrix being mineralized as it is formed. The collagen of lamellar bone is arranged in well-oriented longitudinal bundles, while that of woven bone is arranged in short randomly oriented bundles, as its name implies.

tion. If we accept the fact that the deposition of ^{45}Ca and tetracycline from blood represents bone mineral deposition, then the diffuse deposition of ^{45}Ca means mineral is being deposited diffusely and continuously in woven bone. Constant high mineral content with continuous mineral deposition suggests that mineral must be leaching out at almost the same rate as it is being deposited. This rapid turnover of the mineral in woven bone probably represents an accelerated form of recrystallization in the recently deposited mineral. Thus, it would appear that the fundamental difference between woven and lamellar bone deposition is that woven bone

1.   Mineralizes immediately as matrix is formed
2.   Recrystallizes rapidly over a period of weeks following its deposition

10. Initial Deposition of Mineral as ACP

When the pH of solutions of physiologic concentrations of calcium and phosphate ions is increased, a fluffy white precipitate forms. If left overnight at room temperature, the white precipitate is replaced by colorless long narrow needle crystals. It is well documented that the initial mineral precipitate is ACP which is subsequently hydrolyzed to HA

$$Ca_9(PO_4)_6 \xrightarrow{H_2O} Ca_{10}(PO_4)_6(OH)_2$$

In vitro under physiological concentrations of calcium and phosphate ions, pH, and ion strength the conversion of ACP to HA proceeds with a half-time of about one hour.[20]

The same process has been demonstrated to occur in mineral deposition in injured cells[11,28] in electron microscopy studies. ACP is initially deposited in mitochondria and is subsequently converted to HA: The same sequence has been observed in depositing bone in electron microscopic studies where freeze dry fixation was used.[29,30]

Numerous reports of electron microscopic studies of bone formation have appeared which have reported only HA crystal deposition. These studies generally utilized buffering at higher than physiologic pHs or high phosphate concentrations, either of which will force the conversion of ACP to HA. Thus, it appears increasingly evident that in all biologic systems Ca PO_4 mineral is initially deposited as ACP which is then slowly converted to HA.[31] The conversion of ACP to HA is strongly inhibited by the presence of pyrophosphate which exists at a concentration of about 3.5×10^{-6} *mM* in most biologic fluids.[32] In vitro, the inhibition of ACP conversion to HA by physiologic concentration of pyrophosphate is thought to be due to both inhibition of ACP dissolution as well as to poisoning of the growth surfaces of HA crystals.[23]

11. ACP Deposition in Bone Formation

In lamellar bone formation the osteoblasts deposit osteoid which is uncalcified bone matrix composed of about 80% collagen fibers and 20% mucoprotein on a dry-weight basis. Microradiographic studies as well as electron probe studies have demonstrated that osteoid is not actually free of calcium phosphate mineral, but contains about 5% of the concentration that is attained when mineralization is complete. Electron microscopic studies using "freeze dry" fixation have revealed numerous evenly spaced round deposits of mineral measuring about 25 Å in diameter.[29] The presence of membrane vesicles containing mineral have been reported in osteoid[33] comparable to those extensively documented to be present in calcifying cartilage.

After about 4 to 8 days following deposition, the osteoid matrix suddenly calcifies. This front of calcification, located 5 to 10 μm beneath the plane of osteoblasts is called the "mineralization front". The sequence of events that lead to the massive precipitation of mineral has been the subject of active investigation for the past 25 years and is still controversial. Just before, or at the time of, the mineral deposition, there is a loss of about 50% of the chondroitin sulfate polysaccharide.[34] Simultaneously, acid groups are exposed in the organic matrix, which has been recognized by histologists as the basophilic line separating osteoid from mineralized bone. Electron microscopy studies of freeze-dry fixed tissue reveal closely packed grains of mineral which fuse with extremely dense mineral containing both grains and needle crystals. Is this process like that in vitro where the initial precipitation of mineral is as ACP followed by conversion to HA? Because of the extreme density of the mineral to electron microscopic examination, no realistic assessment of the relative amounts of mineral deposited as ACP and HA have been possible, and examination of the structure of the mineral has been restricted to the junctional edges of the mineralization front and artifactual breaks in sections.

12. ACP to HA Conversion

As mentioned above, metastatic extraosseous mineralization is initially deposited as ACP, and usually does not progress beyond that stage, presumably because the conversion of ACP to HA is inhibited by the presence of pyrophosphate and ATP.

In areas of lamellar bone formation, the mineral is only transiently present as ACP at the "mineralization front" and is rapidly hydrolyzed and converted to HA. As pointed out by Fleisch,[23] the purpose of the large quantity of phosphatase in osteoblasts is probably to destroy all pyrophosphate in the area of mineralization (mineralization front) in order to facilitate the conversion of ACP to HA (Figure 6).

E. Radio-Alkaline Earth and Transuranic Elements Deposition in Bone (Volume and Surface Seekers)

Much information about the deposition of bone seeking radionuclides was obtained between 1950 and the early 1960s as part of an extensive study of the radiotoxicity of long-lived bone seeking radionuclides in order to be able to evaluate the risk to health imposed by accidental exposure. The deposition patterns of radioactive calcium, phosphate, radium, plutonium, thorium, yttrium, and strontium were all studied both radiochemically and radioautographically over the life cycle of several species of animals. The author played an active role in this work, and as such is aware of the basic findings

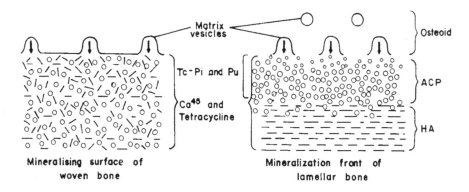

FIGURE 6. A schematic diagram proposed arrangement of the mineralization front of lamellar bone (on the right) contrasted with the mineralizing surface of woven bone (on the left). Above the large circles represent matrix vesicles which may play a role in seeding the matrix with calcium phosphate as the rupture indicated by the small arrows. The small circles represent ACP, which at a lower level, is transformed to HA crystals whose long axes are oriented with the collagen fibers. It is proposed that the Tc-phosphate imaging agents are deposited on the immature mineral (ACP) and are released at the time of its conversion to HA. This conversion may not occur in woven bone, which would account for the greater retention of imaging agents at 24 hr.

which are directly applicable to our current dilemma in understanding the mechanism of the deposition of bone imaging agents.

Radioautographic studies demonstrated that the bone seeking elements were divided rather sharply into two classes. One is the "surface seekers" that plate out on bone surfaces in contact with the perfusing blood and fluids, but do not penetrate to the interior of the bone. The other class is the "volume seekers" that penetrate bone and follow the metabolic path of calcium metabolism.

1. Surface Seekers

In radioautographic studies of plutonium,[35] thorium, yttrium, and uranium,[36] all were found to be exclusively deposited on the surfaces of bone tissue which are in contact with the perfusing blood and its extravascular fluids. These surfaces were those of the trabeculae, the endosteal surface of the medullary cavity, the periosteal surface, and the inner surfaces of the Haversion canals of cortical bone. No element penetrated beneath the bone surface into the mineralized bone. These materials appear to be deposited at the first mineralized surface contacted as the material moves from blood to bone. The "surface seekers" as should be noted are all foreign to bone and not normal constituents, being large hydrated multivalent ions. It is as if some unknown barrier filtered them out from the salt solution which percolates into the interior of mineralized bone[37] and deposited them at the surface.

It should be noted that the only high quality radioautographic study of diphosphonate (tritiated HEDP) revealed that it was deposited at the bone surface and did not penetrate into the mineralized bone.[38] On the basis of this study at 24 hr following injection, it is concluded that diphosphonates are probably "surface seekers". Using ^{96}Tc-pyrophosphate Guillemart et al.,[39] have shown that this complex is also deposited on trabecular surfaces at bone formation sites, but does not penetrate into the intermineralized bone and is therefore a "surface seeker".

2. Volume Seekers

Radioautographs of radio-calcium,[40] radium, strontium, and phosphate,[41] revealed that they all penetrated deeply and almost uniformly throughout the bone tissue, and appeared in these deep locations as early as 5 to 10 min after injection. It was as if they were swept into the deepest recesses of densely mineralized bone along with the water and nutrients. Once deep in dense, old bone, most of the activity was retained, being lost only very slowly with a half-time of five or more years. Sodium isotopes, on the other hand, were seen to be swept into the dense bone, but then washed back out again within a few days as the blood levels dropped.

The process that irreversibly binds the alkaline earth isotopes in densely mineralized old bone was called the "paradox of old bone"[40] by the author, referring to the fact that there was evidence of continued acquisition of new mineral in bone which was not increasing in mineral content. The author postulated that the process was one of slow continual recrystallization of bone mineral. Several years later Marshall et al.,[42] rediscovered the process and even entitled it "the paradox of diffuse activity" and interpreted the process as due to intracrystalline diffusion.[42] The fact that Marshall's Ph.D. thesis was on diffusion problems may well have played a role in his interpretation.

Roland et al.[43] pointed out that the volume seekers, calcium and radium, are to a great extent deposited on bone surfaces, much as the surface seekers, immediately after injection. These surface deposits wash away over a period of 12 to 24 hr as the diffuse deposition, and hot spots in forming bone increase. This suggests that the great bulk of the injected dose that rapidly enters bone within the first hour after injection is taken up by ion exchange at the first mineralized surface contacted. As the blood levels decrease, it exchanges back off of these surfaces and is redistributed in the slower processes of deposition of new mineral, both in the area of bone formation and recrystallization of old mineral deposits.

The concept of the surface mineral being the principal source of exchangeable calcium is indirectly supported by the finding that in vitro about 5% of the calcium in powdered micron-size bone is exchangeable, while in vivo the maximum fraction of all bone calcium which is exchangeable is about 0.2%. Thus, it would appear that only 1/25 or less of the bone volume is engaged in exchange with blood calcium. This would correspond to the volume of bone contained in the first 5 μm beneath all bone surfaces in contact with blood vessels.

It should be noted that the volume seekers are all major chemical constituents of bone mineral or their chemical analogs in the same family of elements. This raises the question as to whether bone has a chemical barrier that excludes multivalent ions that might interfere with mineral metabolism, and thus restricts them to surface deposition.

3. Bone Formation and "Surface Seekers"

Bone-seeking elements, without exception, have been found to concentrate in forming bone of growing animals and healing fractures. The surface seekers, such as plutonium, concentrate in forming bone to a less extent than alkaline earth volume seekers. Uranium ions, unlike the other surface-seeking elements, are intensely concentrated in forming bone.[36] The alkaline earth elements, such as calcium, fit into the crystal lattice of the forming bone mineral and become actual building material. The surface seekers probably adsorb to one or more of the mineral phases and are trapped by the physical packing of mineral crystals as mineralization is completed. Large organic molecules, such as alizarin red dye and fluorescent tetracycline, have been extensively used as agents to permanently label bone which was formed at the time of administration to animals and man. Presumably, they are fixed in bone by the process of being adsorbed to phosphate groups of crystal surfaces, and subsequently locally trapped as the water environment is replaced by ever-increasing numbers of new forming crystals.

The diphosphonates[38] and ^{96}Tc-pyrophosphates[39] have been shown to be "surface seekers", and to concentrate on bone-forming surfaces. It is very probable that the

Tc-diphosphonates behave similarly. Tritiated HEDP has been further shown to concentrate at the "mineralizing front" beneath osteoid.[38] Guillemart's [96]Tc-pyrophosphate radioautograms were not sufficiently well resolved to be able to distinguish between localization in the osteoid and the underlying mineralization front.[39] Unfortunately, they interpreted their results as being consistent with deposition in the collagen of osteoid. The basis for this suspicion is the questionable results of Kaye et al.,[44] suggesting collagen deposition of Tc-pyrophosphate in vitro. It is a well-documented fact that collagen chemically separated from its mucoprotein associate, as in purified rat tail collagen, will readily calcify in vitro if the solutions contain calcium and phosphate at biologic concentrations and pH.[45] Thus Kaye's experiments with purified collagen were probably complicated with calcification of the collagen. Without precise knowledge of whether Kaye's preparations were completely free of ACP and HA, the collagen concentration of Tc-phosphates should not be accepted without reservation. All of the claims of soft tissue and bone uptake of Tc-phosphates in collagen are subject to the uncertainty of whether traces of ACP were present.

Another interesting aspect of the "surface seekers" is their concentration in osteoclasts. The author and Jee[35] first showed that [239]Pu concentrated in osteoclasts at 48 hr, presumably obtaining it from the bone being resorbed. Guillemart et al.,[39,46] demonstrated that [96]Tc-pyrophosphate is also concentrated in osteoclasts at 48 hr, but not at 3 hr, just as was the case with plutonium. This finding further suggests the parallel of metabolism of the Tc-diphosphonates and the heavy metal "surface seekers". The question of how the [96]Tc-pyrophosphate is bound in the osteoclasts arises. Is the Tc free from the pyrophosphate and behaving like a 4+ valence heavy metal? Or is the Tc bound to pyrophosphate and concentrating by binding to the abundant phosphatase enzyme in the osteoclasts cytoplasm? These questions are fundamental to building a rational model of Tc-phosphate metabolism.

F. Kinetic Studies with ^{99m}Tc-Diphosphonates

Kinetic studies with ^{99m}Tc-diphosphonates in bone were undertaken in our laboratory in an attempt to differentiate between different types of bone lesions. In particular, we were hoping that the bone uptake curves would be demonstrably different for metastatic tumor lesions and benign reactive lesions (e.g., arthritis and unrecognized trauma). Our second goal was to be able to evaluate physiologic parameters involved in the concentration of bone imaging agents in bone and bone lesions, as well as to be able to objectively compare one imaging agent with another.

Most of bone imaging agents concentrate in bone during the first 60 min after injection. During this period, the soft tissue background is very high and the agent in bone can only barely be differentiated from the background. In order to be able to quantitatively evaluate the bone activity during this critical early period, we developed a technique for subtraction of both the blood and soft tissue background from digital images of the bone.

1. Method of Kinetic Study

The technique we developed for data collection was to place a large field of view (LFOV) Anger camera over the skeletal area of the patient to be studied (usually the posterior pelvis). A dose of 500 μCi of ^{99m}Tc-labeled human serum albumin was injected, and an image made between 5 and 10 min after injection and stored in a 96 × 96 matrix on the magnetic disc of a minicomputer. On completion of the blood pool image, the patient was given a dose of 10 mCi of ^{99m}Tc-labeled bone imaging agent intravenously, and sequential 15-sec digital images recorded for 50 min and stored on magnetic disc.

The patient was reimaged at 2,4,8, and 24 hr following injection, in as close to the same position under the camera as possible. Following each imaging procedure, a standard was imaged at the surface of the collimator and recorded. This standard consisted of an 8 × 8 × 4 inch box containing about 1/10 of the administered dose of bone imaging agent and filled with water. This count was used to correct the camera images for detection sensitivity, and for physical decay of the ^{99m}Tc. No flood field correction was made because it was determined that this procedure introduced up to 15% error in the count values in small areas of interest.

Venous blood samples were collected at the time of blood pool imaging, and at 5-min intervals throughout the 50-min kinetic study of the bone imaging agent. In addition, venous blood samples were collected at the times of delayed imaging. The activity in the blood samples was calculated in each case as percent of the administered dose per mℓ of blood using an appropriate fraction of the injection dose counted in a well counter.

Rectangular areas of interest (AOIs) were placed over the various bone locations to be measured, as well as the soft tissue background (see Figure 7). The counts within the AOIs were measured in each clinical image and corrected for camera sensitivity and physical decay of the ^{99m}Tc.

2. Background Subtraction

The background activity is composed of agent in the blood as well as the soft tissue in front of and behind the bone. The correction procedure used was to first subtract the activity from the AOIs contributed by the ^{99m}Tc albumin, then the agent activity in the blood, and lastly, the agent activity in the extravascular soft tissue compartment. This left the agent activity derived from bone in the AOI.

A curve of the blood activity was constructed from the measured blood values by a least squares fit of the sum of three exponential functions to the data. At the time of injection, 100% of the administered activity was assumed to be present in the blood.

Blood background in the individual AOIs over bone and soft tissue was calculated for the individual bone images utilizing (1) the blood activity of the bone imaging agent at the time of imaging, and (2) the ratio of AOI ^{99m}Tc-albumin counts to venous blood activity counts as determined from the initial blood pool image and blood sample. The product of (1) and (2) yields agent counts in the blood contained within the various AOIs of the individual bone images.

The soft tissue AOI when corrected for the contained blood yields the extravascular soft tissue background. It was invariably found that the extravascular soft tissue background rose quite rapidly during the first 2 min after injection. When the extravascular soft tissue background was subtracted on a per pixel basis from bone, it yielded negative values for the bone AOIs during the first 5 min, indicating that there was less extravascular soft tissue contribution over bone than elsewhere. It was arbitrarily decided to subtract 80% of the extravascular soft tissue background from bone AOIs to yield the net bone AOI values.

The effect of making the blood and extravascular soft tissue corrections on bone AOI curves is illustrated in Figure 8. The relative magnitude of the contribution by blood, extravascular soft tissue background, and net bone activity is illustrated for the normal femoral head in a study using ^{99m}Tc-HMDP over a 24-hr period. The background correction procedure has converted the gross bone count curve from a rapidly rising and slowly falling curve to a slowly rising and plateauing curve. The importance of background correction is immediately apparent. No meaningful idea of what is taking place in the bone during the first hour is possible without background correction.

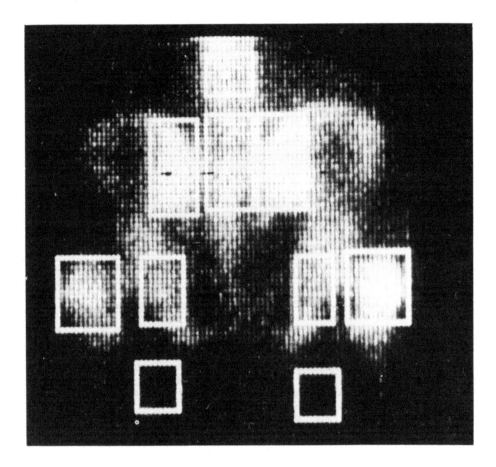

FIGURE 7. Photograph of display of a digital bone image of the posterior pelvis of a patient at 24 hr following a dose of 10 mCi of ^{99m}Tc-MDP. Rectangular areas of interest (AOIs) are seen from above, downward over the lumbar vertebrae, sacroiliac joints, sacrum, femoral heads, ischia, and background of the medial thighs.

3. HEDP, MDP, and HMDP Uptake Curves (Normal Bone)

When net bone activity curves are plotted over a 24-hr period as shown in Figure 9, they are seen to rise during the first two hours and then plateau between 2 and 8 hr. At 24 hr, the counting rate of normal bone shows a drop to almost half of its 4-hr value in the case of HEDP. The curve of HEDP tends to rise slightly less rapidly than that of MDP, and plateaus at a lower level. The decrease in bone activity between 8 and 24 hr occurs to a comparable extent with both HEDP and MDP. The uptake curves of HMDP are comparable to those of MDP except that less loss occurs between 8 and 24 hr. This suggests that the tightly bound component has a higher binding energy in the case of HMDP than MDP.

From the normal uptake curves, it is apparent that the process of fixation is reversible to a great extent. Differences in uptake between the agents are due, not only to differences in the initial rate of uptake, but also in the rates at which the accumulated activity is lost.

4. HEDP, MDP, HMDP Uptake Curves (Bone Lesions)

In order to compare the uptake of HEDP and MDP in bone lesions, we performed kinetic studies on four patients with both agents. One patient had metastatic carcinoma of the prostate in a sacroiliac joint, a second had acute osteomyelitis of the femoral

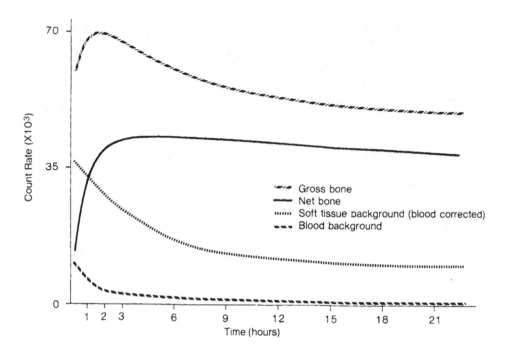

FIGURE 8.   The gross bone count rate (counts per minute) for an AOI over a sacroiliac joint of normal patient receiving 10 mCi of ^{99m}Tc-HMDP. The magnitude of the blood background (lowest curve) and the soft tissue background (next higher curve) are depicted. The subtraction of the two backgrounds from the gross bone curve results in solid net bone uptake curve.

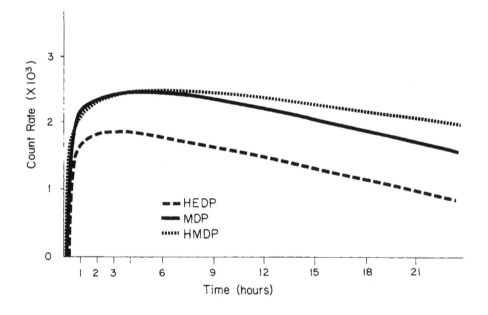

FIGURE 9.   The net bone uptake curves for the femoral heads of normal patients are compared for ^{99m}TC-HMDP, MDP and HEDP over a 24 hr period. During the first three hr HMDP and MDP uptake is comparable. The HEDP curve is lower throughout. Between 8 and 24 hr the MDP and HEDP curve fall to almost half their maximal values while the HMDP curve during the same period falls only slightly.

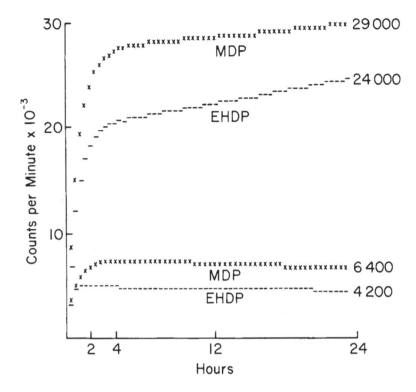

FIGURE 10. The net bone uptake curves for the sacroiliac joints of a patient given 10 mCi of ⁹⁹ᵐTC-MDP (Xs) and repeated with the same dose of HEDP 3 days later (dashes). The two higher curves are of the sacroiliac joint involved with metastatic carcinoma of the prostate. The two lower curves are of the opposite normal sacroiliac joint. The curves of MDP are higher than those of HEDP in both the lesion and control bone. At 2 hr each agent is 45 times as concentrated in the lesion as in the control normal bone. The uptake curves of the lesion with both agents continue to rise between 4 and 24 hr, while those of normal bone progressively fall during this same period suggesting a difference in metabolic handling of the agents in woven bone of lesions and normal bone.

head, and a third had a sacroiliac strain because of an old lower leg amputation on the opposite side.

The uptake curves for lesion due to metastatic Ca of the prostate as well as the control opposite sacroiliac joint are shown in Figure 10. It is apparent that the curves of the bone lesion rise faster and to a higher level than the control values with both agents. The ⁹⁹ᵐTc-MDP curves are higher than those of HEDP in all cases. The striking difference between the tumor lesion and control bone with both agents is the difference in the behavior of the curves between 8 and 24 hr. It is apparent that the curves of the bone lesion with both agents continue to rise, while those of the control normal bone fall during this same period. This suggests that a difference in the metabolic handling of the agent is occurring in the metastatic lesion as compared to the control bone, which is comparable in both the HEDP and MDP studies. Though not depicted, the uptake curves of ⁹⁹ᵐTc-HMDP in tumor lesions is comparable to that of MDP; however, as depicted in Figures 8 and 9, the curves of normal bone tend to plateau between 8 and 24 hr instead of falling as those of MDP and HEDP.

G. Compartmental Analysis

1. General Aspects

In compartmental analysis, it is assumed that a biological system consists of several subsystems, called compartments, each of which is homogeneous and well-mixed. The compartments exchange material with each other, the rate of transfer of material from one compartment to another being proportional to the amount of material in the compartment. The constants of proportionality that determine the rate of transfer between compartments are termed transfer (or rate) constants. A set of differential equations represents the compartmental model. With the aid of a computer, transfer constants are determined that best fit the model to the experimental data.

An important piece of information obtainable from curve fitting procedures is that if the best fitted theoretical curve of the model does not have the same shape as that of the actual data, the model is probably wrong. The model is then changed by rearranging, adding or subtracting compartments in a pattern that is consistent with the known or perceived physiology of the system. When a model is found that produces a good fit of the theoretical to the actual bone curve, it indicates that the model is consistent with the data. It does not indicate that it is the only model which will fit the data, or that it is the correct model.

The actual bone curve from an area of interest is composed of the sum of the contributions from many small areas of bone where different processes are occurring at different rates. Two bone compartments might divide the population of small areas into those with processes equilibrating with blood rapidly, and those equilibrating slowly. The model assumes that we have represented the different processes with separate compartments, and that the sum of the contributions from the contained bone engaged in a single process can be reflected by a single compartment. The problem arises that, as we add more compartments, multiple solutions will fit the actual data equally well, and the capacity of the analysis procedure to determine whether one model is better than another is lost. Thus, in general, the best model is that which has the least number of compartments and still produces good fits to actual experimental data.

It is important to keep in mind that the model should fit data over a period of time when the physiologic changes are thought to be operative. A single bone compartment with a single transfer constant from blood to bone (Figure 11a) will adequately fit bone uptake curves during the first half hour. By adding a return component (Figure 11b), the data can be nicely fitted for 60 min, but will not fit the data at four hours. In order to fit the data for 4 hr, a second compartment is required as illustrated in Figure 11c. Thus, the longer the period of observation, the greater the number of physiologic processes which must be included as separate compartments in the model. In order for the computer program to be able to distinguish between these added compartments, an increase in both the number and accuracy of the late data points is required.

Because of the 6-hr half-life of ^{99m}Tc, we were only able to collect meaningful bone kinetic data out to 24 hr. For this reason, a model was developed which would fit data to this point. This model is depicted in Figure 12a where two parallel compartments are depicted as reversibly communicating with blood which represents loosely bound and tightly bound activity in bone.

Probably a better model for depicting data over a 24-hr period would be that depicted in Figure 12b, where an additional irreversibly bound compartment has been added. To date, we have made our analyses using the model in Figure 12a because of the practical problem of not having enough measured data points between 8 hr and 24 hr (when the patient is asleep) to be able to distinguish between the two models. The problem is to determine whether the bone activity has plateaued at 24 hr, or is still falling, and, if so, at what rate. Without data points between 8 and 24 hr, or at 36 hr, this cannot be determined.

Charkes et al.[21] have proposed a five compartmental model for describing ^{18}F me-

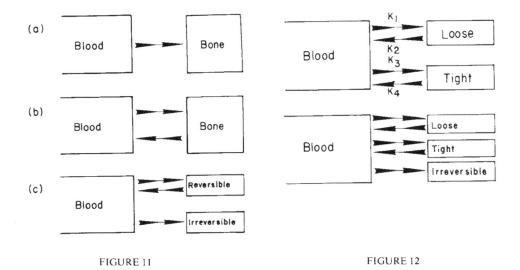

FIGURE 11   FIGURE 12

FIGURE 11.   Three preliminary compartmental models tested for describing the movement of bone imaging agents between blood and bone. All three models can be adequately fitted to experimental blood and bone data for limited periods of time following injection; (a) fits data for ½ hr (b) for 1 hr, and (c) for 4 hr. All fail to fit data over a 24 hr period.

FIGURE 12.   (a) is the compartmental model used in analyzing the kinetic blood and bone data reported in the present studies. It consists of two bone compartments, each of which reversibly communicates with blood. The transfer constants K_3 and K_4 describe uptake and release of imaging agent by tight binding sites in bone. The transfer constant K_1 and K_2 describe the uptake and release of loose or weak binding sites of bone. Functionally, the curve of the tight binding compartment is determined by the blood curve and the bone curve after 4 hr. The loose binding compartment is required to account for the additional agent initial concentrated and rapidly lost from bone. (b) is a model much like (a) with an additional compartment consisting of irreversible binding sites. Additional delayed bone measurements are required to be able to determine which model would best fit the true bone activity curves.

tabolism. The part of their model which is germane to our study is the interaction between blood and bone. They represent bone as two reversible compartments connected in series. The first of these is ascribed to bone ECF space which, as depicted in Figure 1, is located between blood and bone. In Charkes usage this includes the ECF space of the bone marrow. The question arises as to why we do not have a bone ECF space in our model. We did not include it because such a space would be equilibrated within a few minutes, or less, with capillary blood, and as such would thereafter be a simple extension of the blood compartment and be kinetically invisible. Just because a physiologic space exists is not a justification for insisting on its identification with a kinetic compartment, particularly if the activity in such a compartment could not possibly be separated from blood and soft tissue background.

2. Results of Compartmental Analysis

The net bone activity curves are seen to rise progressively during the first 4 hr following injection in Figure 9. During the first 5 min, the rise in bone activity is quite steep. The activity accumulating in the bone must be derived from the blood perfusing it. If we fit the initial rise of the bone activity during the first 5 min with the integral of the blood curve during this time, we can approximate the amount of activity that has to be delivered by blood to the bone. In Figure 13, the integral of the activity delivered to the femoral head in a ^{99m}Tc HEDP study is compared to the observed bone uptake curve over a 4-hr period. The difference between the rather high curve of the integral

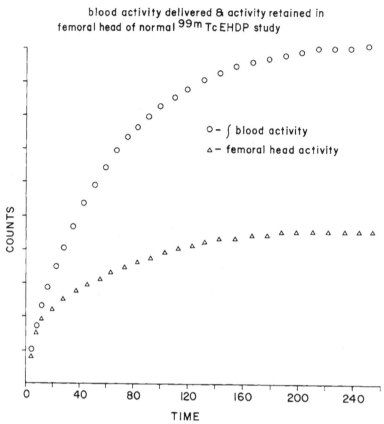

FIGURE 13.   The net uptake of ^{99m}Tc-HEDP in a normal femoral head is plotted as triangles in the lower curve. During the first 5 min when the bone content is rapidly increasing the bone content can be considered as being proportioned to the integral of the blood activity curve. The curve formed of circles is the integral of the measured blood curve fitted to the first 5 min of the bone uptake data. The integral of the blood activity would represent the total activity transferred from blood to bone, while the bone uptake curve is small part retained by bone. The difference is the total activity which was released back to blood from loose binding sites. (From Arnold, J. S., in *Principles of Radiopharmacology,* Vol. 3, Colombetti, L. G., CRC Press, Boca Raton, Fla., 1979, 215.)

of activity extracted from blood, and the lower bone uptake curve, represents the integral of the activity which was lost back to blood during this same period. The blood and bone count data for the first 4 hr of study were analyzed with the model shown in Figure 11c, consisting of an exchangeable loosely bound, as well as irreversibly bound, compartments. The results of fitting the model to the data is graphically depicted in Figure 14. The fit of the theoretical curve and the actual bone data are excellent. The curve of the activity in the loosely bound (exchangeable) and tightly bound (nonexchangeable) compartments is plotted to illustrate their relative contributions as a function of time following injection. It is apparent that the loosely bound component rises rapidly and falls rapidly, becoming negligible at 2 hr. This indicates that the loosely bound activity is of no consequence in clinical bone imaging because its activity has been lost by the time such images would normally be undertaken.

a. Comparison of HEDP, MDP, AND HMDP

The kinetic comparison of ^{99m}Tc-HEDP and MDP was undertaken in five patients

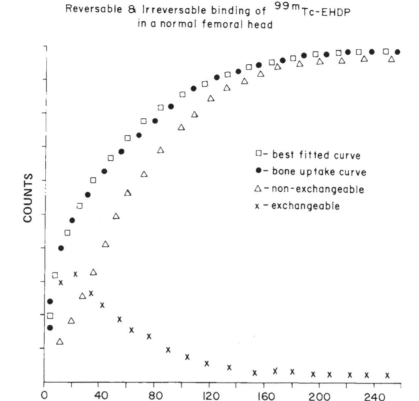

Reversable & Irreversable binding of ^{99m}Tc-EHDP in a normal femoral head

□ – best fitted curve
● – bone uptake curve
△ – non-exchangeable
x – exchangeable

COUNTS

TIME

FIGURE 14. The net uptake curve of a normal femoral head in a ^{99m}Tc-HEDP study. There is good fit to the bone of curve using the compartmental model depicted in Figure 11c. The curves of the activity calculated to be contained in the loosely bound (exchangeable) and tightly bound (non-exchangable) compartments plotted over the 4-hr period of study. The activity in the loosely bound compartment rises and falls rapidly such that it contributes little to bone activity after 2 hr follwoing injection when most clinical images are made. (From Arnold, J. S., in *Principles of Radiopharmacology*, Vol. 3, Colombetti, L. G., Ed., CRC Press, Boca Raton, Fla., 1979, 216.)

who were studied with both agents in separate studies conducted within seven days of each other. In this comparison the transfer constants between blood and the loosely and tightly bound compartments were calculated using the model illustrated in Figure 12a. In the case of HMDP, ten single studies on patients were available for comparison, but unfortunately no multiple studies with an additional agent. In order to compare the agents, only normal femoral head and sacroiliac joint parameters were utilized. Inclusion of bone lesions would have further spread the interpatient variation. The results of this study are presented in Tables 1 and 2.

Of the ten patients studied with HMDP, two were studied in the thoracic area of the spine, and two had extensive metastatic disease of the pelvis, reducing the number suitable for comparative study of normal bone to six cases.

It should be noted that the transfer constants K_1 and K_3 are in units of blood clearance. The values listed are in units of mℓ of blood cleared of agent per minute per cubic centimeter of bone. These units are comparable to those used in renal and hepatic clearance studies. The units in which K_2 and K_4 are expressed are reciprocal minutes or the fraction of the agent which is transferred from the bone binding compartment

Table 1

AVERAGE TRANSFER CONSTANTS FOR
SACROILIAC JOINTS IN PAIRED HEDP(E)
AND MDPP(M) AS WELL AS UNPAIRED
HMDP(H) STUDIES

| | Average | σ | Significance of difference | | |
			E—M	M—H	E—H
K_1^d E	0.218	0.116	—	a	—
M	0.232	0.095			
H	0.132	0.058			
K_2 E	0.106	0.048	—	c	a
M	0.106	0.012			
H	0.042	0.013			

| | Average | σ | Significance of difference | | |
			E—M	M—H	E—H
K_3^d E	0.085	0.039	c^e	—	c
M	0.196	0.098			
H	0.180	0.030			
K_4 E	0.0052	0.0022	—	b	a
M	0.0043	0.0011			
H	0.0020	0.0012			

Note: Five patients and eight sacroiliac joints for HEDP and
MDP studies. Six patients and twelve sacroiliac joints for
HMDP studies.

a   Significant difference between means at level of $p < 0.05$.
b   Significant difference between means at level of $p < 0.005$.
c   Significant difference between means at level of $p < 0.001$.
d   In ml of blood cleared per minute per cubic centimeter of
    bone.
e   Fraction present transferred per minute.

back to blood per minute. K_1 is the blood cleared by the loose binding sites per minute.
The chief determinant of this constant is the rate of the initial rise of the bone curve
and the early values of the blood curve. The high background and low bone activity
during the first 10 min of study contributed to the error of K_1. K_3 is well measured
and depends, as does K_4, on the accuracy of the delayed bone imaging AOI counts,
camera sensitivity, etc. In the paired studies where both ^{99m}Tc-HEDP and MDP measurements were made in the same patients, the correlation of the four transfer constants in the seven bone AOIs was extremely good between the two studies. This suggests that if studies were to be repeatedly performed in an individual patient with the
same agent, the derived transfer constants would be expected to fall within 20% for
the different studies. The transfer coefficients for the sacroiliac joints (SI) and femoral
heads (FH) are quite similar. This K_1 and K_3 being larger for the SI is probably due to
greater blood flow and remodeling of trabecular bone. Slightly less spread in the SI
data is probably the result of somewhat higher counting rates, and thus better counting
statistics than that of the FHs. Since the ^{99m}Tc-HEDP (E) and MDP(M) studies were

Table 2
AVERAGE TRANSFER CONSTANTS FOR NORMAL FEMORAL HEADS IN PAIRED STUDIES WITH HEDP(E) AND MDP(M) AS WELL AS UNPAIRED HMDP(H) STUDIES

| | Average | σ | Significance of difference | | |
			E—M	M—H	E—H
K_1 E	0.093	0.047	—	a	—
M	0.132	0.066			
H	0.0619	0.0271			
K_2 E	0.092	0.017	—	a	b
M	0.076	0.017			
H	0.047	0.017			

| | Average | σ | Significance of difference | | |
			E—M	M—H	E—H
K_3 E	0.038	0.016	c	—	c
M	0.092	0.027			
H	0.081	0.015			
K_4 E	0.0057	0.0021	—	b	a
M	0.0054	0.0014			
H	0.0027	0.0009			

[a] Significant difference between means at level of $p < 0.05$.
[b] Significant difference between means at level of $p < 0.005$.
[c] Significant difference between means at level of $p < 0.001$.
[d] In ml of blood cleared per minute per cubic centimeter of bone.
[e] Fraction present transferred per minute.

paired, each having been made in the five individual patients, the statistical significance of the interagent comparison is greatly enhanced, which is recorded in Tables 1 and 2 in columns E-M as the significance of the difference between means by the student's "t" test.

There appears to be no difference in the magnitude of the constant K_1 or K_2 in HEDP and MDP studies. Thus, the metabolism of the loosely bound component is comparable for both agents. The values of K_1 for HMDP are significantly lower than those of MDP, even considering the two groups of patients studied. The greatest difference noted is that the values of K_2 are lower for HMDP than MDP. The low values for K_1 for HMDP mean that less agent in the blood is being transferred to loose binding sites. The low values for K_2 mean that the binding energy at the loose binding sites is somewhat greater.

The important difference between the behavior of the three agents appears to be in their values of K_4, which is a measure of the energy of binding to tight binding sites. The tighter the agent is bound to the bone binding site, the smaller the value of K_4. The data suggest that the K_4 values for HMDP are lower than for MDP and probably HEDP as well.

In Table 3, three algebraic relations of transfer constants for the sacroiliac joints are presented for the three ^{99m}Tc diphosphate agents. $K_1 + K_3$ gives the total blood

Table 3

CALCULATED PARAMETERS OF AVERAGE
TRANSFER CONSTANTS OF NORMAL
SACROILIAC JOINTS IN HEDP, MDP, AND
HMDP STUDIES

	Average	σ	E—M	M—H	E—H
			Significance of difference		
$K_1 + K_3$ E	0.303	0.121	a	—	—
M	0.428	0.151			
H	0.312	0.075			
$K_3/K_1 + K_3$ E	0.302	0.138	a	a	b
M	0.454	0.124			
H	0.588	0.095			
K_3/K_4 E	17.3	8.8	c	c	c
M	41.6	14.7			
H	119.1	60.4			

Note: $K_1 + K_3$ Blood clearance in ml/min/cc of bone.
   $K_3/(K_1 + K_3)$ Fraction of activity cleared from blood
            which is bound to tight binding sites.
   K_3/K_4 Ratio of influx to egress from the
            tightly bound compartment.

[a]   Significant difference between means at level of $p < 0.05$.
[b]   Significant difference between means at level of $p < 0.005$.
[c]   Significant difference between means at level of $p < 0.001$.
[d]   In ml of blood cleared per minute per cubic centimeter of
   bone.
[e]   Fraction present transferred per minute.

clearance of the agent by a cubic centimeter of bone, and is therefore a measure of the product of extraction efficiency and blood flow to the bone. The blood flow to the same normal bone areas from patient to patient can be considered constant, particularly in paired studies in the same patient. Thus, in the comparison of agents in normal bone, the values of $K_1 + k_3$ are measures of the ability of bone to extract the agents from blood. The quotient, $K_3/(K_1 + K_3)$ is the fraction of agent extracted from blood which is bound to tight binding sites. The ratio K_3/K_4 is related to the ability of the tight binding sites to extract blood activity and the ability of the same site to retain the activity once it is bound. Since an increase in K_3 or a decrease in K_4 acts to increase retained bone activity, the ratio K_3/K_4 is a measure of the combined long term retention of the activity in bone.

In Table 3, it is apparent that there is no remarkable difference in $K_1 + K_3$ or blood clearance between the three bone agents. The paired studies suggest that clearance of MDP may be 25 or 30% greater than that of HEDP which is the magnitude of difference reported for the difference in diffusion through capillaries measured for Tc-pyrophosphate and HEDP by Hughes et al.[46]

In Table 3, it is apparent that $K_3/(K_1 + K_3)$, the fraction of the extracted activity bound by tight binding sites, is least for HEDP, 50% greater for MDP, and almost 100% greater for HMDP. The value of 0.302 for HEDP indicates that 30% of the ^{99m}Tc HEDP extracted from blood is bound by tight binding sites, while 59% of the ^{99m}Tc HMDP is bound by tight binding sites. The parameter $K_3/K_1 + K_3$ can be considered a measure of partitioning of the agent (which arrives in the ECF space above bone surfaces) between loose and tight binding sites. It appears from these data that of the agent arriving at bone surfaces 1½ times as much MDP, and 2 times as much HMDP are bound at tight binding sites than HEDP. We can assume that at the tracer levels involved, there is a vastly greater number of both types of binding sites than molecules of agent. The partitioning between loose and tight binding sites must therefore be based on the average energy or avidity of binding of the agent to the variety of different types of binding sites that are equally available to each agent. Clearly, HMDP must bind tightly to some sites where HEDP and MDP are only loosely bound. Conversely, it would appear that HEDP binds tightly to few sites, and is thus potentially capable of being more selective in detecting bone formation sites where tight binding occurs.

The ratio K_3/K_4, the index of long term retention of bone activity, is seen to be markedly and significantly different for the three diphosphonates. HEDP is the least, MDP about 140% greater, and HMDP about 500% greater than HEDP. There is almost no overlap in the data, and as such, the ratio is an excellent method of clearly separating the agents on the basis of their kinetic behavior. From the practical standpoint of skeletal imaging, the longer imaging is delayed the better the HMDP image will be, relative to MDP and HEDP. As will be discussed later, this may be a disadvantage from the standpoint of lesion imaging as contrasted with normal skeletal imaging.

In summary, the difference in kinetic behavior of ^{99m}Tc HEDP, MDP, and HMDP as measured by their transfer constants between blood and the loose and tight binding compartments of bone are as follows: MDP is extracted from blood most efficiently into both loose and tight binding compartments. HEDP is extracted from blood slightly less efficiently than MDP and the loosely bound fraction behaves kinetically, almost identically to that of MDP. The extraction of blood activity by the tight binding compartment in the case of HEDP is half that of either MDP or HMDP. This suggests that tight binding of HEDP is restricted to fewer chemical surface configurations than the other agents. HMDP appears to extract less activity from blood on loose binding sites, and the binding energy at these fewer sites is greater. The extraction of HMDP from blood by tight binding sites is almost equal to that of MDP. However, once bound, it is released at only half the rate that MDP and HEDP are released. The rate of release of tightly bound activity (K_4) has been found to be decreased in conditions where bone apposition rates are decreased, suggesting some biological regulation of this parameter. Since the K_4 values of HMDP are much smaller than those of HEDP and MDP, it is apparent that the biologic process which appears to release tightly bound HEDP and MDP does not work to the same extent in the case of HMDP. This suggests that in normal remodeling bone, tightly bound HMDP is not released in the bone apposition processes where ACP is converted to HA. HEDP and MDP may be more extensively bound to ACP, which is rapidly being renewed in the bone formation process. Alternatively, HMDP may have a sufficiently greater binding affinitive for forming HA that it is rebound at the mineralization front when released from ACP before it can be lost from the bone surface.

b. Bone Lesion Analysis

Our experience with the kinetic behavior of the ^{99m}Tc-diphosphonates in bone lesions is based on the studies in five patients of a few hours duration, in four lesions in three patients with paired HEDP-MDP studies, and in two patients with HMDP alone.

The data for the four lesions studied in three patients with both HEDP and MDP will be presented because they represent four different types of lesions which are all unilateral. Each lesion can be directly compared to its normal control bone area in the same patient, and two comparable studies make it possible to compare intertest variability as well as relative agent effectiveness in characterizing the lesions.

Table 4

RATIO OF LESION TO NORMAL BONE TRANSFER CONSTANTS FOR
FOUR LESIONS IN THREE PATIENTS STUDIED WITH BOTH
^{99m}TC-HE DP AND MDP

	Osteomyelitis		Acute fracture		SI strain		Metastatic CA	
	E(1)	M(2)	E(3)	M(4)	E(5)	M(6)	E(7)	M(8)
K_1	3.04	3.36	2.58	3.57	1.19	0.84	1.0	1.57
K_2	1.61	6.09	0.59	0.60	0.94	1.01	0.29	0.47
K_3	1.78	1.86	3.83	3.45	1.57	1.21	4.0	3.7
K_4	1.48	1.52	0.33	0.47	1.10	1.02	0.45	0.73
$K_1 + K_3$	2.69	2.68	3.09	3.50	1.37	1.35	1.44	2.28
$K_3/K_1 + K_3$	0.662	0.694	1.24	0.988	1.15	0.90	2.78	1.65
K_3/K_4	1.20	1.22	11.7	7.39	1.43	1.19	8.9	5.10

In Table 4, the data for the ratios of the transfer constants for lesion to normal bone are depicted for the four lesions, each studied with both HEDP and MDP. The columns are numbered in parentheses. The first two data columns labeled E(1) and M(2) for HEDP and MDP, respectively, are the ratios of lesion to normal transfer constants for the case of acute osteomyelitis of a femoral head. Columns 3 and 4 are comparable data for the acute femoral fracture. Columns 5 and 6 for a case of sacro-iliac strain. Columns 7 and 8 are for a case of carcinoma of the prostate with metastasis to a sacroiliac joint.

The striking thing about the data is that the values of the ratios of the lesion to control bone transfer constants for HMDP and MDP are quite comparable. This indicates that HEDP and MDP are measuring about the same thing, and that the results of repeated studies are almost identical. The high value for the K_2 L/N ratio in Column 2 is believed to be an artifact due to a low detection sensitivity of one side of the crystal detector in the 2-hr delay image since a spuriously high K_2 value for the iliac crest of the same side was noted, while the rest of the bone values were in the expected range.

In Table 4, Columns 1 and 2 for the osteomyelitis case, the K_1 values are increased over normal by 200% while the K_2, K_3, and K_4 values are modestly increased from 50 to 80% above normal. The values of $K_1 + K_3$ indicate that blood clearance was increased by 170% with both agents, which is probably accounted for on the basis of increased blood flow alone, and is consistent with the 200% increase in clearance by the loose binding sites, and only an 80% increase in clearance by the tight binding sites. The index of long term retention (K_3/K_4) is only increased by 20% above normal for both agents, which is consistent with the rather faint increase in uptake in the clinical images for this patient. Low lesion concentration is, of course, the usual finding in clinical imaging in acute osteomyelitis. The claims that increased blood flow is responsible for the imageability of bone lesions generally are not borne out in these studies. Here we have a case where there was a 170% increase in blood flow which resulted in only a 20% increase in retained activity at late imaging times. The problem is that increased activity delivered by the increased blood flow is leaked back to blood at a normal rate, because it is principally bound to loose binding sites.

In the case of the acute fracture (Table 4, Columns 3 and 4), there is marked increase in blood clearance by the lesion above that of normal bone by about 200% for both agents. The increase in blood clearance is almost equally shared by loose and tight binding sites. The histological lesion anticipated in this case is woven bone of callus tissue. The decrease in the K_4 value by 60% of that in normal bone is consistent with the concept of trapping of the tightly bound agents in woven bone. The proportion of the activity cleared from blood which is bound to tight binding sites ($K_3/K_1 + K_3$) in the lesion is almost the same as in normal bone for both agents. For purposes of clinical imaging, the increase in K_3 above normal accounts for an increased amount of agent being delivered from blood to the lesion. But probably of equal importance is the fact that the tightly bound activity is retained in the lesion better than it is in normal bone since the values of K_4 are decreased by 60% of that of normal bone. The ratio of K_3/K_4 reflects the combined effect in late clinical images which indicates a 12-fold increase over normal bone in HEDP and 7-fold increase in MDP.

In the case of sacroiliac (SI) strain (Table 4, Columns 5 and 6), the only apparent change in transfer constants from the normal control bone in this lesion is the 35% increase in blood clearance $(K_1 + K_3)$ by the lesion. The clearance by tight binding sites accounts for the increase in blood clearance since clearance by the loose binding sites, K_1, is that of normal bone. This increase in clearance by tight binding sites, with no increase in loose binding site clearance, suggests that the blood flow may not have been increased and that the increased clearance by tight binding sites must have resulted from an increased extraction efficiency of the agent from blood. The index of delayed retention, K_3/K_4, is slightly less than the increase in K_3 because the values of K_4 are slightly greater than in normal bone. Thus, the concentration of bone imaging agents in reactive bone lesions (due to increased bone remodeling) appears kinetically to be accounted for on the basis of an increase in extraction efficiency by tight binding sites in bone.

In the case of metastatic carcinoma of the prostate involving the sacroiliac joint (Table 4, Columns 7 and 8) the transfer constants of the lesion compared to normal bone indicate a 300% increase in blood clearance by tight binding sites. Unfortunately, there are some technical problems associated with the magnitude of the K_1 values as indicated by the unusual spread of values. However, there appears to be little if any change in K_1 values from normal with either agent. This suggests, as reasoned previously, that there probably is little if any increase in blood flow, as a marked increase in clearance by loose binding sites is seen in conditions where increased blood flow is known to be a clinical feature of the disease process. The increase in blood clearance by tight binding sites without an appreciable increase in blood flow, must again indicate that the mechanism of concentration is increased efficiency of extracting agent from the perfusing blood. In addition to the increased blood clearance by tight binding sites (K_3) there is a marked decrease in rate of loss of the tightly bound agent back to blood, which is 65% less than normal in the HEDP study. The combined effect of the increase in blood clearance by tight binding sites and its slow rate of loss from these sites results in an increase in the index of agent concentration at delayed times (K_3/K_4) of ninefold for HEDP and fivefold for MDP. In the case of metastatic bone tumor, the histologic lesion concentrating the agent is woven bone,[16] as all pathologists know. This is comparable to the bone present in callus of early healing fractures. The increase in K_3 and the decrease in K_4 are characteristics of these lesions. The surprising thing to the author is that there was not a substantial increase in K_1 reflecting an increase in blood flow. Because tumors generally have a rich blood supply a definite increase in blood flow was anticipated.

In reviewing Table 4, it is apparent that in each lesion which revealed a decrease in the rate of loss of tightly bound agent (K_4) there was also a comparable decrease in the rate of loss of loosely bound agent (K_2). K_2 seems to be behaving comparably to K_4. It is apparent that decreases in these two constants are seen in the two cases where woven bone was present: the healing fracture and the metastatic carcinoma. In our model we have made the assumption that there is a clean division between loose binding and tight binding sites in bone. This is, of course, an oversimplification. More probably there is a spectrum of binding sites with binding energies ranging from very loose to very tight or irreversible. The separation of the spectrum into two arbitrary compartments results in considerable overlap. The finding that K_2 is behaving like K_4

values in woven bone lesions suggests that some firmly binding sites in woven bone are being included in the loosely bound compartment.

In comparing HEDP and MDP in bone lesions it is apparent that in all cases except the acute osteomyelitis, there was a greater concentration of HEDP at tight binding sites compared to normal bone than was seen in the MDP study. This is indicated by the greater values of K_3 L/N ratio, and usually the smaller values of K_4 L/N ratio for HEDP than MDP. Why was this the case when our comparison of normal bone with the same agents showed a decrease in K_3 blood clearance and an increased K_4 loss from tight binding sites for HEDP as compared to MDP? In bone lesions HEDP behaves almost the same as MDP. In normal bone opposite the lesion, less HEDP is cleared from the blood by tight binding sites than MDP, and this is lost back to blood at a slightly greater rate than that of MDP. The ratio of activity in the lesion to that in normal bone is greater in HEDP than in MDP, not because the lesion has more activity but because the normal bone has less activity.

The clinical problem of differentiating between benign reactive bone lesions and tumors is important and frequent. Our kinetic studies were undertaken in order to be able to approach a solution to this problem by nuclear medicine techniques. Based on our work to date, it appears that metastatic bone lesions as woven bone containing lesions, are kinetically characterized by having increased K_3 values, decreased K_4 values, and probably slightly increased K_1 values. Benign reactive bone lesions being formed by increased amounts of lamellar bone formation are kinetically characterized by having an increase in K_3 values, normal or slightly increased K_4 values, and probably normal K_1 values. The mechanism of concentrating the activity in the bone lesions, in each case, is by an increase in the extraction efficiency of blood borne agent by tight binding sites in bone. The difference in metabolism of the agents in the two forms of lesion bone is in the rate of release of the tightly bound bone imaging agents. In tumors there is a decrease in the rate of loss of the agents from the tight binding sites, producing a flat or slowly rising late retention curve. In reactive lesions there is a normal, slightly increased, rate of loss of tightly bound agent (half-time of 2 hr) producing a rather steeply falling late retention curve between 8 and 24 hr following injection.

III. SUMMARY

The mechanism of fixation of ^{99m}Tc-phosphate bone imaging agents has been explored using compartmental analysis of kinetic studies in normal and lesion bone in patients. Based on these studies, it is concluded that the agents are both loosely and tightly bound to bone. The clearance of blood by loose binding sites is correlated with the blood flow to the bone. The clearance of blood by tight binding sites is correlated with bone formation. The mechanism by which the increase in blood clearance occurs in bone formation sites is due, principally, to an increase in the efficiency of extracting the agent from the perfusing blood. The rate of loss of the agents from tight binding sites is under biologic control, and is thought to be due to the dissolution of amorphous calcium phosphate (ACP) on which initial binding takes place. The rate of loss from tight binding sites normally occurs with a half-time of 2 hr, the rate being decreased and the half-time increased in woven bone of bone tumors and callus of healing fractures as well as in osteomalacia. ^{99m}Tc-diphosphonate uptake in bone is completely unrelated to the rate of mineral deposition because it is increased in both high and low rates of bone apposition. The kinetic behavior of the Tc-phosphate compounds as well as the radioautographic evidence indicates that the compounds are deposited on bone surfaces and do not penetrate the mineralized bone tissue and, as such, belong to the "surface seeking" group of radionuclides which include the rare earth and transuranic elements. They adsorb to resting bone surfaces where they probably bind to hydroxyapatite (HA). They are more intensely concentrated in forming bone where they may be principally adsorbed to ACP.

The mechanisms of fixation of Tc-phosphate agents in metastatic calcification sites is probably adsorption on ACP rather than on HA, as is generally assumed. This is suggested not only by electron microscopy findings, but also because of the large avidity for binding by a small mass of mineral that is kinetically characteristic of bone formation in which a comparatively small mass of ACP is thought to play a principal role.

REFERENCES

1. Siegel, B. A., Donovan, R. L., Alderson, P. O., and Mack, G. R., Skeletal uptake of ^{99m}Tc-diphosphonate in relation to local bone blood flow, *Radiology*, 120, 121, 1976.
2. Sagar, V. V., Piccone, J. M., and Charkes, N. D., Studies of skeletal kinetics. III. Tc-^{99m} (Sn) methylenediphosphonate uptake in the canine tibia as a function of blood flow, *J. Nucl. Med.*, 20, 1257, 1979.
3. Fogelman, I., Bassent, R. G., Turner, J. G., Citrin, D. L., Boyle, I. T., and Greig, W. R., The use of whole-body retention of Tc^{99m} diphosphonate in the diagnosis of metabolic bone disease, *J. Nucl. Med.*, 19, 27, 1978.
4. Frost, H. M., *Bone Remodeling and its Relationship to Metabolic Bone Disease*, Charles C Thomas, Springfield, Ill., 1973, 176.
5. Bonte, F. J., Parkey, R. W., Grahana, K. D., Moore, J., and Stokey, E. M., A new method of radionuclide imaging of myocardial infarcts, *Radiology*, 110, 473, 1974.
6. Pugh, P. R., Buja, L. M., Parkey, R. W., et al. Cardioversion and "false positive" Tc^{99m} stannous pyrophosphate myocardial scintigrams, *Circulation*, 54, 399, 1976.
7. Metsui, K., Yamada, H., and Chiba, K., Visualization of soft tissue malignancies by using ^{99m}Tc-polyphosphate pyrophosphate and diphosphonate, *J. Nucl. Med.*, 14, 632, 1973.
8. Oren, V. O. and Uszler, J. M., Liver metastases of oat cell carcinoma of lung detected on ^{99m}Tc-diphosphonate bone scan, *Clin. Nucl. Med.*, 3, 355, 1978.
9. Vanek, J. A., Cook, S. A., and Bukowski, R. M., Hepatic uptake of ^{99m}Tc labeled diphosphonate in amyloidosis; case report, *J. Nucl. Med.*, 18, 1086, 1977.
10. Richards, A. G., Metastatic calcification detected through scanning with ^{99m}Tc-polyphosphate, *J. Nucl. Med.*, 15, 1057, 1974.
11. D'Agostino, A. N., An electron microscopic study of cardiac necrosis produced by 9 alpha-fluorocortisol and sodium phosphate, submitted, 1980.
12. Matthews, J. L., Martin, J. H., and Carson, F. L., Ultrastructure of calciphylaxis in skin, *Metab. Bone Dis. Rel. Res.*, 1, 219, 1978.
13. Kim, K. M., Calcification of matirx vesicles in human aortic valve and aortic media, *Fed. Proc.*, 35, 156, 1976.
14. Sagar, V. V., Piccone, J. M., and Charkes, N. D., Studies of skeletal tracer kinetics. III. Tc-^{99m}(Sn) methylenediphosphonate uptake in the canine tibia as a function of blood flow, *J. Nucl. Med.*, 20 1257, 1979.
15. Charkes, N. D., Mechanisms of skeletal tracer uptake, *J. Nucl. Med.*, 20, 794, 1979.
16. Galasko, C. S. B., The pathological basis for skeletal scintigraphy, *J. Bone Jt. Surg.*, 57B, 353, 1975.
17. Hughes, S., Khan, R., Davis, R., and Lavender, P., The uptake by the canine tibia of the bone-scanning agent ^{99m}Tc-MDP before and after an osteotomy, *J. Bone Jt. Surg.*, 60B, 579, 1978.
18. Lavender, J. P., Khan, R., and Hughes, S., Blood flow and tracer uptake in normal and abnormal canine bone: comparisons with Sr-85 microspheres, K_r-81m, and ^{99m}Tc-MDP, *J. Nucl. Med.*, 20, 413, 1979.
19. Jung, A., Bisaz, S., and Fleisch, H., The binding of pyrophosphate and two diphosphonates crystals, *Calcif. Tissue Res.*, 11, 269, 1973.
20. Eanes, E. D. and Meyer, J. L., The maturation of crystalline calcium phosphates in aqueous suspensions at physiologic pH, *Calcif. Tissue Res.*, 23, 259, 1977.
21. Charkes, N. D., Makler, P. T., and Philips, C., Studies of skeletal tracer kinetics. 1. Digitial-computer solution of a five-compartment model of ^{18}F fluoride kinetics in humans, *J. Nucl. Med.*, 19, 1301, 1978.

22. Robertson, W. G. and Morgan, D. B., Effect of pyrophosphate on the exchangeable calcium pool of hydroxyapatite crystals, *Biochem. Biophys. Acta*, 230, 495, 1971.
23. Fleisch, H., Russell, R. G. G., and Bisaz, S., Influence of pyrophosphate on the transformation of amorphous to crystalline calcium phosphate, *Calcif. Tissue Res.*, 2, 49, 1968.
24. Francis, M. D., Russell, R. G. G., and Fleisch, H., Diphosphonates inhibit formation of calcium phosphate crystals in vitro and pathological calcifications in vivo, *Science*, 165, 1264, 1969.
25. Hosain, P., Spencer, R. P., Ahlquist, K. J., and Sripada, P. K., Bone accumulation of the Tc[99m] complex of carbamyl phosphate and its analogs, *J. Nucl. Med.*, 19, 530, 1978.
26. Spencer, R. P., personal communication, 1979.
27. Van Langevelde, A., Driessen, O. M. J., Pauwels, E. K. J., et al., Aspects of Tc[99m] binding from an ethane-1-hydroxy-1, 1-diphosphonate-Tc[99m] complex to bone, *Eur. J. Nucl. Med.*, 2, 47, 1977.
28. Gerlach, U., Hohling J. H., and Themann, H., Metabolism and structure of connective tissue during extraosseous calcification, *Clin. Orthop. Res.*, 69, 118, 1970.
29. Molnar, Z., Development of the parietal bone of young mice-crystals of bone mineral in frozen-dried preparations, *J. Ultrastruct. Res.*, 3, 39, 1959.
30. Gay, C. V., Schraer, H., and Hargest, T. E., Ultrastructure of matrix vessicles and mineral in unfixed embryonic bone, *Metab. Bone Dis. Related Res.*, 1, 105, 1978.
31. Posner, A. S., Betts, F., and Blumenthal, N. C., Properties of nucleating systems, *Metab. Bone Dis. Rel. Res.*, 1, 179, 1978.
32. Russell, R. G., Bisaz, S., Doneth, A., Morgan, D. B., and Fleisch, H., Inorganic pyrophosphate in plasma, in normal persons and in patients, *J. Clin. Invest.*, 50, 961, 1971.
33. Anderson H. C., Calcium-accumulating vessicles in the intercellular matrix of bone, *Ciba Found. Symp.*, 11 (New Series), 213, 1973.
34. Baylink, D., Wergedal, J., and Thompson, F., Loss of protein polysaccharides at sites where bone mineralization is initiated, *J. Histochem. Cytochem.*, 20, 279, 1972.
35. Arnold, J. S. and Jee, W. S. S., Bone growth and osteoclastic activity as indicated by radioautographic distribution of plutonium, *Am. J. Anat.*, 101, 367, 1957.
36. Kisieleski, W. E., Faraghan, W. G., Norris, W. P., and Arnold J. S., The metabolism of uranium[233] in mice, *J. Pharm. Exp. Ther.*, 104, 459, 1952.
37. Arnold, J. S., Frost, H. M., and Buss, R. O., The osteocyte as a bone pump, *Clin. Ortho.*, 78, 47, 1971.
38. Jones, A. G., Francis, M. D., and Davis, M. A., Bone scanning: radionuclide reaction mechanisms *Semin. Nucl. Med.*, 6, 3, 1976.
39. Guillemart, A., LePapa, A., Galey, G., and Besnard, J. C., Bone kinetics of Ca[45] and pyrophosphate labeled with technetium-96: an autoradiographic evaluation, *J. Nucl. Med.*, 21, 466, 1980.
40. Arnold, J. S., Jee, W. S. S., and Johnson, K. D., Observations and quantitative radioautographic studies of Ca[45] deposited in vivo in forming Haversion systems and old bone of the rabbit, *Am. J. Anat.*, 99, 291, 1956.
41. Leblond, C. P., Wilkinson, G. W., Belanger, L. F., and Robichon, J., Radioautographic visualization of bone formation in the rat, *Am. J. Anat.*, 86, 289, 1957.
42. Marshall, J. H., Rowland, R. E., and Jawsey, J., Microscopic metabolism of calcium in bone. V. The paradox of diffuse activity and long term exchange, *Radiat. Res.*, 10, 259, 1959.
43. Roland, R. E., Exchangable bone calcium, *Clin. Ortho. Rel. Res.*, 49, 233, 1966.
44. Kaye, M., Silverton, S., and Rosenthall, L., Technetium-99m-pyrophosphate: studies in vivo and in vitro, *J. Nucl. Med.*, 16, 40, 1975.
45. Glimcher, M. J., Composition, structure, and organization of bone and other mineralized tissues and the mechanism of calcification, in *Handbook of Physiology-Endocrinology VII*, Aurbach, G. D., Ed., Williams & Wilkins, Baltimore, 1976, 25.
46. Guillemart, A., Besnard, J. C., and LaPape, A., Skeletal uptake of pyrophosphate labeled with technetium-96 as evaluated autoradiography, *J. Nucl. Med.*, 19, 895, 1978.
47. Hughes, S. P. F., Wahner, H. W., Kelley, P. J., Bassingthwaight, J. B., and Davies, D. R., Transcapillary extraction by the canine tibia of 99m-Tc pyrophosphate and HEDP, *J. Nucl. Med.*, 18, 605, 1977.
48. Jones, J. L., Davis, W. L., Jones, R. G., Miller, G. W., and Matthews, J. L., The effect of cytochalasin B on the endosteal lining cells of mammalian bone, *Calcif. Tissue Res.*, 24, 1, 1977.
49. Kung, H., Ackerholt R., and Blau, M., Are two phosphate groups necessary for bone localization of Tc-99m complexes? *J. Nucl. Med.*, 18, 624, 1977.
50. Arnold, J. S., Kinetic analysis of bone imaging agents, in *Principles of Radiopharmacology*, Vol. 3, Colombetti, L. G., Ed., CRC Press, Boca Raton, Fla., 1979, 205.

Chapter 6

RENAL SECRETION AND FILTRATION STUDIES

L. Rao Chervu and M. Donald Blaufox

TABLE OF CONTENTS

I. INTRODUCTION

The mammalian kidney is a plasma-processing organ of remarkable efficiency. Approximately 20 to 25% of the cardiac output is delivered to the kidneys each minute, and every 30 min the equivalent of the plasma volume is filtered through the glomerular capillaries and flows past the lining membranes of the tubular epithelial cells. The tubular reabsorption and secretion adding to or subtracting from this filtrate continuously renews the internal environment of the body. The constant composition of the blood and the extracellular fluid depends to a large extent therefore on the functional integrity of the kidney. The body depends principally upon the kidney to maintain a constant osmolarity and concentration of the principal cations and anions. It also helps to maintain a constant pH, as well as to excrete a number of products of metabolism. While the kidney has both endocrine and excretory functions, radiopharmaceuticals have not been developed for the in vivo assessment of the former, while in vitro measurements by RIA are routinely available. On the other hand, noninvasive studies of the excretory function have been fully developed with the use of radiopharmaceuticals, and these have become valuable alternatives to constant infusion and/or catheterization procedures which are routinely used for the evaluation of the status of renal function. The diagnosis and management of the patient with renal disease at times require, besides measurement of the serum electrolytes, the acquisition of functional information which may include:

1.   Total renal blood flow
2.   Effective renal plasma flow
3.   Glomerular filtration rate
4.   Morphology of the kidney
6.   Postvoiding residual urine volume
6.   Ureterovesical reflux
7.   General evaluation of the integrity of the urinary collecting system

The traditional method of assessing renal structure radiographically, and more recently using ultrasound and computed tomography, although morphologically superior, are not as accurate as radionuclidic techniques for the functional evaluation of the diseases of the urinary tract. Nuclear medicine procedures may be used in conjunction with the radiographic procedures for estimating renal size or shape, and functional integrity in the evaluation of potentially surgically correctable disorders of renal perfusion and urodynamics.[1-5] In renal transplant cases, nuclear medicine has contributed to the early evaluation of the vascular patency of the anastomosis, as well as differentiation of the progressive onset of rejection from acute tubular necrosis and other sources of reduced renal function.

The qualitative and quantitative aspects of the radionuclide renal function studies have evolved over a period of the past two decades, a very short time in terms of modern medicine. Radiopharmaceuticals incorporating short-lived, high photon yield isotopes have been made increasingly available for renal functional assessment[6] which, in turn, has helped to replace the techniques which were based on external probe counting techniques and the quantification of radioactivity in blood samples. A reduced radiation dose to the patient, especially in pediatric cases, which is of great concern in many of the radionuclide studies, has facilitated serial evaluation wherever necessary, within reasonably short intervals with minimal discomfort and risk to the patient. High resolution gamma camera-imaging devices in conjunction with on-line data processing computer systems are now available at a large number of centers, and these have helped to quantitate with great precision and speed many renal function studies. The agents currently available, though widely used, do not yet fully satisfy all the criteria for ideal utilization in several renal function studies, and a continuing search for better radio-pharmaceuticals is definitely in order.

II. PHYSIOLOGICAL CONSIDERATIONS

A. Factors Determining Urinary Excretion

The major routes of excretion of most physiologically inert compounds occur by one or both of two different pathways, either via the hepatobiliary system or via the urinary tract. The mechanisms by which the major route of excretion of some compounds in the bile and others in urine occur are not fully clarified.[7] While a large number of compounds which are in the molecular weight range of 355 to 465 are efficiently excreted in both the urine and bile, a loss of functional integrity in one route of excretion generally results in a compensatory excretion via the other. Compounds of molecular weights lower than 400 are usually preferentially excreted in the urine. The molecular size selectivity for urinary and biliary excretion is a property of the excretory organ itself. The properties of polarity and molecular weight, strength of protein binding, and the functional status of the principal excretory organ decide the fate of excretion of any particular compound. Under these circumstances it is essential to examine the different classes of radiopharmaceuticals that have been used for radionuclidic studies from the viewpoint of their mechanism of renal handling.

B. Physiology of Renal Excretory Function

The structural and functional unit of the kidney is the nephron which consists of a single glomerulus composed of a capillary network which is intimately attached to the renal tubular system. There are approximately one million nephrons in a single human kidney. The interconnecting capillaries which form the glomerular tuft are enclosed by an epithelial capsule (Bowman's capsule). The second major component of the nephron is the renal tubule or tubular exchanger which is composed of a sequence of proximal tubule, loop of Henle, distal tubule, and the collecting tubule (Figure 1). The glomerulus filters about 180 ℓ a day of essentially protein-free filtrate that has approximately the same osmolar concentration of all the crystalloids as that in the aqueous phase of the plasma, and the tubules selectively process this filtrate. About 99% of the filtered water is reabsorbed so that only 1 to 2 ℓ/day are excreted in the urine. The crystalloids also are conserved or excreted selectively by the two tubular exchange processes of reabsorption and secretion. Since each of these components of the nephron handles materials introduced in the bloodstream in a distinctive and specific manner, evaluation of the pattern of clearance and the degree and duration of retention on the binding sites of the renal tubular cells is quite complicated. A brief description of the four main processes, namely glomerular filtration, tubular reabsorption, tubular secretion, and tubular fixation is presented below.

1. Glomerular Filtration

Arterial blood entering the kidney passes through the glomerular capillaries where it is subjected to ultrafiltration that results in the passage of water and water-soluble low and medium molecular weight materials from the bloodstream to the lumen of the nephron. The glomerular filtration is a purely physical process in which hydrostatic pressure at the glomerular capillary level provides a net positive pressure to overcome the opposing frictional resistance in the membrane and colloid oncotic pressure and results in a protein-free filtrate of plasma entering Bowman's space. The process of filtration is one that does not require a local expenditure of metabolic energy by the cells of the filtering surface. Characteristics that distinguish glomerular filtration from

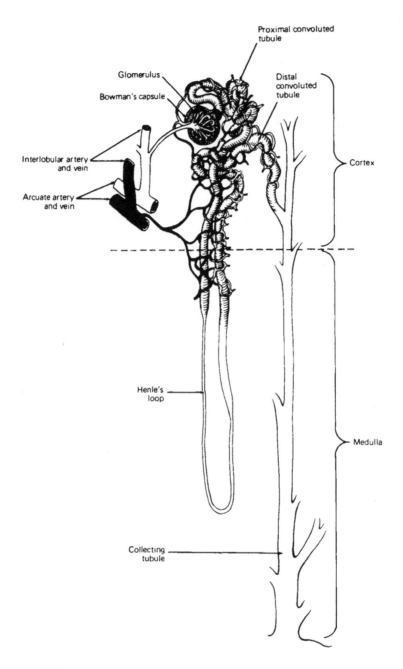

FIGURE 1.   Topology of nephron. (Reproduced with permission from Junqueira, L. C., Carneiro, J., and Contopoulos, A. N., *Basic Histology,* 2nd ed., Lange, Los Altos, Calif., 1977.)

transcapillary exchange in other organs are that the glomerulus almost completely excludes the plasma proteins of the size of albumin and larger from its filtrate; and secondly, the glomerular capillary exhibits an unusually high permeability to water and small molecules. The ultrafiltrate formed through the filtration process contains all the freely diffusible substances in the same concentration that occur in the plasma. Diffusibility requires dimensions below a certain size consistent with the pore dimensions of the membrane and subject to Gibbs-Donnan distribution in the case of electrolytes. Morphological and functional studies of glomerular permeability indicate that in addition to molecular size, molecular charge and probably molecular shape play an important role as determinants in the filtration processes of macromolecules. Molecules of virtually identical size and configuration have been shown to have distinctly different filtration properties as they pass from negative to positive or neutrally charged states.[8,9] The passive process of glomerular filtration may be quantitated by the measurement of the rate of renal clearance of a particular substance in plasma provided the same substance meets certain criteria that will be discussed later.

The absolute GFR depends on both the pressure and the rate at which the blood is delivered to the kidneys via the renal arteries, although changes in arterial pressure and renal blood flow may not result in changes in GFR over a wide range (autoregulation). GFR may be lowered, however, in response to other factors and only increases in a few situations. In healthy young men, GFR averages about 125 mℓ/min/1.73 m^2 and remains relatively stable except for transient variations in response to specific stimuli. The GFR is lower in infants and increases to adult values (adjusted for surface area) by about 2 years of age. It is reduced with advancing age as a result of loss of functioning nephrons. The GFR is an important determinant of the rates of subsequent renal transport processes and ultimately of urine excretion. The glomerular filtration rate is thus a critical parameter in a number of clinical situations.

2. The Tubular Reabsorption

Through this mechanism a number of solutes are transported from the glomerular filtrate in the tubular lumen to the peritubular capillary fluid. Very discrete active or passive mechanisms which are sometimes interdependent are involved in the reabsorption of various components of the filtrate. Several of these reabsorption mechanisms are located in the proximal segment of the renal tubules. These include the transport mechanism for a number of compounds including glucose, many amino acids and low molecular weight proteins, and they are remarkably efficient in operation. Some inert molecules such as inulin are cleared from the blood and not subject to significant net tubular secretion or reabsorption. The mechanism of action of most diuretic drugs is a result of the inhibition of active transport of a number of solutes in different segments of tubule.

3. Tubular Secretory Process

This involves the transport of materials from the peritubular fluid to the tubular lumen. Blood leaving the glomerular capillaries perfuses the renal tubules where diffusion or cellular transport may move molecules from the capillary blood to the lumen of the nephron, or conversely materials in the lumen of the nephron may diffuse or be transported across the renal tubule from the lumen of the nephron back to the blood. The factors involved in this process include active secretory mechanisms that exhibit an absolute limitation of transport capacity, and passive secretory mechanisms involving diffusion of materials down concentration gradients or gradients of electric potential. These mechanisms adequately explain the tubular secretion of a number of organic acids, strong organic bases, and various other organic compounds. The active site of secretion of many of these compounds is the proximal tubule and a continuous supply of energy is necessary to move the transported materials from blood to urine either directly or indirectly. These energy requiring systems are very susceptible to metabolic inhibition.

4. Renal Fixation

Certain materials show prolonged retention in the kidneys and have high affinity for unidentified binding sites in the renal tubules. These materials gain access to the renal tubular binding sites either by diffusion or transport from the lumen or capillary. The nature of binding sites in the renal tubules is quite diverse, and may include thiol or disulfide moieties that have very high affinity for heavy metal cations. For the most part they are not well defined or understood.

Because of the many possible modes of renal handling of radiopharmaceuticals, it is essential that the detailed mechanism of renal handling of any agent be established for consideration of application for renal function and proper interpretation of studies.

III. RENAL CLEARANCE; ASSOCIATED CONCEPTS

An assessment of renal function by clearance measurements using radionuclidic techniques may involve measurement of total renal function by the measurement of Glomerular Filtration Rate (GFR), or Effective Renal Plasma Flow (ERPF), or individual renal function measurement through relative uptake, parenchymal transit, transfer through renal pelvis, and pelvic transit functions. Renal clearance can be defined in terms of the amount of the plasma that must be cleared of the solute in a given period of time to provide a measured quantity of the same solute in the urine. Normal blood constituents that are maintained in a reasonably steady state, such as creatinine, may be used for the measurement of renal clearance.[10] If the plasma concentration of any compound is maintained constant, the minimum amount of plasma passing through the kidney can be derived from the classic Fick equation: $C = UV/P$ where: C = clearance (mℓ/min), U = urine concentration (mg/mℓ), V = the urine flow rate (mℓ/min), and P = the plasma concentration (mg/mℓ). If the plasma concentration of a given substance that is totally cleared in a single passage through the kidneys is kept constant at a level of X mg/mℓ by continuous intravenous infusion, and if the kidneys were then found to excrete each minute Y mg of the substance, the volume of the plasma cleared completely each minute would be Y/X mℓ and the plasma flow would be equal to Y/X mℓ per minute. If the material is totally cleared, the minimum plasma value must equal the total flow. Thus, the resultant value is the plasma clearance expressed as mℓ per minute. The assumption made in this technique is that a steady state exists, and the arterial and plasma concentrations are equal. If a substance is almost totally removed from plasma by either glomerular filtration and/or tubular secretion, its clearance yields an estimate of renal plasma flow. There is no ideal substance that is extracted totally from the plasma by the kidney, since most of the materials have a certain degree of plasma protein binding or red cell binding, and filtration and reabsorption may occur to a certain degree. Even paraaminohippuric acid (PAH) is extracted only about 90% and orthoiodohippurate (OIH) is at best 80% extracted. Realistically, one can only calculate effective renal plasma flow with these nonideal substances instead of total plasma flow.

The clearance equation may be applied to any substance whose rate of excretion is measurable, from a very small rate of clearance to a maximum which obviously cannot exceed the total renal plasma flow. The measurement of the concentration of a particular solute in urine and plasma can be either by chemical analysis or by quantifying its radioactive label. The plasma flow or the clearance of substance as such provides no information about the mechanisms by which the kidney removes that substance from the plasma. The substance may be removed by ultrafiltration or tubular secretion or a combination of these. In order to assess the magnitude of each of these processes, it is necessary to determine the GFR and renal plasma flow. If the substance is filtered at the glomerulus and appears in the ultrafiltrate in a concentration equal to its concentration in plasma and is neither secreted nor reabsorbed by the tubules, it may be used as a measure of the glomerular filtration rate. Many materials, however, are not only filtered by the kidney, but are also reabsorbed or secreted. The magnitude of the tubular clearance of these compounds depends upon their plasma concentration and the tubular transport capacity for absorption or secretion. Several of these substances are of limited clinical importance for the quantitative evaluation of renal function since the limits of their normal clearance are difficult to define.

The kidney has a definable maximum tubular transport capacity to reabsorb or secrete some materials in the plasma. The maximum tubular transport capacity, T_m, implies that there is certain threshold concentration in the plasma which saturates the transfer process. All the filtered quantity below that maximum of a given substance may be absorbed or secreted. At a high plasma concentration, the total clearance of a substance which is secreted (PAH) or absorbed (glucose) asymptomatically approaches the glomerular filtration rate at which point the amount secreted or absorbed is negligible compared to the total blood concentration. Radiopharmaceuticals differ from the nonradioactive compounds with respect to their maximum tubular transport capacity since their concentration in the plasma is far below the amount usually required to reach the tubular maximum. Therefore, this does not present a practical problem. A discussion of the procedures for evaluating renal function with radiopharmaceuticals is presented in later sections.

IV. RADIOPHARMACEUTICALS FOR RENAL FUNCTION MEASUREMENT

A large number of radiopharmaceuticals have been suggested for measuring renal function. In the majority of cases, the specific parameter of renal function which could be measured by each of these materials has not been clearly defined. In Figure 2 the topography of the nephron and the excretion mechanisms of the radiopharmaceuticals used in these function studies are indicated. Compounds as shown in the figure are only a representative sampling of the large number of agents that have been utilized. The four categories of agents which are shown in the figure are the agents for measurement of renal blood flow and the intrarenal distribution of blood flow, GFR agents, renal plasma flow agents, and agents that show prolonged retention in the kidneys and that have high affinity for binding sites in the renal tubules. The pathways of different substances from the glomerular and peritubular capillaries to various parts of the nephron are shown in Figure 3. Several other agents including ^{99m}Tc-pertechnetate and other ^{99m}Tc-labeled compounds are available for renal perfusion and imaging purposes but their excretion mechanisms are not well defined. A complex interaction of glomerular filtration, tubular reabsorption, and/or tubular secretion is involved in their renal handling. Due to this inherent difficulty, it is obvious that the clearance values obtained with these radiopharmaceuticals are not of any significant importance in patients with renal disease since they cannot be properly standardized or interpreted.

In studying renal function with a gamma-emitting nuclide, it is necessary to clearly define not only its mechanism of excretion by the kidney, but also its kinetics in the body. In many of these cases, a detailed study of the kinetics has not been available except in a few cases, as for radiohippuran for which detailed compartmental analysis has been carried out (Figure 4).[11-15]

A. Radiopharmaceuticals for the Measurement of Glomerular Filtration Rate

The measurement of glomerular filtration rate to follow the evolution of renal disease has been stressed by many workers long before the availability of radioactive tracers. These have an important role in the clinical or investigative evaluation of renal function as mentioned earlier. GFR measurements were performed in man, using a wide variety of nonradioactive substances such as inulin, hyposulfite, mannitol, or

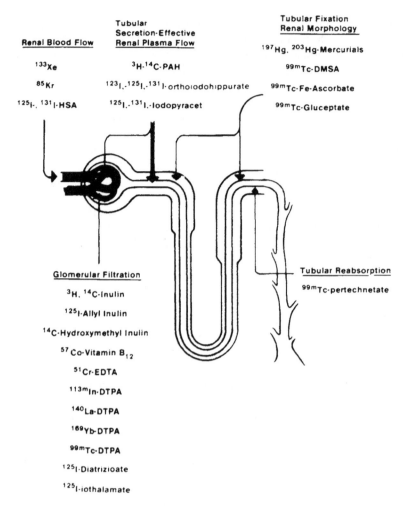

FIGURE 2.   Renal function agents. Radiopharmaceuticals which are in vogue for
renal function measurement are indicated above with different nephron segments
with different excretion mechanisms. Agents which remain fixed in the renal tubules
for relatively long periods of time are useful for evaluation of renal morphology
and to delineate comparative individual renal function. Glomerular filtration rate
can be assessed using agents which are filtered and subsequently excreted in urine
without any further tubular absorption or secretion providing there is little or no
protein binding of the agent. Renal plasma flow measurement is obtained with
agents which are partially filtered by the glomerulus and in addition secreted actively
by the proximal tubules and transported with minimal or no tubular reabsorption
into the bladder urine.

creatinine.[16] In order for any of these agents to be used for the measurement of GFR,
the following criteria must be fulfilled:

1.   The substance must not be bound to the plasma proteins, nor sieved in the ul-
      trafiltration process, and it must be freely filterable through the glomerular cap-
      illary membranes.
2.   It must be physiologically inert and not metabolized by the kidneys.
3.   It must be neither secreted nor absorbed by the renal tubules.
4.   It must not have any toxic effects nor exert any effect on the renal function when
      infused in large quantities over a wide range of concentrations.
5.   It must be quantifiable with a high degree of accuracy under routine laboratory
      conditions in both plasma and urine.

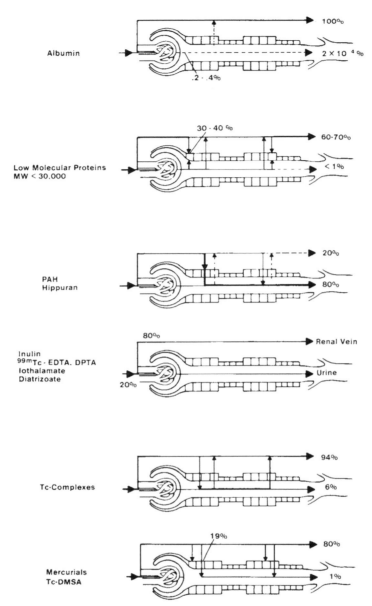

FIGURE 3. Intrarenal pathways of different renal agents. Percentages shown are relative to renal arterial concentration. All of the numbers given must be considered to be estimates which vary under a wide range of physiologic and pathophysiologic conditions. Large molecules like albumin remain completely in circulation while low molecular weight proteins appear to some extent in the glomerular filtrate and are reabsorbed and retained for a variable time in the proximal tubular cells. Agents actively secreted by the tubules such as PAH and hippuran are filtered partially at the glomerulus with very limited reabsorption. Passively filtered agents are excreted into the urine and are not reabsorbed. ^{99m}Tc-Complexes are filtered and secreted with varying degrees and reabsorption occurs for most of them with some retention in the proximal and distal tubular cells. The exact mechanism of handling of these compounds which are highly protein bound requires considerable further investigation. Mercurials bound to plasma proteins are fixed in tubular cells following tubular secretion with only 1% of the plasma content appearing in the urine acutely. (Adapted from McAfee, J. G. and Subramanian, G., in *Clinical Scintillation Imaging*, Freeman, L. M. and Johnson, P. M., Eds., Grune & Stratton, New York, 1975.)

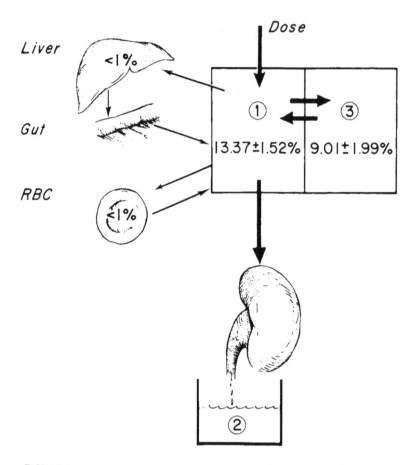

FIGURE 4. Compartmental analysis of hippuran distribution. Detailed studies of
the compartmental distribution and excretion of renal agents handled by the kidney
are limited to very few agents. Compartmental analysis of hippuran distribution
and excretion in man shown in the above figure indicates that hippuran is distributed
in a volume approximating 22% of the body weight with negligible enterohepatic
cycle and without significant uptake in red cell. Urinary secretion almost completely
accounted for the injected dose.

Inulin, a starchlike polymer of fructose, of molecular weight about 5000 and molecular diameter of 12 Å , fulfills all the necessary criteria for use as a measure of glomerular filtration rate and was originally proposed for GFR measurements by Richards et al.,[17] and Shannon and Smith[18] in 1935. It is completely filterable, is neither reabsorbed nor secreted by the renal tubules, it is inert and nontoxic, and its administration does not influence renal function directly. Its concentration in biologic fluids can be measured with adequate reliability. Since inulin is a foreign substance, measurements of its clearance require that it be administered intravenously during the timed collection of urine. Another disadvantage is that its analytical determination is somewhat cumbersome for routine clinical use. However, since the inulin molecule seems to satisfy all the criteria for the measurement of GFR, as proven by ample experimentation on a number of animal species and also in man, this has been used extensively.[19] The clinical use of inulin clearance as a measure of the GFR has been accepted as a standard, and inulin clearances provide the basis for a standard of reference against which the routes or mechanism of excretion of other substances can be ascertained. The inulin preparation is administered in a concentrated priming dose in order to raise the plasma concentration to about 10 to 20 mg/100 mℓ, and it is infused at a rate to maintain the plasma level constant. The blood samples and urine samples are collected to measure their inulin content by chemical analysis. Endogenous creatinine or urea clearances also have been widely used for clinical purposes as an index of glomerular filtration rate. Creatinine can be quantified with reasonable accuracy, and its administration is not required since it is a normal body constituent with relatively constant plasma concentration. Its excretion is almost entirely dependent upon glomerular filtration though partial excretion by the kidney tubules has been reported.[20]

The disadvantages associated with the accurate chemical analysis techniques which are quite demanding, and the relative inconveniences involved in using nonradioactive substances and bladder catheterization for the measurement of this important parameter in the study of renal function and evaluation of patients with renal disease, have encouraged a search for radioactive compounds that have the same renal excretory behavior as inulin. The agents that have been mainly investigated are radiolabeled inulin derivatives, radioiodinated urological contrast agents (hypaque, renografin, conray, etc.), radiocyanocobalamin (vitamin B_{12}), and several metal chelates. Several of these radiopharmaceuticals evaluated as GFR agents are discussed below.

1. Inulin and Derivatives

Inulin labeled with ^{14}C at the carboxyl group and the hydroxyl methyl ^{14}C inulin have both been suggested as agents for the measurement of GFR.[21-23] The former has been reported to differ from inactive inulin in its physiologic behavior and shows a highly significant deviation of the clearance from that of native inulin.[22] The hydroxy methyl ^{14}C-inulin clearances show a high degree of correlation with native inulin.[23] However, these preparations have not gained widespread clinical use owing to the fact that the beta activity of the isotope requires the use of liquid scintillation counter for measurement of the samples and does not permit external counting. An ^{131}I-labeled inulin derivative has been introduced by Brooks et al.,[24] by attaching an allyl ether group to the inulin molecule followed by the addition of radioiodine at the double bond. Concannon et al.,[25,26] have successfully purified this compound from free iodine by repeated passage through anion exchange columns, and established a high degree of correlation between the chemical inulin method and ^{125}I-allyl inulin method for the determination of GFR. Though evidence to date has not shown that the iodinated inulin derivatives have different biologic behavior from that of inactive inulin, clearance values of several inulin derivatives[27-31] including ^{131}I-chloroiodopropyl inulin and ^{131}I-propargyl inulin yield clearance values on an average slightly lower than the chemical values simultaneously determined, though a significant correlation exists between the two series of measurements. The stability of these molecules is suspect, and the gradual liberation of free iodine in the preparation and the dissociation of iodine in vivo after the injection, as evidenced by the thyroid uptake, are two important factors that might lead to a falsely low estimate of GFR because of the markedly lower clearance rate of inorganic iodide.[32] A stable preparation of ^{51}Cr-inulin has been reported by Johnson et al., and this agent has also been used for determination of GFR in dog and in man.[33-35]

2. Vitamin B_{12}

Studies in rabbit[36] and in man by Watkins et al.,[37] have shown that cyanocobalamin and hydroxocobalamin labeled with ^{57}Co or ^{58}Co are eliminated by the kidney essentially by means of glomerular filtration similar to the renal clearance of inactive vitamin B_{12}. The rationale for application of these agents[38,39] for the measurement of GFR is as follows: The endogenous B_{12} saturates the serum protein, transcobalamin I (TC I) and any radioactive B_{12} added in vitro or in vivo is recovered in the transcobalamin II (TC II) fraction. The transcobalamin II fraction acts as a B_{12} transport protein

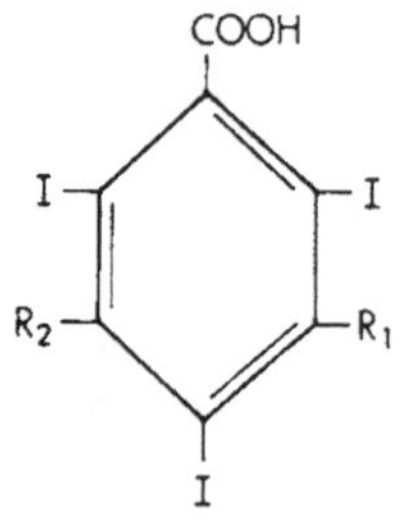

GENERAL STRUCTURAL FORMULA OF IODINATED CONTRAST AGENTS

FIGURE 5.   General structural formula of iodinated contrast agents. General configuration of labeled urological contrast agents which are generally excreted by GFR. I = site of iodine label and R = organic side chain.

in vivo and the tissue sites, especially the liver, remove the vitamin with constant increase in the amount bound to TC I. With the administration of large amounts of stable vitamin B_{12}, the binding capacity of the tissue and the plasma binding sites in vivo is exceeded, and the circulating free nonprotein-bound vitamin will be excreted solely by glomerular filtration. The radioactive vitamin B_{12} given shortly thereafter will be largely unbound, and its clearance is a measure of the GFR.

The renal clearance of vitamin B_{12} labeled with ^{57}Co or ^{58}Co has been compared with that of inactive inulin.[40-48] Serum presaturation is carried out using cyanocobalamin or hydroxocobalamin, the latter being more firmly bound to the plasma proteins.[49,50] This method yields clearance values lower than inulin clearance values, the ratio of U/P total B_{12} to U/P inulin ranging from 0.43 to 0.99. This wide discrepancy has been attributed mainly to the variability of the binding of the labeled agent to the plasma proteins owing to the partial exchange of radioactive B_{12} with inactive vitamin during the clearance procedure[45,48,51] even in the same patient under different conditions. This can be corrected to some extent by separating the bound and free radioactive vitamin B_{12} in each plasma sample using a complex and tedious procedure. Too many variables, such as the temperature of the incubating mixture, the temperature of the system during the measurement of binding, and plasma dilution influence the determination.[48] Significant uptake of ^{57}Co B_{12} in liver,[44] besides other major tissue and body fluid distributions, indicates a multicompartmental distribution of the cyanocobalamin, and estimation of GFR via the single injection techniques tends to be inaccurate. Thus, though the renal clearance of unbound radioactive B_{12} is practically the same as that of inulin, the ^{57}Co B_{12} procedure for measurement of GFR does not appear to be useful as a routine clinical procedure.

3. Labeled Urological Contrast Agents

Iodinated benzoic acid derivatives with more than two or three atoms of iodine per molecule have been used extensively as contrast media for intravenous urography since iodine is a heavy atom with a high X-ray absorption. The renal handling of these contrast media with respect to the degree of glomerular filtration, tubular reabsorption, or tubular secretion, and their extrarenal pathways differ considerably. Agents that have been labeled with ^{125}I or ^{131}I and investigated for renal function studies (GFR) are illustrated in the general chemical structure shown in Figure 5, with the substituents of the various agents shown in Table 1. In case of urokon, proximal tubular secretion at lower plasma levels and a self-inhibiting effect on the proximal tubular secretion at higher levels,[11] and considerable binding to serum protein[52] preclude its use for further renal function studies. Diatrizoate and iothalamate have been studied extensively for renal function testing.

Table 1
UROLOGICAL CONTRAST AGENTS

Generic name	Compound	R$_1$ (Position 3)	R$_2$ (Position 5)
Hypaque®, Renografin®	Diatrizoate	NHCOCH$_3$	NHCOCH$_3$
Conray®	Iothalamate	NHCOCH$_3$	CONH CH$_3$
Miokon	Diprotrizoate	NHCOCH$_2$CH$_3$	NHCOCH$_2$CH$_3$
Urokon®, Triopac	Acetrizoate	H	NHCOCH$_3$

Note: Refer to figure 5.

a. Diatrizoate

[125]I or ^{131}I-labeled diatrizoate is commercially available in the form of a neutral aqueous solution of the sodium salt (Hypaque®), or the methyl glucamine salt (Renografin®). Kimbel and Borner[53] described the use of [131]I Hypaque® for renal function studies in 1955, and Winter and Taplin[54] applied this tracer in renography. The renal clearance and body distribution in animals and in man have been studied by a number of investigators.[55-67] Free iodine is liberated from labeled Hypaque® gradually during storage as a result of autoradiolysis, and the amount of iodine released in dependent upon the specific activity of the preparation.[32] The presence of the free iodine is reflected in lower clearance values[66] and in thyroid uptake which, though small, points to the necessity of rigid quality control in the preparations. Plasma protein binding of this agent is low.[52,67] Tubular reabsorption and tubular secretion are ruled out, the former on the basis of constancy in clearance values with wide variations of urine flow,[63] and the latter with tubular blockade experiments and on grounds of constancy of clearance values with increasing plasma loading of the stable urographic agent or PAH loading.[59,68] The clearance of the diatrizoate (Hypaque® [125]I or ^{131}I) may be used on the strength of the experiments cited above for the measurement of GFR in man by the constant infusion procedure employing the UV/P relationship for renal clearance. Correlation between Hypaque® or Renografin® clearance values and those of inulin is significant with a mean clearance ratio of unity in normal and pathological conditions in adults and in children. With increasing reductions in GFR, the extrarenal pathway for radioiodinated Hypaque® becomes more important as shown in detailed studies on the total clearance.[69,70]

b. Iothalamate

[125]I-labeled iothalamate is commercially available in pure form with minimal contamination of free iodide (<2%) even after storage up to 60 days.[71] The clearance properties are quite similar to diatrizoate. The nature of its plasma binding, its renal handling, and its dependence on carrier doses need detailed investigation. Plasma binding of 8 to 27% has been reported with ultrafiltration techniques[67] while others report only 3% binding in man.[71] Tubular excretion of the tracer is absent, as indicated in experiments with dogs[72] and judging from the constancy of clearance values at high and low plasma concentrations of the unlabeled agent in man.[73] Clearance values obtained with iothalamate have been validated with those of inulin in animals[74-76] and in a large number of patient series in adults and in children.[71,73,77-86] Extrarenal elimination has been reported for the tracer in the anephric rat.[87] The characterization of distribution and elimination of the tracer in adult man in different stages of renal disease is awaited.

4. Metal Chelates

Foreman et al.[88,89] observed that ^{14}C-labeled ethylenediaminetetraacetic acid (EDTA), when administered parenterally in rats and in man, is eliminated essentially

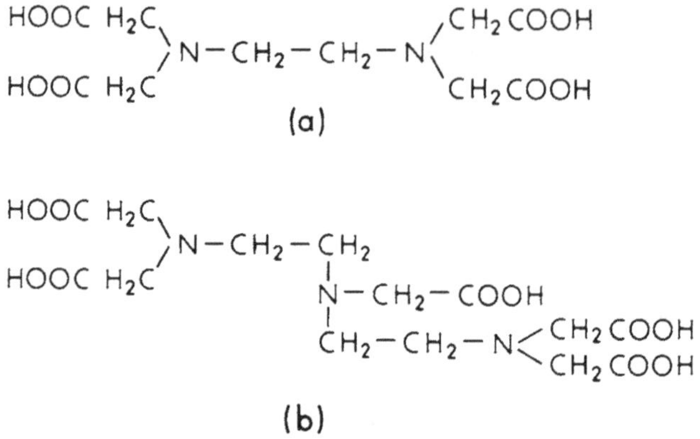

FIGURE 6. Structural formula of (a) ethylenediaminetetraacetic acid (EDTA) and (b) diethylenetriaminepentaacetic acid (DTPA).

unchanged by GFR and tubular secretion to the extent of 95 to 98% in about 6 hr. Clearances of EDTA in the rat have been found to be much higher than that of inulin.[90] EDTA and its analogs (Figure 6) form very strong complexes with a number of metals, and they have been used for therapy for radioactive and stable metal poisoning in humans.[91] The in vitro stability constants of these complexes are high; however, their stability may be affected in vivo. Chelates that enter into a cellular space may be exposed to an increased mass of binding agents and to systems with extremely high affinities for specific metals when transchelation might occur. Thus, in vivo, the chelates may be bound to different organs (e.g., ^{203}Hg-EDTA is detectable in the kidneys for several days). A large number of physiologically stable metal chelates tagged to radionuclides have been proposed as renal agents during the last few years, of which ^{51}Cr-EDTA, ^{99m}Tc-DTPA, ^{140}La-DTPA, ^{169}-Yb-DTPA, and ^{113m}In-DTPA received somewhat greater attention for application for GFR measurements.[92-97] Several other complexes of EDTA, DTPA, and citrate with ^{57}Co, ^{58}Co, ^{68}Ga, ^{99m}Tc, ^{111}In, ^{113m}In, ^{114m}In, ^{115m}In, ^{169}Yb, and ^{197}Hg may also be excreted by glomerular filtration.[97-100] The studies with these complexes are quite limited, and physical properties in many cases are not satisfactory for human application. Adequate validation of the true renal handling is available only for a few agents discussed below.

a. ^{51}Cr-EDTA

This complex was originally suggested by Myers et al.,[101,102] for renography and was reinvestigated by Stacy and Thorburn as a GFR agent.[92] It has a high degree of radiochemical stability and high specific activity. Preparations (60 mCi/mg) are available in very pure form commercially. A very small fraction is bound to the serum proteins (<2%).[51,92,103] Red cell binding occurs to a small extent.[103,104] No tubular secretion for this complex has been demonstrated.[92] The extrarenal elimination of the tracer is insignificant,[103,105] and whole-body retention of activity after several days is less than 1%, as determined by whole-body counting.[106] Slight retention of activity of ^{51}Cr-EDTA has been reported in the kidney after 72 hr.[105]

These data at least partially support that the labeled ^{51}Cr-EDTA is excreted mainly by glomerular filtration. Continuous infusion and single-injection techniques have shown a high degree of correlation between the clearance of this agent and classical inulin clearances.[103-105,107-111] The absolute value of the clearance measured with ^{51}Cr-EDTA is lower than that of inulin, by as much as 5 to 20% in animals and in man. The reason for the difference between the clearances is not known with certainty and awaits a more detailed study of its distribution. However, this shortcoming must be considered in applying this agent, since an unpredictable underestimate of the GFR will usually result.

b. DTPA Complexes

[140]La, [113m]In, and [169]Yb-DTPA[63,94-96,112-115] have been studied in some detail for the measurement of GFR by several workers, and it appears that these complexes are eliminated by the kidney essentially by glomerular filtration. The effects of increasing plasma concentration of the agent, varied urine flow rates, and tubular blockade in animals result in no significant differences in the plasma clearances.[63] [140]La-DTPA causes a high radiation dose because of its associated beta decay, and hence it is not very suitable for human application. Comparisons of [169]Yb-DTPA, [113m]In-DTPA and ^{14}C-inulin have been made in a large number of patient series by Sziklas et al.,[114] and they report a high degree of correlation between the clearances obtained simultaneously. Gel filtration characteristics of the plasma and urinary radioactivity indicate that the complex is stable in vivo and that the radioactivity is not bound to a protein fraction.[113] Reasonably stable preparations of ^{99m}Tc-DTPA[116-119] have now become commercially available, and clearance studies with this agent have been reported in animals and in a limited number of patient series.[75,93,120-127] In dogs, using constant infusion techniques, no change in ^{99m}Tc-DTPA clearance was noted at different urine flow rates or during tubular blockade. Slightly lower clearance values have been obtained compared to iothalamate clearances in patients. About 5% of the activity is reported to be protein-bound and insignificant diffusion of activity into cellular blood fractions.[93,119] Biological distribution and excretion of ^{99m}Tc-DTPA has been studied in detail in rodent and nonrodent species and in man.[128,129] The renal handling of the agent under different degrees of renal impairment and the extrarenal pathways must be evaluated in greater detail, besides further study on the suitability and stability of the preparation from different suppliers for routine GFR determination with this agent.[127] Although the agent may be expected from the data noted above to underestimate GFR, it has several advantages. Clinically and in animals the underestimate is only about 5% during single injection clearance. The ideal physical properties of ^{99m}Tc and the simplicity and readily available DTPA labeling outweigh this disadvantage. Since most clinical situations can tolerate a 5% error, and since combined clearance and imaging are readily carried out with ^{99m}Tc-DTPA, this agent comes closest to being the current radiopharmaceutical of choice for estimating GFR on a routine basis.

B. Renal Plasma Flow Agents

The clearance of a tubular-secreted compound yields a measure of the renal plasma flow, provided it is completely extracted from the renal blood. The measurement of this parameter is based on the application of the Fick principle as stated previously. If the compound were as close to being totally extracted as possible in a single pass during the course of perfusion of the kidney, the calculated clearance would be somewhat less than the total renal plasma flow, whence a measure of the maximum effective renal plasma flow would be obtained. Paraaminohippuric acid (PAH) is eliminated by the kidneys partially through glomerular filtration (20%) and partially by tubular secretion (80%). During one passage through the normal kidney about 90% of the PAH in the blood is extracted by the tubule and the rest is returned to the general circulation. This compound has been chosen as the standard reference compound[12] for measuring the effective renal plasma flow (ERPF), since no other substance has been shown to achieve such a high extraction.

The extraction ratio is the fraction of renal arterial blood cleared of an agent and is expressed by the equation:

$$ER = \frac{A - V}{A} \qquad (1)$$

where ER = extraction ratio, A = arterial concentration of the agent, and V = venous concentration (renal) of the agent.

If the clearance of a substance and its extraction ratio are known, the total renal plasma flow can be calculated:

$$RPF = Cl \times \frac{1}{ER} \qquad (2)$$

where RPF = renal plasma flow and Cl = clearance. Further, if the renal arterial hematocrit is known, the renal blood flow can be calculated:

$$RBF = Cl \times \frac{1}{ER} \times \frac{1}{1 - HCt} \qquad (3)$$

where RBF = renal blood flow and HCt = hematocrit (renal arterial). Many organic acids of diverse structure and properties are subject to active tubular secretion.[12,130] Organic bases, too, are excreted in similar but separate mechanisms. Members of both classes of compounds undergo passive tubular reabsorption to a certain extent, which can partially be predicted on the basis of their physical properties and physiological variables, some of which are lipid solubility and pK_a. The tubular fluid pH and volume influence the degree of reabsorption or secretion. Several organic acids and bases secreted by the renal tubules have been studied, and the chemical structural requirement has been examined in detail with no clear definition of substrate specificity.[131] A compound that is rapidly excreted unchanged, and as a consequence not extensively metabolized, is ideally suited for the measurement of the effective renal plasma flow. Iodopyracet was the first substance used for the measurement of ERPF, but due to the complicated chemical method of estimation and a large extrarenal pathway, it has given way to *p*-aminohippuric acid (PAH). While the chemical method of estimation of ERPF using PAH is precise, a radioisotopic method is preferable in some situations, particularly if external measurements with rapid processing of clinical results are needed. PAH molecular structure is such that it does not appear to be possible to label it with a suitable gamma-emitting nuclide. Smith et al.,[12] suggested that OIH (orthoiodohippuric acid) might replace PAH for the measurement of the effective renal plasma flow. The preparation of ^{131}I-labeled OIH has added a valuable radiopharmaceutical for the measurement of ERPF.[132] The chemical structural formulas of PAH, OIH, and iodopyracet are shown in Figure 7.

1. Iodopyracet (Diodrast®)

This radiolabeled agent is considered for kidney function by Billion et al.[133,134] Clearance of tracer doses of iodopyracet ^{131}I are consistently 20 to 25% lower than that of PAH.[134] Significant biliary excretion precludes its use for renogram or single injection studies.[135,136] The measurement of ERPF with carrier iodopyracet administration[77,137] gave clearance ratios of ^{131}I iodopyracet to PAH of 0.96 to 1.03, which is quite close to the values obtained by using inactive compound.[138] The exact role of the carrier in the elevation of the clearances is not known, and awaits a more detailed understanding of the plasma binding of the compound and the renal handling of the bound form during the excretory process.

2. Orthoiodohippurate (Hippuran®)

The radiolabeling of this agent with ^{131}I, ^{125}I, or ^{123}I is accomplished by simple exchange reactions which were described first by Tubis et al.,[132] and later modified by other workers.[139-145] The most common and troublesome radiochemical impurity of radiohippuran is inorganic iodide, the content of which increases during storage and is influenced by a variety of factors including light, temperature, and specific

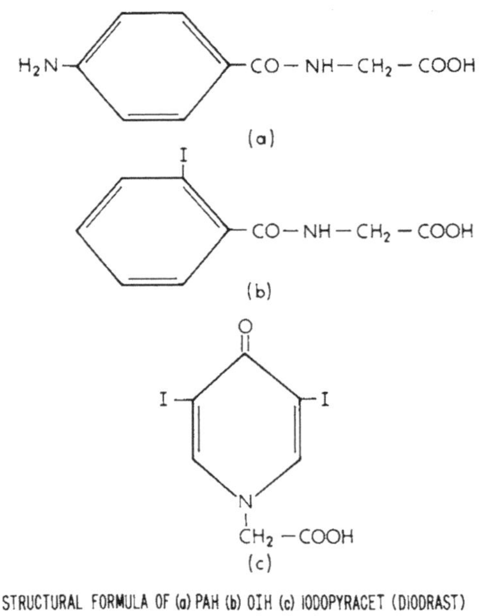

STRUCTURAL FORMULA OF (a) PAH (b) OIH (c) IODOPYRACET (DIODRAST)

FIGURE 7. Renal plasma flow agents in vogue. The major agents that are suitable for estimating renal plasma flow (a) paraaminohippuric acid, (b) orthoiodohippurate, and (c) iodopyracet.

activity.[146-148] No more than 2% of free iodide is acceptable in Hippuran® preparations used for clinical purposes. It is necessary to maintain adequate quality control of the preparations and to check free-iodine content at the time of use if potentially erroneous clearance values are to be avoided.

The application of the compound as a measure of ERPF was first studied in detail by Burbank et al.,[149] and by Schwarz and Madeloff,[150] who determined simultaneous renal clearances of Hippuran® and PAH in man. The clearance values of OIH were lower than those of PAH, with mean clearance ratios of OIH/PAH being 0.85. These data have been corroborated by several other workers in studies in both animals and in man. Tabulations of these extensive data are given by Mailloux and Gannon.[151]

Many reasons for the lower clearances of hippuran compared to PAH have been advanced. These include (1) presence of free radioiodine in the preparation, (2) plasma protein binding, and (3) difference in tubular transport. The literature data are somewhat conflicting in arriving at a plausible explanation for the discrepancy. However, a combination of all three factors in affecting the clearance values probably provides the full explanation, since the effect of any one of these factors is small. It is interesting that since Hippuran® is a weak acid, and tends to be in nonionized form in the distal tubule, a small amount of back diffusion may occur here. A detailed compartmental analysis of the distribution and mode of excretion of radioiodinated Hippuran® has been made by Blaufox et al.[13-15] (Figure 4 and Figure 8). It was reported that the injected dose is almost completely excreted from the body in the urine, with a negligible fraction in the enterohepatic cycle and in the red cell.

Currently the ^{131}I derivative is in widespread use.[1-4,125,152] This agent unfortunately cannot be used in large doses and the images it provides are of rather poor quality. Its use for clearance determinations have also been extended to children.[83] In this situation the risk of infection and and trauma of bladder catheterization may be less desirable than the small radiation burden incurred with the use of a radiopharmaceutical.

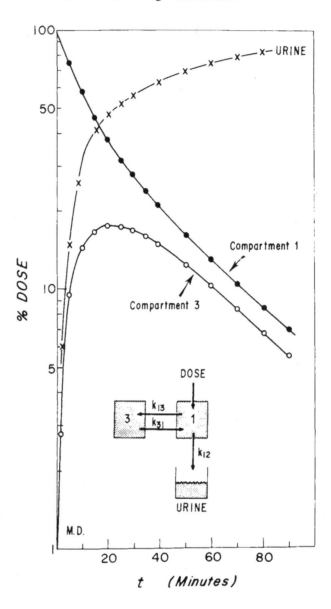

FIGURE 8.   Events occurring during the performance of a renogram. The renogram which traces time course of activity in the
kidney with an external detector reflects the kinetics of disposition
of the dose in three body compartments, blood (Compartment 1),
urine (Compartment 2) and extravascular fluid (Compartment 3),
each of which affects the renogram profile. The visual interpretation of this complex function to derive any clinical parameter is
fraught with too many errors and the general model shown here
displays the relative distribution of a dose of radiohippuran at any
point in time after a bolus injection. (From Blaufox, M. D. and
Freeman, L. M., *Semin. Nucl. Med.*, 3(1), 35, 1973. With permission.)

Increasing numbers of centers are using ^{123}I-labeled Hippuran®[145,153,154] which can
be used in much larger quantities. Renal clearance and extraction parameters of [123]I-
Hippuran® have been compared with ^{131}I-Hippuran® and PAH.[155] This agent is not
readily available because of the short half-life of ^{123}I and its great expense. It should

be noted that it has excellent properties and can be used for flow, clearance, and imaging studies.

C. Agents for Measurement of Renal Blood Flow

Clearances of agents which measure renal function can vary independent of renal blood flow and consequently the clearance data from these compounds cannot reliably substitute for the direct measurement of renal blood flow. The minimum renal plasma flow can be estimated by Hippuran® and the blood flow calculated using the relationship RBF = RPF (1/1 −HCt) which is discussed above. The renal hematocrit differs from the peripheral hematocrit by several percent and this affects the values thus obtained for the renal blood flow if the renal HCt is not known. In renal disease where the renal clearance is affected by reduced extraction, any estimate of renal blood flow may be quite misleading if it is calculated from clearance values. Thus functional tests like the renogram or clearances are unsuitable for accurate evaluation of the renal blood flow.

In order to obtain an accurate measurement of the renal blood flow, the extraction ratio of any substance which is excreted by the kidney can be determined by renal venous sampling and the renal blood flow for each kidney can be calculated. The extraction ratio is a measure of the efficiency of the kidney is removing a given substance from the circulation. Arterial samples or peripheral vein samples and renal vein samples must be obtained for its calculation.

The available techniques for the measurement of renal blood flow may be broadly classified under three categories:[156,157]

1. Methods based on the application of nondiffusible intravascular indicators, ^{131}I-labeled albumin, ^{99m}Tc-albumin and ^{32}P — or ^{51}Cr-labeled erythrocytes
2. Methods based on a very high extraction of an indicator or a test substance from blood involving use of agents excreted largely by tubular secretion, PAH or Hippuran®. Other agents including ^{86}Rb or ^{42}K, or radiolabeled microspheres and ^{131}I-labeled albumin macroaggregates have also been used
3. Use of inert diffusible indicators including ^{85}Kr and ^{133}Xe by gas washout techniques

These sophisticated methods require renal arterial and/or venous catheterization and in some cases ureteral catheterization. They are warranted only when an accurate estimate of individual kidney perfusion is urgently needed as in cases of renovascular hypertension and other severe renal disease states. Generally they have not proven to be of great value in clinical practice although they have provided a great deal of useful, basic physiologic information.

D. Agents for Quantitation of Functional Renal Mass

An accurate means of measuring unilateral renal function mass may be very important in the evaluation of the pathophysiology of renal disease and consequent therapeutic intervention. Measurements based on Hippuran® or DTPA clearances are semiquantitative and show significantly greater variations over a wide range of relatively uncontrollable circumstances. ^{131}I Hippuran® has been utilized for measuring the clearances of each kidney with Bianchi's technique based on ureteral compression.[158] In addition, there are some data which suggest that the relative slope of the second portion of the renogram obtained over each kidney may provide a reliable ratio of the relative fractional contribution of each kidney to total renal function.[159]

1. Mercurials

The quantitation of total renal mass has been achieved by radiotracer fixation of

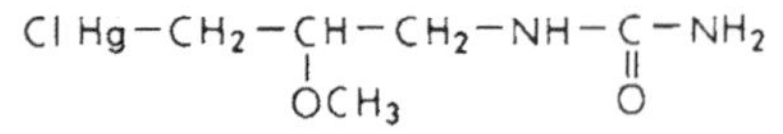

$$Cl\ Hg-CH_2-\underset{\underset{OCH_3}{|}}{CH}-CH_2-NH-\underset{\underset{O}{\parallel}}{C}-NH_2$$

FIGURE 9. Structural formula of chlormerodrin.

mercurials[160-163] (e.g., chlormerodrin, Figure 9) and in particular $^{197}HgCl_2$ by Raynaud.[164-166] These are bound to plasma proteins (or thiols) while in the circulation and reach the tubular fluid following secretion by the proximal tubules or filtration of the nonprotein-bound fraction. A large portion of the injected dose is fixed to the cells of the proximal tubule. The basic principle underlying this methodology is that the delivery of the radionuclide to the kidney is dependent upon renal blood flow and the functional renal mass. The absolute amount of radiopharmaceutical trapped by the kidneys can then provide an estimate of the relative renal blood flow between the two kidneys, and since under most circumstances this is proportional to the renal mass, an estimate of the relative renal mass results. Implicit here is the assumption that there is an accurate correlation between extraction and mass which is supported by the data showing that PAH extraction falls only late in the course of renal disease. In certain unilateral renal diseases such as pyelonephritis, this may not be completely true and the presence of these conditions would represent an inherent potential source of error with an underestimate of function of the pathologic kidney. If the extraction of both kidneys fall proportionately the basic method remains valid. An estimate of the functional renal mass may be quite accurate if one is correct in assuming that the extraction ability of normal tissue remains normal throughout the course of renal disease, and that only those tubules which are destroyed contribute to the fall in extraction, then the extraction would be directly proportional to the overall functional renal mass. The amount of radioactivity taken up by each kidney should bear a direct relationship to the functional tissue and should facilitate detection of changes in renal function and renal growth patterns.

Raynaud has evolved an elaborate quantitative technique[165] with $^{197}HgCl_2$ to estimate the relative renal uptake. When radio mercury is injected in soluble form, the renal radioactivity increases slowly to about 19% of the injected dose within 24 hr. It remains at this level until the 15th day, after which it decreases slowly during a period of the next 100 to 120 days. Daily urinary excretion of ^{197}Hg is small and practically constant representing 0.5% of the injected dose except for the first day on which it is approximately 1%. Fecal excretion is high during the first and second day and decreases rapidly thereafter. It can reach 5% of the injected dose during the first three days.[166] The rate of renal uptake from the second to the fifth day after injection is constant; it depends on the functional activity of the renal parenchyma. The renal uptake of mercury as a functional test along with clearance of inulin and PAH has been well-established by Raynaud.[164] Knowledge of the cellular behavior of mercury uptake is somewhat fragmentary. Mercury crosses the proximal tubular cell membrane either at the apical pole or at the basal pole. It is then fixed on metallothionein, a protein of small molecular weight which has approximately 50 mercaptan groups per mole.[167] The slow urinary excretion indicates that mercury associated with this molecule remains in the cell. Renal uptake of bismuth has also been proposed for evaluating individual renal function but it has not been studied in such great detail as mercury.[166]

The shortcoming of ^{197}Hg or ^{203}Hg is the high radiation dose, which prohibits its use in the U.S., and its high liver uptake, which poses a technical complication.

2. ^{99m}Tc-Complexes

Several radiopharmaceuticals that have been used to define renal anatomy are cur-

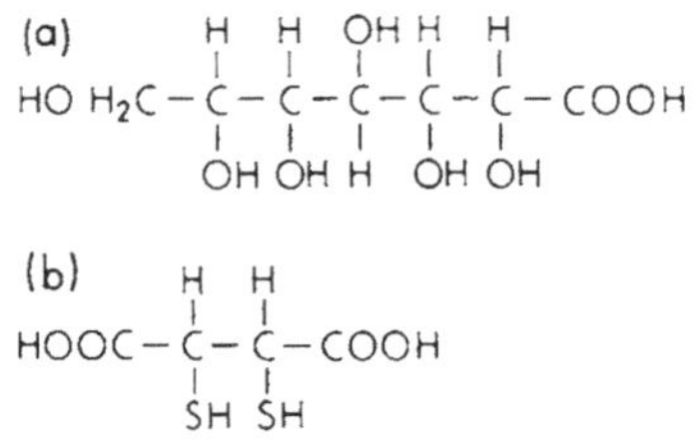

(a)

$$HO\ H_2C - \underset{\underset{OH}{|}}{\overset{\overset{H}{|}}{C}} - \underset{\underset{OH}{|}}{\overset{\overset{H}{|}}{C}} - \underset{\underset{H}{|}}{\overset{\overset{OH}{|}}{C}} - \underset{\underset{OH}{|}}{\overset{\overset{H}{|}}{C}} - \underset{\underset{OH}{|}}{\overset{\overset{H}{|}}{C}} - COOH$$

(b)

$$HOOC - \underset{\underset{SH}{|}}{\overset{\overset{H}{|}}{C}} - \underset{\underset{SH}{|}}{\overset{\overset{H}{|}}{C}} - COOH$$

FIGURE 10. Structural formula of a) glucoheptonate and b) dimercapto succinic acid.

rently available labeled with ^{99m}Tc including iron ascorbate, pencillamine acetazolamide, DTPA, glucoheptonate, and DMSA.[117,128,168-179] ^{99m}Tc-glucoheptonate and DMSA (dimercaptosuccinic acid) labeled with ^{99m}Tc (Figure 10) are currently widely used for renal morphology.[180-183] ^{99m}Tc-DMSA achieves higher renal concentration than any other renal agent.[177,184,185] The biological behavior of ^{99m}Tc-DMSA is similar to that of mercury with predominant cortical localization and very slight urinary excretion. Studies have shown good correlation of the uptake of ^{99m}Tc-DMSA with the relative blood flow and the relative creatinine clearance of each kidney.[186,187] The advantages of ^{99m}Tc-DMSA over ^{197}HgCl$_2$ include superior energy characteristics and dosimetry and the substantial reduction in the delay necessary before imaging. The only drawback in the utilization of ^{99m}Tc-DMSA has been the variation in the composition of the preparation which might interfere in the absolute uptake determinations.[184,185] Variations in the kidney depth, configuration, and rotation that will affect the measurements independent of actual organ functions have to be taken note of in order to utilize this agent as a renal functional agent. Additionally, the short half-life of ^{99m}Tc may necessitate imaging in patients at a time when the renal collecting system still contains some activity. Although liver uptake is low, it does present another potential problem.

V. METHODS OF RADIONUCLIDIC EVALUATION OF RENAL FUNCTION

As outlined in previous sections several radionuclides and radiopharmaceuticals have been developed for GFR, ERPF, and renal blood flow measurements. The experimental methodology utilized for these determinations is varied. Clearance methods either with continuous infusion or single injection techniques involving collection of urine and/or blood samples, and in vitro measurement of radioactivity have been developed with a great degree of sophistication to measure renal function parameters.[188] The use of continuous infusion clearances for measurement of renal function has been replaced clinically by single injection clearance measurements that provide a workable substitute. Radiorenography in conjunction with serial scintiphotography and precisely timed blood and urine sampling with a suitably labeled radiopharmaceutical yields a wealth of information on renal functional status. The clinical procedures for radiorenography and the analysis of renograms have become sophisticated with the interfacing of imaging equipment with advanced computer facilities. Pharmacokinetic modeling for radionuclide distribution and excretion in man has enabled more accurate and detailed analysis of the renogram.

The principles underlying the determination of each of these parameters are given along with the clinical method currently adopted at our institution.

A. Single Injection Clearances

Standard clearance methods for estimating GFR and EPRF are often tedious and

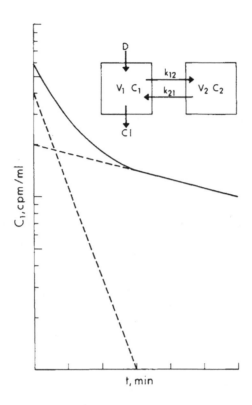

FIGURE 11. Standard two-compartment model and plasma disappearance curve. The standard two-component model and a typical plasma disappearance curve for a radiopharmaceutical excreted solely by the kidney is shown in Figure. The dose D is injected into the plasma, distributes in the central compartment in the Volume V_1 at a changing concentration C_1 owing to excretion by renal clearance (C1) or by diffusion into a second compartment V_2 at rates K_{12} or K_{21} which are dependent upon the relative concentrations in the two compartments. The plasma disappearance curve is a composite of two exponentials and satisfies the equation $X = A e^{-at} + B e^{-bt}$ where A and B are the relative intercepts and a and b are the slopes of the two components.

impractical in many chronic experiments since these methods require bladder catheterization and accurate collection of urine samples, as well as continuous intravenous infusion. The emergence of the single injection clearance technique, however, obviates some of these problems and this is one of the most important contributions of radionuclides to nephrology. The two most widely used single injection clearance methods are the two-compartment model and the simplified clearance method. The concepts of the theory and its application[13,108,189,190] for the typical two-compartment model and the single compartment model are given in Figure 11 and Figure 12. The dose injected into the central compartment with volume V_1 diffuses into and out of this compartment into a second compartment V_2 at a constant rate depending upon the relative concentrations between the two compartments. At the same time the injected dose would also be cleared by the kidney at a rate equal to its clearance. Analysis of the plasma disappearance curve plotted as counts/min/ml of plasma with time gives two

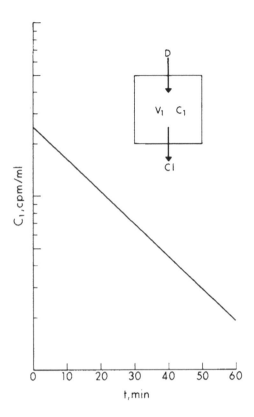

FIGURE 12. Single compartment model. A simplified single compartment model of radiopharmaceutical kinetics is shown in Figure. The dose is injected into a theoretical single compartment V_1 and is excreted by renal clearance (Cl). A predetermined portion of the clearance curve yields a single component curve from which clearance is calculated as: (Dose injected/Intercept at t_o) × slope of line.

slopes and intercepts of the two components from which the clearance is calculated as shown in Figure 13. In case one assumes that the material is distributed in a single compartment and is excreted only by the kidney as shown in Figure 12, the clearance is derived from the slope of the plasma disappearance curve and the zero time intercept concentration of the activity. The simplified clearance method is not quite accurate and may be applied in selected clinical situations only.

As mentioned earlier, the models assumed are a simplification of the events that occur after the intravenous injection of the radiopharmaceutical into the body. In the cases where an expanded extracellular fluid volume or ascites is present, distribution of the indicator in the extracellular fluid is slow and the plasma disappearance curve does not fit a simple two-component exponential function. In these instances these methods of estimation of the clearance tend to be overestimated.

1. Measurement of GFR

[51]Cr EDTA has been widely used for routine determination of glomerular filtration rate using single injection method.[111] At our institution ^{125}I-iothalamate or ^{99m}Tc-DTPA are routinely used. With the latter, one has to be aware of the problems and pitfalls for an accurate estimation of the GFR. ^{99m}Tc-DTPA preparation was made as

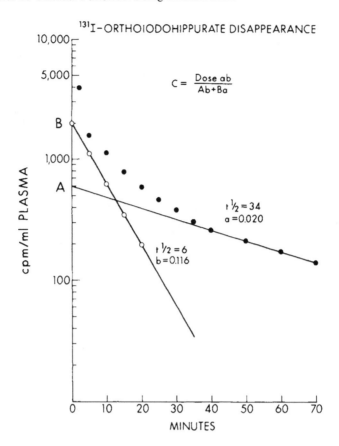

FIGURE 13.   Plasma disappearance curve for hippuran. Intravenous injection of ^{131}I-orthoiodohippurate is given and several timed blood samples are drawn during 1 hr post injection. The plasma radioactivity plotted against time on a semilog paper yields a curve which is resolved graphically into slow and fast components with intercepts A and B and slopes a and b, respectively. The clearance, C, is calculated from the expression shown in the insert.

per the instructions of the manufacturer using freshly milked ^{99m}Tc-pertechnetate from a generator system. Approximately 100 to 200 μCi of technetium DTPA is injected into the patient in an anticubital vein in a volume of 0.5 mℓ. A corresponding volume is drawn to serve as a standard to calibrate the dose. Immediately after injection, timed blood samples are drawn and the plasma radioactivity measured on a scintillation counter. Plasma radioactivity is then plotted on a semi-log graph paper and the two components resolved from the resulting curve and GFR calculated as shown in Figure 13. The GFR determinations measured using the technetium DTPA show a good degree of correlation with simultaneous iothalamate clearance estimations in animals, and the ratio approaches 1.

2. Measurement of Effective Renal Plasma Flow

The clearance of ^{131}I-iodohippurate by the single injection method has been shown[15,191] to have a good correlation with the urinary clearance of paraamino hippurate (PAH). The plasma disappearance curve as shown in Figure 13 is drawn by plotting the activity of multiple time blood samples or external monitoring of the disappearance of radionuclide from which the effective renal plasma flow is calculated by resolving the two components of the curve and applying the derived values in the equation with the injected dose. Approximately about 20 μCi of ^{131}I-orthoiodohippuric acid activity is injected into the patient and timed blood samples are drawn over a period of 1 hr at intervals of every 10 min, plasma activities measured from which the clearance is calculated as shown in Figure 13. Plasma radioactivity can also be estimated with the help of a scintillation counter placed over a vascular pool such as the temporal region or over the pericardium though this may not lead to as accurate results as the blood sample collection. Thus, the measurement of the effective renal plasma flow and the GFR by the above methods indicated would provide a means of assessing the relative tubular function and the glomerular filtration rate which is diagnostically quite informative.

B. Radiorenography

The radiorenogram was introduced into clinical nuclear medicine by Taplin et al.[54,92] to demonstrate the variation with time of the activity of a radionuclide within each kidney with the help of an externally placed detector after intravenous injection of a suitable radioactive material. With the introduction of ^{131}I-orthoiodohippurate by Tubis et al.,[132] the procedure has become considerably more informative in the evaluation of renal function.[159,193,194] The radiation probes placed over each kidney provide a curve indicative of the nature of the handling by the kidney of a given gamma-emitting radiopharmaceutical.

The events that occur during the performance of the renogram are indicated in Figure 8. Within 10 to 12 sec after the injection of 25 to 30 μCi of ^{131}I-orthoiodohippurate intravenously, and with the two detectors one placed over each kidney prompt increase in radioactivity is seen in the view of the radiation detector. Within a short interval of time, in about 30 to 40 sec, the rapidly rising count rate reaches a peak as the rate of uptake continues quite rapidly as the activity bolus is seen within the aorta and kidneys and major blood vessels. At this point in time, there is an inflexion of the curve after which the rise in the activity is less rapid and reaches a peak in 3 to 5 min which is a phase in which the bolus disperses into the body fluids, extracted by the kidney and transit to the renal tubules. During the renal transit time, the curve gradually declines owing to the removal of the activity in the renal pelvis followed by the pelvic transit into the lower urinary tract and accumulation in the bladder. The fall is quite rapid and exponential since successive entry of plasma into the kidney contains diminishing activity within the view of the detector. Thus the whole curve consists of the uptake, parenchymal transit or transfer form the tubules, removal, and the pelvic transit. Since the curve represents a composite of all these overlapping functions of several phases of the renal function, any individual functional evaluation from the overall curve is subject to error. Minor renal dysfunction cannot be diagnosed with any degree of confidence. At best the function of one kidney against the other can be compared in these studies, and these techniques have been useful and particularly sensitive in the detection of asymmetrical disease. A comparison of the curves thus generated over the two kidneys helps to differentiate patients with unilateral involvement from those with bilateral disease. This is shown in Figures 14, 15, and 16 indicating the marked asymmetry in the renogram which is a characteristic finding in unilateral renal disease and urinary obstruction of outflow from the left kidney.

With the advent of the scintillation camera coupled with data storage and processing systems, the precise outlines of the kidneys can be determined and time activity curves can be generated prospectively by flagging various areas of interest (Figure 17). Comprehensive evaluation of the renal function can be achieved by analysis of the radioisotope study using this technique.

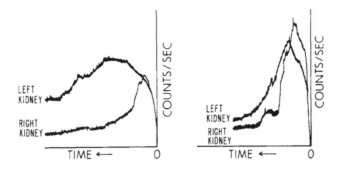

FIGURE 14.   Patient renogram pre- and post-surgical relief of ureteral vesical obstruction. Patient renogram pre- and post-surgical relief of ureteral vesical obstruction is shown. In the renogram at left
the clearance of the activity from the left kidney is sluggish whereas
the renograms at right are normal for both kidneys.

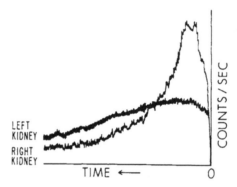

FIGURE 15.   Renogram in unilateral renal disease. Renogram in unilateral renal disease is
shown. Reduced uptake by the left kidney is
sharply in contrast to the normal right kidney
curve which is a characteristic finding in unilateral renal disease, as in renovascular hypertension. However, although a clear cut difference in
function exists between the two kidneys, the
highly sensitive study has only a modest specificity and specific etiologic diagnoses should not be
made without supporting information.

VI. SUMMARY

Pathophysiological assessment of renal disorders by radionuclides has become accepted in clinical nephrology and urology as a valuable noninvasive technique for the
evaluation of the urinary tract and for the diagnosis of renal disease. Relatively simple
determinations of serial clearances using radiopharmaceuticals for the assessment of
renal function have been established as valuable alternatives to the classical chemical
procedures that are time consuming and cumbersome for routine applications. Though
in many cases, the renal and extrarenal handling of the radiopharmaceutical by the
body and the distribution of the agent in the different body compartments under different pathological conditions have not been clearly established, a few agents like ^{125}I-
sodium iothalamate, ^{99m}Tc-DTPA are useful for routine clinical application for the

^{131}I ORTHOIODOHIPPURATE RENOGRAMS

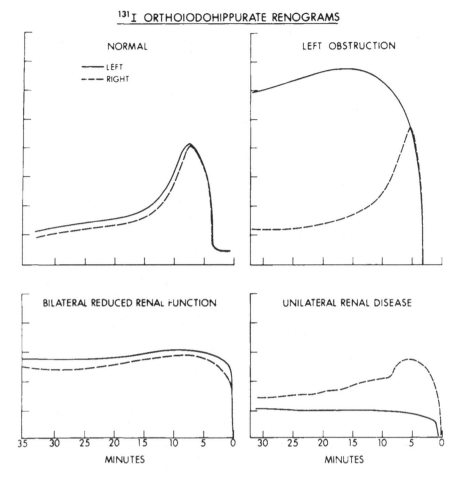

FIGURE 16. Typical renogram patterns in hippuran renography. Normal variations do occur in the normal renogram curves owing to variation in hydration, probe position and background activity. (Reproduced with permission from *PDR for Radiology and Nuclear Medicine*, Litton Industries, Oradell, N.J., 1972.)

measurement of GFR. Similarly, ^{131}I-hippuran is applicable for the measurement of the effective renal plasma flow. Among the agents currently available for clinical evaluation of the morphology of the kidney, ^{197}Hg has been utilized for the quantitative measurement of the functional mass of each kidney. An alternative agent which delivers a much lower radiation dose is ^{99m}Tc-DMSA, and this agent is currently being evaluated as a substitute for ^{197}Hg for functional mass determination of each kidney. Dynamic imaging studies coupled with data processing equipment using ^{99m}Tc-labeled radiopharmaceuticals including ^{99m}Tc-pertechnetate, ^{99m}Tc-glucoheptonate, and ^{131}I-Hippuran® aid in the evaluation of disorders of renal perfusion and other sources of reduced renal function.

ACKNOWLEDGMENTS

This work was supported in part by the American Cancer Society Grant VC-172 awarded to Elliott Robbins, M. D. and from Departmental Funds from the Division of Nuclear Medicine. Our thanks are due to Saqui Huq for assistance during the preparation and to Elsie Corvi for her excellent secretarial assistance.

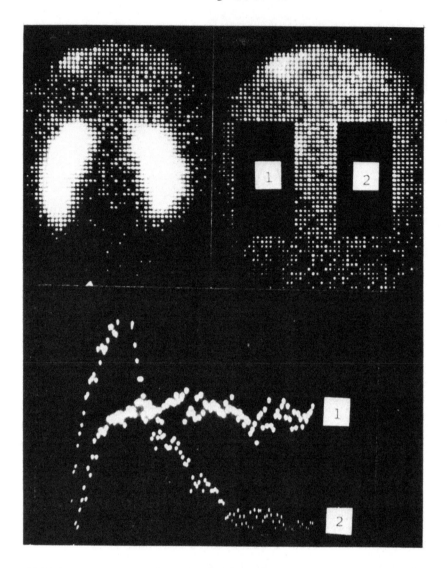

FIGURE 17.   Dynamic scintigraphic renal study. Following the dynamic Anger Camera study of the distribution of radioactivity in each kidney a specific area of interest on each kidney is selected and time activity curves are generated. The activity curves show that the curve over left renal area (1) is abnormal, with a diminished rate of uptake after the initial rise and reduced rate of outflow, compatible with urinary tract obstruction and/or a moderate reduction in renal function. In contrast for the right kidney (2) a typical normal renogram results in a prompt uptake of activity followed by a rapid decline. (From Blaufox, M. D. and Freeman, L. M., *Semin. Nucl. Med.*, 3(1), 33, 1973. With permission.)

REFERENCES

1. Blaufox, M. D., Ed., *Evaluation of Renal Function and Diseases with Radionuclides*, S. Karger, Basel, 1972.
2. Freeman, L. M. and Blaufox, M. D., Eds., *Radionuclide Studies of the Genitourinary System*, Grune & Stratton, New York, 1975.
3. Atkins, H. L. and Freeman, L. M., The investigation of renal disease using radionuclides, *Postgrad. Med. J.*, 49, 503, 1973.

4. Zum Winkel, K., Jost, H., Motzkus, F., and Golde, G., Renal function studies with radioisotopes, in *Dynamic Studies with Radioisotopes in Medicine*, International Atomic Energy Agency, Vienna, 1971.

5. Zum Winkel, K., Kidney secretion and filtration studies using radiotracers, in *Principles of Radiopharmacology*, Vol. 3, Colombetti, L. G., Ed., CRC Press, Boca Raton, 1979.

6. Chervu, L. R., Freeman, L. M., and Blaufox, M. D., Radiopharmaceuticals for renal studies, in *Radionuclide Studies of the Genitourinary System*, Freeman, L. M. and Blaufox, M. D., Eds., Grune & Stratton, New York, 1975.

7. Levine, W. G., Biliary excretion of drugs and other xenobiotics, *Annu. Rev. Pharmacol. Toxicol.*, 18, 81, 1978.

8. Brenner, B. M., Hostetter, T. H., and Humes, H. D., Molecular basis of proteinuria of glomerular origin, *N. Engl. J. Med.*, 298, 826, 1978.

9. Venkatachalam, M. A. and Rennke, H. G., The structural and molecular basis of glomerular filtration, *Circ. Res.*, 43, 337, 1978.

10. Bennett, W. M. and Porter, G. A., Endogenous creatinine clearance as a clinical measure of glomerular filtration rate, *Br. Med. J.*, 4, 84, 1971.

11. Woodruff, M. and Malvin, R., Localization of renal contrast media excretion by stop flow analysis, *J. Urol.*, 84, 677, 1960.

12. Smith, H., Finkelstein, N., Alminosa, M., Crawford, B., and Graber, M., The renal clearance of substituted hippuric acid derivatives and other aromatic acids in dogs and man, *J. Clin. Invest.*, 24, 388, 1945.

13. Blaufox, M. D., Orvis, A., and Owen, C. A., Jr., Compartmental analysis of the radiorenogram and distribution of ^{131}I-hippuran in dogs, *Am. J. Physiol.*, 204, 1059, 1963.

14. Blaufox, M. D. and Merril, J. P., Simplified hippuran clearance measurement of renal function in man with simplified hippuran clearances, *Nephron*, 3, 274, 1966.

15. Blaufox, M. D., Potchen, E. J., and Merril, J. P., Measurement of effective renal plasma flow in man by external counting methods, *J. Nucl. Med.*, 8, 77, 1967.

16. Pitts, R. F., *Physiology of the Kidney and Body Fluids*, Year Book Medical Publishing, Chicago, 1972.

17. Richards, A. N., Westfall, B. B., and Bott, P. A., Renal excretion of inulin, creatinine and xylose in normal dogs, *Proc. Soc. Exp. Biol. Med.*, 32, 73, 1934.

18. Shannon, J. A. and Smith, H. W., The excretion of inulin, xylose and urea by normal and phlorizinized man, *J. Clin. Invest.*, 14, 93, 1935.

19. Smith, H. W., *Principles of Renal Physiology*, Oxford University Press, New York, 1956.

20. Berlyne, G. M., Varley, H., Nilwarangkur, S., and Hoerni, M., Endogenous creatinine clearance and glomerular filtration rate, *Lancet*, 2, 874, 1964.

21. Cotlove, E., C-14 carboxyl-labeled inulin as a tracer for inulin, *Fed. Proc.*, 14, 32, 1955.

22. Marlow, C. G. and Sheppard, G., Labeled tracers of inulin for physiological measurements, *Clin. Chim. Acta*, 28, 469, 1970.

23. Marlow, C. G. and Sheppard, G., ^{51}Cr-EDTA, (hydroxymethyl-^{14}C) inulin and inulin-T for the determination of glomerular filtration rate, *Clin. Chim. Acta*, 28, 479, 1970.

24. Brooks, S. A., Davies, J. W. L., Graber, I. G., and Ricketts, C. R., Labeling of inulin with radioactive iodine, *Nature (London)*, 188, 675, 1960.

25. Concannon, J. P., Summers, R. E., Brewer, R., Cole, C., Weil, C., and Foster, W. D., ^{125}I allyl inulin for the determination of glomerular filtration rate, *Am. J. Roentgenol., Radium Ther., Nucl. Med.*, 92, 302, 1964.

26. Summers, R. E., Concannon, J. P., Weil, C., and Cole, C., Determination of simultaneous effective renal plasma flow and glomerular filtration rate with ^{131}I-o-iodohippurate and ^{125}I-allyl inulin, *J. Lab. Clin. Med.*, 69, 919, 1967.

27. Schmidt, H. A. E., Untersuchungen uber die Verwendbarkeit von Inulin-I-131 zur funktions-diagnostik der niere. II. Blutspiegel und Ausscheidung von inulin-I-131 Nach simultaner applikation. Eine Nuklearnudizinische method zur berechnung der inulin-I-131 clearance, *Klin. Wochenschr.*, 42, 967, 1964.

28. Hör, G., Steinhoff, H., Heinze, H. G., Pabst, H. W., and Hadid, D., ^{131}I-inulin for quantitative determination of glomerular filtration rate, *Acta Radiol.*, 6, 579, 1967.

29. Haas, J. P. and Prellwitz, W., Die Bestimmung der renalen und totalen Clearance mit jod-131-markiertem chlorjodpropyl-Inulin, *Radioaktive Isotope in Klinik und Forschung*, Vol. 7, Urban & Schwarzenberg, Munich, 1967, 462.

30. Adam, W. E., Hardt, H., Bonatz, K. G., and Bettge, S., Untersuchungen zur Bestimmung des glomerulusfiltrats mit radioaktivem inulin, *Klin. Wochenschr.*, 45, 818, 1967.

31. Tubis, M., Parsons, K., Rawalay, S. S., and Crandall, P. H., The preparation of labeled carbohydrates for biochemical uses, *J. Nucl. Med.*, 7, 338, 1966.

32. Bianchi, C., Hegesippe, E., Meozzi, A., Rosa, V., and Sossi, S., Effect of autoradiolysis on the renal clearance of ^{131}I labeled Hypaque and Hippuran, *Minerva Nucleare, 9*, 152, 1965.

33. Johnson, H. E., Hartley, B.,and Gollan, F., Preparation and properties of Cr-51 labeled inulin, *J. Nucl. Med., 8*, 97, 1967.

34. Johnson, A. E. and Gollan, F., Determination of glomerular filtration rate by external monitoring of Cr-51 labeled inulin, *Int. J. Appl. Rad. Isot., 19*, 43, 1968.

35. Materson, B. J., Johnson, A. E., and Perez-Stable, E. C., Inulin labeled with chromium-51 for determination of glomerular filtration rate, *JAMA, 207*, 94, 1969.

36. Cresseri, A. and Marro, F., Clearance renale della vitamina B_{12} nel coniglio, *Boll. Soc. Ital. Biol. Sper., 33*, 1662, 1957.

37. Watkin, D. M., Barrows, C. H. Jr., Chow, B. F., and Shock, N. W., Renal clearance of intravenously administered vitamin B_{12}, *Proc. Soc. Exp. Biol. Med., 107*, 219, 1961.

38. Hall, C. A. and Finkler, A. E., Measurement of the amounts of the individual vitamin B_{12} binding proteins in plasma, *Blood, 27*, 611, 1966.

39. Hall, C. A. and Finkler, A. E., Function of transcobalamin. II. A B_{12} binding protein in human serum, *Br. J. Haematol., 12*, 529, 1966.

40. Nelp, W. B., Wagner, H. N., Jr., and Reba, R. C., Renal excretion of vitamin B_{12} and its use in measurement of glomerular filtration rate in man, *J. Lab. Clin. Med., 63*, 480, 1964.

41. Cutler, R. E. and Glatte, H., Simultaneous measurement of glomerular filtration and effective renal plasma flow with [57]Co-cyano-cobalamin and ^{125}I-hippuran, *J. Lab. Clin. Med., 65*, 1041, 1965.

42. Shearman, D. J. C., Calvert, J. A., Ala, F. A., and Girdwood, R. H., Renal excretion of hydroxycobalamin in man, *Lancet, 2*, 1328, 1965.

43. Breckenbridge, A. and Metcalfe-Gibson, A., Methods of measuring glomerular filtration rate. A comparison of inulin, vitamin B_{12} and creatinine clearances, *Lancet, 2*, 265, 1965.

44. Jeremy, D. and McIver, M., Inulin, [57]Co-labeled vitamin B_{12} and endogenous creatinine clearances in the measurement of glomerular filtration rate in man, *Aust. Ann. Med., 15*, 346, 1966.

45. Foley, T. H., Jones, N. F., and Clapham, W. F., Renal clearance of [57]Co-cyanocobalamine: importance of plasma protein binding, *Lancet, 2*, 86, 1966.

46. Ekins, R. P., Nashat, F. S., Portal, R. W., and Sgherzi, A. M,. The use of labeled vitamin B_{12} in the measurement of glomerular filtration rate, *J. Physiol. (London), 186*, 347, 1966.

47. Weeke, E., [57]Co-cyanocobalamin in the detection of the glomerular filtration rate, *Scand. J. Clin. Lab. Invest., 21*, 139, 1968.

48. Ogilvie, R. I. and Ruedy, J., Determination of glomerular filtration rate with [57]Co-B_{12} and plasma protein binding of cyanocobalamin, *Can. J. Physiol. Pharmacol., 47*, 349, 1969.

49. Skeggs, H. R., Hanus, E. J., McCanley, A. B., and Rizzo, V. J., Hydroxocobalamin; Physiological retention in the dog, *Proc. Soc. Exp. Biol. Med., 105*, 518, 1960.

50. Withey, J. L. and Kilpatrick, G. S., Hydroxocobalamin and cyanocobalamin in Addisonian anaemia, *Lancet, 1*, 16, 1964.

51. Knapp, M. S. and Walker, W. H. C., Glomerular filtration rate, *Br. Med. J., 2*, 836, 1967.

52. Lasser, E. C., Farr, R. S., Fujimagari, T., and Tripp, W. N., The significance of protein binding of contrast media in roentgen diagnosis, *Am. J. Roentgenol., Radium Ther., Nucl. Med., 87*, 338, 1962.

53. Kimbel, K. H. and Borner, W., Uber den Verbleib von ^{131}I markiertem Urograffin im Korper, *Naunyn-Schmiedebergs Arch. Pharmacol. Exp. Pathol., 226*, 262, 1955.

54. Winter, C. C. and Taplin, G. V., A clinical comparison and analysis of radioactive diodrast, hypaque, miokon and urokon renograms as tests of kidney function, *J. Urol., 79*, 573, 1958.

55. McChesney, E. W. and Hoppe, J. O., Studies of the tissue distribution and excretion of sodium diatrizoate in laboratory animals, *Am. J. Roentgenol., Radium Ther., Nucl. Med., 78*, 137, 1957.

56. Blaufox, M. D., Sanderson, D. R., Tauxe, W. N., Wakim, K. G., Orvis, A. L., and Owen, C. A., Jr., Plasmatic diatrizoate-^{131}I disappearance and glomerular filtration in the dog, *Am. J. Physiol., 204*, 536, 1963.

57. Meschan, I., Deyton, W. E., Schmid, H. E., and Watts, F. C., The utilization of ^{131}I labeled renografin as an inulin substitute for renal clearance determination, *Radiology, 81*, 974, 1963.

58. Tauxe, W. N., Burbank, M. K., Maher, F. T., and Hunt, J. C., Renal clearances of radioactive orthoiodohippurate and diatrizoate, *Mayo Clin. Proc., 39*, 761, 1964.

59. Morris, A. M., Elwood, C., Sigman, E. M., and Catanzaro, A., The renal clearance of ^{131}I labeled meglumine diatrizoate (Renografin) in man, *J. Nucl. Med., 6*, 183, 1965.

60. Bianchi, C., Meozzi, A., and Zampieri, A., Glomerular filtration rate measured by the urinary clearance of hyposulfite and radioiodine labeled hypaque in subjects with normal and impaired renal function, *Minerva Nucleare, 9*, 157, 1965.

61. Denneberg, T., Clinical studies on kidney function with radioactive diatrizoate (Hypaque), *Acta Med. Scand. (Suppl.), 175*, 442, 1965.

62. Bianchi, C., Coli, A., Gallucci, L., Paci, A., Palla, R., and Rindi, P., The measurement of glomerular filtration rate in children by ^{131}I-Hypaque and external counting, *J. Nucl. Biol. Med.*, 11, 144, 1967.

63. Bianchi, C. and Blaufox, M. D., ^{131}I-hypaque and [140]La-DTPA for the measurement of glomerular filtration in dog, *J. Nucl. Biol. Med.*, 12, 117, 1968.

64. Schmid, H. E., Hutchins, P. M., and Muelbaecher, C. A., The continuous determination of the extraction and clearance of radioiodinated (^{131}I) diatrizoate, an inulin substitute, *J. Appl. Physiol.*, 25, 294, 1968.

65. Ram, M. D., Holroyd, M., and Chisholm, G. D., Measurement of glomerular filtration rate using ^{131}I-diatrizoate, *Lancet*, 1, 397, 1969.

66. Schmid, H. E., Meschan, I., Watts, F. G., Hosick, T., and Muelbaecher, C. A., Effect of free ^{131}I on renal excretion of diatrizoate ^{131}I, an inulin substitute, *Am. J. Physiol.*, 218, 903, 1970.

67. Maher, F. T. and Tauxe, W. N., Renal clearance in man of pharmaceuticals containing radioactive iodine: influence of plasma binding, *JAMA*, 207, 97, 1969.

68. Bianchi, C. and Toni, P., La determinazione delle clearances renali mediante traccianti radioattive, *Minerva Nefrol.*, 10, 116, 1963.

69. Chamberlain, M. J. and Sherwood, T., The extrarenal excretion of diatrizoate in renal failure, *Br. J. Radiol.*, 39, 765, 1966.

70. Segall, H. D., Gall bladder visualization following the injection of diatrizoate, *Am. J. Roentgenol.*, 107, 21, 1969.

71. Sigman, E. M., Elwood, C. M., and Knox, F., The measurement of glomerular filtration rate in man with sodium iothalamate-[131]I (Conray), *J. Nucl. Med.*, 7, 60, 1966.

72. Griep, R. J. and Nelp, W. B., Mechanism of excretion of radioiodinated sodium iothalamate, *Radiology*, 93, 807, 1969.

73. Sigman, E. M., Elwood, C., Reagan, M. E., Morris, A. M., and Catanzaro, A., The renal clearance of ^{131}I-labeled sodium iothalamate in man, *Invest. Urol.*, 2, 432, 1965.

74. Oester, A., Wolf, H., and Masten, P. O., Double isotope technique in renal function testing in dogs, *Invest. Urol.*, 6, 387, 1969.

75. Barbour, G. L., Crumb, K., Boyd, M., Reeves, D., Rastogi, P., and Patterson, R. M., Comparison of inulin, iothalamate, and ^{99m}Tc-DTPA for measurement of glomerular filtration rate, *J. Nucl. Med.*, 17, 317, 1976.

76. Hall, J. E., Guyton, A. C., and Farr, B. M., A single-injection method for measuring glomerular filtration rate, *Am. J. Physiol.*, 232, 72, 1977.

77. Elwood, C. M. and Sigman, E. M., The measurement of glomerular filtration and effective renal plasma flow in man by iothalamate ^{131}I and iodopyracet ^{131}I, *Circulation*, 36, 441, 1967.

78. Elwood, C. M., Sigman, E. M., and Treger, C., The measurement of glomerular filtration rate with ^{125}I sodium iothalamate (Conray), *Br. J. Radiol.*, 40, 581, 1967.

79. Houwen B., Donker, A. J. M., and Woldring, M. G., Simultaneous determination of glomerular filtration rate with ^{125}I-iothalamate and effective renal plasma flow with ^{131}I-Hippuran, in *Proc. of Symposium on Dynamic Studies with Radioisotopes in Medicine*, International Atomic Energy Agency, Vienna, 1971.

80. Maher, F. T., Nolan, N. G., and Elveback, L. R., Comparison of simultaneous clearances of ^{125}I-labeled iothalamate (Glofil) and of inulin, *Mayo Clin. Proc.*, 46, 690, 1971.

81. Gagnon, J. A., Schrier, R. W., Wies, T. P., Kokotis, W., and Mailloux, L. U., Clearance of iothalamate-^{125}I as a measure of glomerular filtration rate in the dog, *J. Appl. Physiol.*, 30, 774, 1971.

82. Cohen, M. L., Smith, F. G., Jr., Mindell, R. S., and Vernier, R. L., A simple reliable method of measuring glomerular filtration rate using single, low dose sodium iothalamate ^{131}I, *Pediatrics*, 43, 407, 1969.

83. Silkalns, G. I., Jeck, D., Earon, J., Chervu, L. R., Spitzer, A., Blaufox, M. D., and Edelman, C. M., Jr., Simultaneous measurement of glomerular filtration rate and renal plasma flow using plasma disappearance curves, *J. Pediatr.*, 83, 749, 1973.

84. Israelit, A. H., Long, D. L., White, M. G., and Hull, A. R., Measurement of glomerular filtration rate utilizing a single subcutaneous injection of ^{125}I-iothalamate, *Kidney Int.*, 4, 346, 1973.

85. Brouhard, B. H., Travis, L. B., Cunningham, R. J., III, Berger, M., and Carvajal, H. F., Simultaneous iothalamate, creatinine and urea clearances in children with renal disease, *Pediatrics*, 59, 219, 1977.

86. Tessitore, N., Lo Schiavo, C., Corgnati, A., Previato, G., Valvo, E., Lupo, A., Chiaramonte, S., Messa, P., D'Angelo, A., Zatti, M., and Machio, G., ^{125}I-Iothalamate and creatinine clearances in patients with chronic renal disease, *Nephron*, 24, 41, 1979.

87. Blaufox, M. D. and Cohen, A., Single injection clearances of iothalamate-^{131}I in the rat, *Am. J. Physiol.*, 218, 542, 1970.

88. Foreman, H., Vier, M., and Magee, M., The metabolism of ^{14}C labeled ethylenediaminetetraacetic acid in the rat, *J. Biol. Chem.*, 203, 1045, 1953.

89. Foreman, H. and Trujillo, T. T., The metabolism of ^{14}C labeled ethylenediaminetetraacetic acid in the human being, *J. Lab. Clin. Med.*, 43, 566, 1954.

90. Heller, J. and Vostal, J., Renal excretion of calcium-disodium ethylenediaminetetraacetic acid, *Experientia*, 20, 99, 1964.

91. Levine, W. G., Heavy metal antagonists, in *The Pharmacological Basis of Therapeutics*, Goodman, L. S. and Gilman, A., Eds., MacMillan, New York, 1970.

92. Stacy, B. D. and Thorburn, G. D., Chromium-51 ethylenediaminetetra acetate for estimation of glomerular filtration rate, *Science*, 152, 1076, 1966.

93. Klopper, J. F., Hauser, W., Atkins, H. L., Eckelman, W. C., and Richards, P., Evaluation of ^{99m}Tc-DTPA for the measurement of glomerular filtration rate, *J. Nucl. Med.*, 13, 107, 1972.

94. Funck-Brentano, J. L. and Leski, M., Measurement of glomerular filtration rate by external counting of the decay curve on plasma ^{140}La-DTPA, in *Proc. Int. Congr. Radioisotopes in the Diagnosis of Diseases of the Kidneys and the Urinary Tract*, Liege, 1967, ICS 178, Excerpta Medica, Amsterdam, 1967.

95. Reba, R. C., Hosain, F., and Wagner, H. N., Jr., Indium-113m diethylmetriaminepentaacetic acid (DTPA): A new radiopharmaceutical for study of the kidneys, *Radiology*, 90, 147, 1968.

96. Hosain, F., Reba, R. C., and Wagner, H. N., Jr., Visualization of renal structure and function with chelated radionuclides, *Radiology*, 93, 1135, 1969.

97. Reba, R. C., Poulouse, K. P., and Kirchner, P. R., Radiolabeled chelates for visualization of kidney function and structure with emphasis on their use in renal insufficiency, *Semin. Nucl. Med.*, 4, 151, 1974.

98. Prpic, B., Isotope nephrographie mit ^{68}Ga-EDTA in der ratte, *Nucl. Med. (Stuttgart)*, 6, 357, 1967.

99. Molnar, G., Pal, I., Stutzel, M., and Jaky, L., Determination of glomerular filtration rate with ^{51}Cr, ^{58}Co, ^{114m}In, ^{115m}In and ^{169}Yb labeled EDTA and DTPA complexes, in *Proc. Symp. Dynamic Studies with Radioisotopes in Medicine*, International Atomic Energy Agency, Vienna, 1971.

100. Pfeifer, K. J., Rothe, R., Bull, V., Frey, G., Muller, O. A., and Heinze, H. G., Bestimmung des glomerulum Filtrates mit ^{169}Yb-EDTA und des effectiven nieren plasma Stromes mit ^{131}I-hippuran, *Fortschr. Rontgenstr.*, 117, 456, 1972.

101. Myers, W. G. and Diener, C. F., EDTA-^{51}Cr gamma ray carrier, *J. Nucl. Med.*, 1, 124, 1960.

102. Winter, C. C. and Myers, W. G., Three new testing agents for the radioisotope renogram: DISA-^{131}I: EDTA-^{51}Cr and Hippuran-^{125}I, *J. Nucl. Med.*, 3, 273, 1962.

103. Garnett, E. S., Parsons, V., and Veall, N., Measurement of glomerular filtration rate in man using a ^{51}Cr edetic acid complex, *Lancet*, 1, 818, 1967.

104. Lingardh, G., Renal clearance investigations with ^{51}Cr-EDTA and ^{125}I-hippuran, *Scand. J. Urol. Nephrol.*, 6, 63, 1972.

105. Brochner-Mortensen, J., Giese, J., and Rossing, N., Renal inulin clearance versus total plasma clearance of ^{51}Cr-EDTA, *Scand. J. Clin. Lab. Invest.*, 23, 301, 1969.

106. Zum Winkel, K., Jahns, E., Herzfeld, U., and Georgi, M., Die strahlen Belastung in der nuklearmedizinischen Nierendiagnostik, in *Deutscher Rontgenkongress*, Thieme-Verlag, Stuttgart, 1967, 222.

107. Favre, H. R. and Wing, A. J., Simultaneous ^{51}Cr edetic acid, inulin and endogenous creatinine clearance in 20 patients with renal disease, *Br. Med. J.*, 1, 84, 1968.

108. Chantler, C., Garnett, E. S., Parsons, V., and Veal, N., Glomerular filtration rate measurement in man by the single injection method using ^{51}Cr-EDTA, *Clin. Sci.*, 37, 169, 1969.

109. Ditzel, J., Vestergaard P., and Brinklov, M., Glomerular filtration rate determined by ^{51}Cr-EDTA complex. A practical method based upon the plasma disappearance curve determined from four plasma samples, *Scand. J. Urol. Nephrol.*, 6, 166, 1972.

110. Hagstam, K. E., Nordenfelt, I., Svensson, L., and Svensson, S. E., Comparison of different methods for determination of glomerular filtration rate in renal disease, *Scand. J. Clin. Lab. Invest.*, 34, 31, 1974.

111. Bröchner-Mortensen, J. and Rodbro, P., Selection of routine method for determination of glomerular filtration rate in adult patients, *Scand. J. Clin. Lab. Invest.*, 36, 35, 1976.

112. Funck-Brentano, J. L., Lellouch, J., and Leski, M., Nouvelle methode de measure de la filtration glomerulaire sans prelevement d'urine. Measure de la clearance du DTPA lanthane 140 par enregistrement de la decroissance de la radioactive recueille par detection externe, *Rev. Fr. Etud. Clin. Biol.*, 12, 790, 1967.

113. Hosain, F., Reba, R. C., and Wagner, H. N., Jr., Measurement of glomerular filtration rate using chelated Yb-169, *Int. J. Appl. Radiat. Isot.*, 20, 517, 1969.

114. Sziklas, J. I., Hosain, F., Reba, R. C., and Wagner, H. N., Jr., Comparison of ^{169}Yb-DTPA, ^{113m}In-DTPA, ^{14}C-inulin and endogenous creatinine to estimate glomerular filtration, *J. Nucl. Biol. Med.*, 15, 122, 1971.

115. Bianchi, C., Coli, A., Palla, R., and Lo Moro, A., The reliability of ^{140}La-DTPA for the determination of glomerular filtration rate in man, *J. Nucl. Biol. Med.*, 17, 158, 1973.

116. Eckelman, W. C. and Richards, P., Instant ^{99m}Tc-DTPA, *J. Nucl. Med.*, 11, 761, 1970.

117. Hauser, W., Atkins, H. L., Nelson, K. G., and Richards, P., Technetium-99m DTPA: a new radiopharmaceutical for brain and kidney scanning, *Radiology*, 94, 679, 1970.

118. Atkins, H. L., Cardinale, K. G., Eckelman, W. C., Hauser, W., Klopper, J. F., and Richards, P., Evaluation of ^{99m}Tc-DTPA prepared by three different methods, *Radiology*, 98, 674, 1971.

119. Chervu, L. R., Lee, H. B., Goyal, Q., and Blaufox, M. D., Use of ^{99m}Tc-Cu-DTPA complex as a renal function agent, *J. Nucl. Med.*, 18, 62, 1977.

120. Hosain, F., Quality Control of ^{99m}Tc-DTPA by double tracer clearance technique, *J. Nucl. Med.*, 15, 442, 1974.

121. Kempi, V. and Persson, R. B., ^{99m}Tc-DTPA (Sn) dry-kit preparation: quality control and clearance studies, *Nucl. Med. (Stuttgart)*, 13, 389, 1975.

122. Heidenreich, P., Kreigel, H., Hör, G., Göger, H., Keyl, W., and Schramm, E., ^{99m}Tc-DTPA (Sn), Biologische und klinische Untersuchungen zur verteilung, kinetic und in-vivo-stabilitat einer renin clearance-substanz, *Strahlentherapie*, 74, 174, 1975.

123. Hilson, A. J. W., Mistry, R. D., and Maisey, M. N., ^{99m}Tc-DTPA for the measurement of glomerular filtration rate, *Br. J. Radiol.*, 49, 794, 1976.

124. Rossing, N., Bojsen, J., and Frederiksen, P. L., The glomerular filtration rate determined with ^{99m}Tc-DTPA and a portable cadmium telluride detector, *Scand. J. Clin. Lab. Invest.*, 38, 23, 1978.

125. Schlegal, J. U., Halikiopoulos, H. L., and Prima, R., Determination of filtration fraction using the gamma scintillation camera, *J. Urol.*, 122, 447, 1979.

126. Bianchi, C., Boradio, M., Donadio, C., Tramonti, G., and Figus, S., Measurement of glomerular filtration rate in man using DTPA-^{99m}Tc, *Nephron*, 24, 174, 1979.

127. Carlsen, J. E., Moller, M. L., Lund, J. O., and Trap-Jensen, J., Comparison of four commercial Tc-99m (Sn) DTPA preparations used for the measurement of glomerular filtration rate, *J. Nucl. Med.*, 21, 126, 1980.

128. Arnold, R. W., Subramanian, G., McAfee, J. G., Blair, R. J., and Thomas, F. D., Comparison of ^{99m}Tc-Complexes for renal imaging, *J. Nucl. Med.*, 16, 357, 1975.

129. McAfee, J. G., Gagne, G., Atkins, H. L., Kirchner, P. T., Reba, R. C., Blaufox, M. D., and Smith, E. M., Biological distribution and excretion of DTPA labeled with Tc-99m and In-111, *J. Nucl. Med.*, 20, 1273, 1979.

130. Weiner, I. M. and Mudge, G. H., Renal tubular mechanisms for excretion of organic acids and bases, *Am. J. Med.*, 36, 743, 1964.

131. Despopoulos, A., A Definition of substrate specificity in renal transport of organic anions, *J. Theor. Biol.*, 8, 163, 1965.

132. Tubis, M., Posnick, E., and Nordyke, R. A., Preparation and use of ^{131}I labeled sodium iodohippurate in kidney function tests, *Proc. Soc. Exp. Biol. Med.*, 103, 497, 1960.

133. Oeser, H. and Billion, H., Funktionelle Strahlen diagnostik durch etiketeirte Roentgenkontrastmittle, *Fortschr. Rontgenstr.*, 76, 431, 1952.

134. Billion, H. and Schlungbaum, W., Distribution of radioactive Perabrodil-M in the human organism and its application in renal clearance tests, *Klin. Wochenschr.*, 33, 1089, 1955.

135. Block, J. B. and Burrows, B. A., Hepatic transport of ^{131}I-Diodrast, *Clin. Res.*, 7, 34, 1959.

136. Fozzard, H. A., Diodrast (^{131}I) whole blood clearance as an index of renal blood flow, *Am. J. Physiol.*, 206, 309, 1964.

137. Elwood, C. M., Armenia, J., Orman, D., Morris, A., and Sigman, E. M., Measurement of renal plasma flow by iodopyracet ^{131}I, *JAMA*, 193, 771, 1965.

138. Chasis, H., Redish, J., Goldring, W., Ranges, H. A., and Smith, H. W., Use of sodium-p-aminohippurate for the functional evaluation of the human kidney, *J. Clin. Invest.*, 24, 583, 1945.

139. Scheer, K. E. and Meier-Borst, W., Die Darstellung von ^{131}I-orthoiodohippursaure durch Austauchmarkierung, *Nucl. Med. (Stuttgart)*, 2, 193, 1961.

140. Mitta, A. E. A., Fraga, A., and Veall, N., A simplified method for preparing ^{131}I-labeled hippuran, *Int. J. Appl. Radiat. Isot.*, 12, 146, 1961.

141. Elias, H., Arnold, C. H., and Koss, G., Preparation of ^{131}I-labeled m-iodohippuric acid and its behavior in kidney function studies compared to o-iodohippuric acid, *Int. J. Appl. Radiat. Isot.*, 24, 463, 1973.

142. Gillet, R., Cogneau, M., and Mathy, G., The preparation of I-123 labeled sodium-iodo hippurate for medical research, *Int. J. Appl. Radiat. Isot.*, 27, 61, 1976.

143. Thakur, M. L., Chauser, B. M., and Hudson, R. F., The preparation of iodine-123 labeled sodium orthoiodohippurate and its clearance by the rat kidneys, *Int. J. Appl. Radiat. Isot.*, 26, 319, 1975.

144. Sinn, H., Maier-Borst, W., and Elias, H., An efficient method for routine production of orthoiodohippuric acid labeled with ^{131}I, ^{125}I or ^{123}I, *Int. J. Appl. Radiat. Isot.*, 28, 809, 1977.

145. Zielinski, F. W., Holly, F. E., Robinson, G. D., Jr., and Bennett, L. R., Total and individual kidney function assessment with iodine-123 orthoiodohippurate, *Radiology*, 125, 753, 1977.

146. Anghileri, L. J., A chromatographic study of the stability of iodine-131 labeled sodium o-iodohippurate, *J. Nucl. Med.*, 4, 155, 1963.
147. Medina, F. C., Carrido, D. W. R., and Val Cob , M., Degradation of ^{131}I-hippuran, ^{131}I-lipiodol, ^{131}I-rose bengal and [198]Au-colloidal gold, *Proceedings of a Panel*, STI/PUB/253, International Atomic Energy Agency, Vienna, 1969, 181.
148. Hotte, C. E. and Ice, R. D., The in-vitro stability of ^{131}I-o-iodohippurate, *J. Nucl. Med.*, 20, 441, 1979.
149. Burbank, M. K., Tauxe, W. N., Maher, F., and Hunt, J. C., Evaluation of radioiodinated hippuran for the estimation of renal plasma flow, *Proc. Staff Meet. Mayo Clin.*, 36, 372, 1961.
150. Schwartz, F. D. and Madeloff, M. S., Simultaneous renal clearances of radiohippuran and PAH in man, *Clin. Res.*, 9, 208, 1961.
151. Mailloux, L. and Gagnon, J. A., Measurement of effective renal plasma flow, in, *Progress in Nuclear Medicine*, Vol. 2, Blaufox, M. D., Ed., University Park Press, Baltimore, 1972.
152. Clorius, J. H., Dreikorn, K., Zelt, J., Rapton, E., Weber, D., Rubinstein, K., Dahm, D., and Georgi, P., Renal graft evaluation with pertechnetate and I-131 Hippuran. A comparative clinical study, *J. Nucl. Med.*, 20, 1029, 1979.
153. Wellman, H. N., Berke, R. A., and Robbins, P. J., Dynamic quantitative renal imaging with ^{123}I-hippuran. A possible salvation of the renogram, *J. Nucl. Med.*, 12, 405, 1971.
154. Chisholm, G. D., Short, M. D., and Glass, H. I., The measurement of individual renal plasma flows using ^{123}I-hippuran and the gamma camera, *Br. J. Urol.*, 46, 591, 1974.
155. Stadalnik, R. C., Vogel, J. M., Jansholt, A. L., Krohn, K. A., Matolo, N. M., Lagunas-Solar, M. C., and Zielinski, F. W., Renal clearance and extraction parameters of ortho-iodohippurate (I-123) compared with OIH (I-131) and PAH, *J. Nucl. Med.*, 21, 168, 1980.
156. Blaufox, M. D., Methods for measurement of the renal blood flow, in, *Progress in Nuclear Medicine*, Vol. 2, Blaufox, M. D., Ed., University Park Press, Baltimore, 1972.
157. Grünfeld, J. P., Sabto, J., Bankir, L., and Funck-Brentano, J. L., Methods for measurement of renal blood flow in man, in *Radionuclide Studies of the Genitourinary System*, Freeman, L. M. and Blaufox, M. D., Eds., Grune & Stratton, New York, 1975.
158. Bianchi, C., Coli, A., Palla, R., and Rindi, P., Divided renal plasma flow measurement; the improvement of a new technique, in *Radioisotopes in the Diagnoses of the Kidneys and the Urinary Tract*, Timmermans, L. and Merchie, G., Eds., Excerpta Medica, Amsterdam, 1969, 273.
159. Taplin, G. V., Dore, E. K., and Johnson, D. E., The quantitative radiorenogram for total and differential renal blood flow measurement, *J. Nucl. Med.*, 4, 404, 1963.
160. Reba, R. C., Wagner, H. N., and McAfee, J. G., Measurement of [203]Hg chlormerodrin accumulation by the kidneys for detection of unilateral renal disease, *Radiology*, 79, 134, 1962.
161. Schlegel, J. V., Varela, R., and Stanton, J. J., Individual renal plasma flow determination without ureteral catheterization, *J. Urol.*, 96, 20, 1966.
162. Tabern, D. L., Kearney, J., and Sohn, H., The quantitative measurement of tubular chlormerodrin binding as an index of renal function: a study of 400 cases, *Can. Med. Assoc. J.*, 103, 601, 1970.
163. Gadbois, W. F. and Corriere, J. N., Jr., The use of segmental [197]Hg chlormerodrin delayed scans to evaluate and to follow individual renal function, *J. Urol.*, 112, 420, 1974.
164. Raynaud, C., Ricard, S., Karani, Y., and Kellershohn, C., The use of the renal uptake of [197]Hg as a method for testing the functional value of each kidney, *J. Nucl. Med.*, 11, 125, 1970.
165. Raynaud, C., A technique for the quantitative measurement of the function of each kidney, *Semin. Nucl. Med.*, 4, 51, 1974.
166. Raynaud, C., Study of renal cell function by radiotracer fixation, in *Principles of Radiopharmacology*, Vol. 3, Colombetti, L. G., Ed., CRC Press, Boca Raton, 1979, 89.
167. Pulido, P., Kagi, J. H. R., and Vallee, B. L., Isolation and some properties of human metallothionine, *Biochemistry*, 5, 1768, 1966.
168. Kountz, S. L., Yeh, S. H., and Wood, J., [99m]Tc (V) Citrate complex for estimation of glomerular filtration rate, *Nature (London)*, 215, 1397, 1967.
169. Charamaza, O. and Budikova, M., Method of preparation of ^{99m}Tc-Sn-Complex for renal scintigraphy, *Nucl. Med. (Stuttgart)*, 8, 301, 1969.
170. Winston, M. A., Halpern, S. E., Weiss, E. R., Endow, J. S., and Blahd, H. W., A critical evaluation of ^{99m}Tc-Fe-Ascorbic Acid complex as a renal scanning agent, *J. Nucl. Med.*, 12, 171, 1971.
171. Hauser, W., Atkins, H. L., and Richards, P., Renal uptake of ^{99m}Tc-Fe-Ascorbic acid complex in man, *Radiology*, 101, 637, 1971.
172. Winchell, H. S., Lin, M. S., Shipley, B., Sargent, T., and Katchalsky-Katzir, A., Localization of polypeptide caseidin in the renal cortex: a new radioisotope carrier for renal studies, *J. Nucl. Med.*, 12, 678, 1971.
173. Halpern, S., Tubis, M., Endow, J., Walsh, C., Kunsa, J., and Zwicker, B., ^{99m}Tc-Pencillamine-Acetazolamide complex. A new renal scanning agent, *J. Nucl. Med.*, 13, 45, 1972.

174. Halpern, S. E., Tubis, M., Golden, M., Kunsa, J., Endow, J., and Walsh, C., ^{99m}Tc PAC, A new renal scanning agent. II. Evaluation in humans, *J. Nucl. Med.*, 13, 723, 1972.

175. Stewart, R. D. H., Forster, L., and Ross, I. H., Radionuclide imaging of the kidneys and bladder using technetium-^{99m} gluconate, *Aust. N.Z. J. Med.*, 2, 336, 1972.

176. Boyd, R. E., Robson, J., Hunt, F. C., Sorby, P. J., Murray, I. P. C., and McKay, W. J., ^{99m}Tc-Gluconate complexes for renal scintigraphy, *Br. J. Radiol.*, 46, 604, 1973.

177. Lin, T. H., Khentigan, A., and Winchell, H. S., A ^{99m}Tc-Chelate substitute for organoradiomercurial renal agents, *J. Nucl. Med.*, 15, 34, 1974.

178. Taylor, A., Davis, G., Halpern, S., and Ashburn, W., ^{99m}Tc pencillamine: A renal cortical scanning agent, *J. Urol.*, 117, 418, 1977.

179. Hagan, P. L., Chauncey, D. M., Halpern, S. E., and Ayres, P. R., ^{99m}Tc-thiomalic acid complex: a non stannous chelate for renal scanning, *J. Nucl. Med.*, 18, 353, 1977.

180. Enlander, D., Weber, P. M., and dos Remedios, L. V., Renal cortical imaging in 35 patients. Superior quality with ^{99m}Tc-DMSA, *J. Nucl. Med.*, 15, 743, 1975.

181. Handmaker, H., Young, B. W., and Lowenstein, J. M., Clinical experience with ^{99m}Tc-DMSA, a new renal imaging agent, *J. Nucl. Med.*, 16, 28, 1975.

182. Bingham, J. B. and Maisey, M. N., An evaluation of the use of ^{99m}Tc-DMSA as a static renal imaging agent, *Br. J. Radiol.*, 51, 599, 1978.

183. Daly, M. J., Milutinovic, J., Rudd, T. G., Phillips, L. A., and Fialkow, P. J., The normal ^{99m}Tc-DMSA range, *Radiology*, 128, 701, 1978.

184. Ikeda, I., Inoue, O., and Kurata, K., Preparation of various ^{99m}Tc-DMSA complexes and their evaluation as radiotracers, *J. Nucl. Med.*, 18, 1222, 1977.

185. Razumenic, N. M. V. and Gorkic, D., Studies of chemical and biological properties of ^{99m}Tc-DMSA, *Eur. J. Nucl. Med.*, 1, 235, 1976.

186. Kawamura, J., Hosokawa, S., Yoshida, O., Fujita, T., Ishii, Y., and Torizuka, K., Validity of ^{99m}Tc-DMSA renal uptake for an assessment of individual kidney function, *J. Urol.*, 119, 305, 1978.

187. Daly, J., Jones, W., Rudd, T. G., and Tremann, J., Differential renal function using technetium-99m, dimercaptosuccinic acid (DMSA): in vitro correlation, *J. Nucl. Med.*, 20, 63, 1979.

188. Cohen, M. L., Radionuclide clearance techniques, *Semin. Nucl. Med.*, 4, 23, 1974.

189. Sapirstein, L. A., Vidt, D. G., Mandel, M. J., and Hanusek, G., Volume of distribution and clearances of intravenously injected creatinine in the dog, *Am. J. Physiol.*, 181, 330, 1955.

190. Rosenbaum, J. L., Kramer, M. S., Raja, R. M., Manchanda, R., and Lazaro, N., Determination of inulin and p-aminohippurate clearance without urine collection, *Nephron*, 10, 347, 1973.

191. Stokes, J. M. and Ter-Pogosiian, M. M., Double isotope technique to measure renal functions, *JAMA*, 187, 120, 1964.

192. Taplin, G. V., Meredith, O. M., Jr., Kade, H., and Winter, C. C., The radioisotope renogram, *J. Lab. Clin. Med.*, 48, 886, 1956.

193. Nordyke, R. A., Tubis, M., and Blahd, W. H., Use of radioiodinated hippuran for individual kidney function tests, *J. Lab. Clin. Med.*, 56, 438, 1960.

194. Farmelant, M. H., Lipetz, C. A., Bikerman, V., and Burrows, B. A., Radioisotope renal function studies and surgical findings in 102 hypertensive patients, *Am. J. Surg.*, 107, 50, 1964.

Chapter 7

ENDOCRINE SYSTEM

D. M. Lyster

TABLE OF CONTENTS

I. INTRODUCTION

This section will be limited to a review of radiopharmaceuticals used in nuclear medicine for visualizing the pancreas and prostate. Other organs of the endocrine system will be covered elsewhere in this volume. Unfortunately, the ability of nuclear medicine to develop radiopharmaceuticals that can be used routinely to visualize the pancreas and prostate has, at best, met with limited success. It would appear that the ability of nuclear medicine to demonstrate normal physiology in the body serves as a disadvantage when trying to develop radiopharmaceuticals that will localize in the pancreas or prostate, as no compound that is unique to these organs has, as yet, been found. Interference from activity in other organs or lack of concentration are the problems routinely encountered. In spite of our extensive knowledge of the biochemistry of these organs and ability to synthesize and label very specific molecules for localization, each review on the subject is compelled to conclude that a totally satisfactory radiopharmaceutical has not been found.[1-3]

II. PANCREAS

The pancreas, liver, intestinal mucosa, spleen, and bone marrow are the organs responsible for the majority of protein synthesis in the body. Therefore, they require a constant supply of amino acids. Digestive enzymes are rapidly synthesized in response to the emptying of the pancreas at each meal. As this response can be influenced by hormones or drugs, the investigation of potential pancreas-localizing agents has also concentrated on these areas.

Of all the amino acids, ^{75}Se-L-selenomethionine has been the most successful for pancreas imaging. Blau and Manske[4] found that selenium did not alter the biological behavior of this amino acid when substituted for sulfur. The oxidation state has also been found to be not critical. ^{75}Se-selenomethionine can be incorporated directly into the pancreas[5] as well as the oxidized form.[6] This mechanism is outlined in Figure 1. Support for direct incorporation was shown by Cowie and Cohen, who observed competition between selenomethionine and methionine in bacteria.[7] In whole blood analyzed 3 hr post-injection with ^{75}Se-selenomethionine, the activity was found in the TCA precipitable fraction.[8] It was felt that some ^{75}Se-selenomethionine was converted to ^{75}Se-selenocysteine which may be bound to proteins by a Se—S bond. In humans, the disappearance of ^{75}Se-selenomethionine has three phases.[9] Their biological half-lives are shown in Figure 2.

T_A is 10 hr, and may represent the elimination of unbound or nonpeptide bound ^{75}Se-selenomethionine and the pancreatic recycling of ^{75}Se-selenomethionine. T_B is 8 days, and may represent incorporation into the metabolizing protein pool. T_C is 90 days, and may represent incorporation into red cells, structural proteins, and other stable proteins. Distribution studies of ^{75}Se-selenomethionine have been carried out in mice and dogs,[10] rabbits,[11] and cats.[5] Although percent uptake varies widely, the pancreas-to-liver ratios are usually high enough to make this radiopharmaceutical minimally suitable for use in nuclear medicine. Because of the interference of activity in the liver and duodenum, various techniques are used to improve the pancreas-to-liver ratios.[12-17] Extensive investigation into the effects of pretreatment with various agents has met with limited success.[16,17] Only propylthiouracil significantly increased the pancreas uptake of ^{75}Se-selenomethionine. Many changes were noted, however, and it is suggested that more studies be carried out to determine the influence of endocrine and other factors on pancreas uptake.[17] Although ^{75}Se-selenomethionine is the agent of choice for pancreas scanning, it is far from the ideal radiopharmaceutical. Besides its variable uptake, it delivers a high radiation dose to the patient.[18]

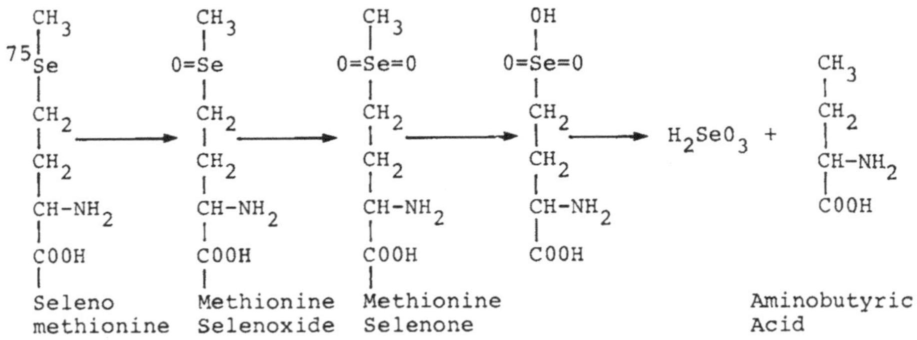

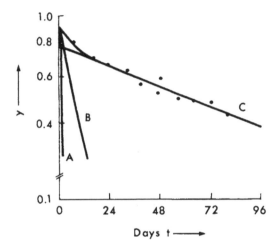

FIGURE 1. Oxidation of selenomethionine for incorporation into the pancreas.

FIGURE 2. ^{75}Se disappearance curve in humans. (From Ben-Porath, M., Case, L., and Kaplan, E., *J. Nucl. Med.*, 9, 168, 1968. With permission.)

Many other amino acids have been investigated for potential use in nuclear medicine.[19] Labeling of the various amino acids has been done with short-lived radionuclides or other halogens. Short-lived radionuclides can be used as peak levels in the pancreas are rapidly obtained. The amino acids, DL-valine,[20] phenylalanine and phenylglycine,[21] DL-tryptophan,[22,23] and methionine[24] have been labeled with carbon-11. The amino acids, L-alanine[25] and L-glutamic acid,[26] have been labeled with nitrogen-13. These have met with limited success depending upon the method of synthesis used. Enzymatic or biosynthetic methods produce the biologically active L-form, but chemical synthesis produces a racemic mixture which decreases the pancreas uptake.[28] Many amino acids have been labeled with the halogens, fluorine, bromine or iodine.[28-31] The iodine-123 labeled material has advantages due to the fact that one does not need to be close to a cyclotron, and its radiation characteristics are much superior to Selenium-75. Its pancreas-to-liver ratio was also very high during the first hour post injection.

As mentioned, many hormones and drugs have been investigated as potential pancreas scanning agents. Attempts to label various sulfonylureas for pancreas scanning have been unsuccessful.[32,33] As the β-cell content of the pancreas is low, this result is not surprising. Another receptor specific radiopharmaceutical, diazoxide, has been studied for pancreas localization.[34] However, in spite of the fact that it causes hyperglycemia by acting directly on the pancreas,[35] no significant localization was noted. The carbon-11[36] or iodo derivatives[37] of diphenylhydantoin, also a hyperglycemic agent, have also been unsuccessful for pancreas scanning.

Radiolabeled enzyme inhibitors only show localization when injected intraperitoneally.[38] However, as observed previously,[2] this material may be on the pancreas rather than in it. Similar localization was also seen for ^{99m}Tc-thioglucose after intraperitoneal injection.[39]

The use of zinc as a potential pancreas scanning agent has been extensively investigated because of the high zinc content of pancreatic tissue. All of these attempts have been unsuccessful and as Blau states,[2] "it is time we concentrated on other areas with a potential for success."

One area that seems to offer great promise for localization of radiopharmaceuticals is the use of labeled antibodies. Despite early failure,[40] many radiopharmaceuticals have been developed using labeled antibodies, and there is justification for the further investigation of these agents as pancreas scanning agents.

III. PROSTATE

Various radiopharmaceuticals have been proposed for prostate localization, and all have been relatively unsuccessful. Radiopharmaceuticals based on organ synthetic activity, hormonal control, and drugs, have all been investigated.

As with the pancreas, zinc has been extensively investigated as a potential prostate scanning agent due to its high content in this organ.[41-45] Due to the physical characteristics of zinc-62 outlined by Yano and Budinger,[46] ^{62}ZnCl$_2$ has been used in a patient with benign prostatic hypertrophy to visualize the prostate.[44] However, no significant accumulation was noted in the period needed to utilize this radionuclide for scanning.

Although the function of polyamines is unknown, they have been found in high concentration in the prostate of humans and rats.[47,48] Interaction of these polycationic substances with polyanionic macromolecules such as RNA and DNA, has been suggested, and they seem to play an important role in cell growth.[49] A proposed transport system for putrescene[50] and spermidine[51] has led to the belief that investigation of various polyamine analogs may lead to the development of a radiopharmaceutical with significant localization in the prostate, as well as the pancreas and rapidly growing tumors.[3] This transport system is shown in Figure 3. Uptake of activity in the prostate and pancreas of rats has been shown at one hour post injection of ^{3}H-putrescine and by six hours, 90% of the activity was in the form of spermidine and spermine.[52] Welsh et al.[53] have investigated N-methylated analogs of various polyamines and were able to visualize the dog prostate using carbon-11 labeled methyl putrescine. Based on these results, Counsell and colleagues[3] have initiated extensive structure activity relationship trials in an attempt to find the best radiopharmaceutical for use in nuclear medicine. Results of the radioiodinated putrescine and spermine analogues have, however, been disappointing as potential prostate scanning agents.

Androgens and estrogens have been shown to have specific binding sites in the prostate.[54-56] However, the total number of binding sites available in the prostate are limited, and radiopharmaceuticals with specific activities of 20 Ci/mmol are required, not to saturate the sites and allow visualization.[2,57] A radioiodinated analog of estradiol[58] and radioiodinated estradiol phosphate[59] have been used to visualize the prostate, but other investigators have not been as successful.[60] Specific activity may account for the difference. Skinner et al.[61] have had encouraging results using a selenium-75 labeled androgen analogue. This analogue was found to bind in the same protein fraction as ^{3}H-dihydro-testosterone (DHT). The usefulness of this will be limited due to the selenium-75.

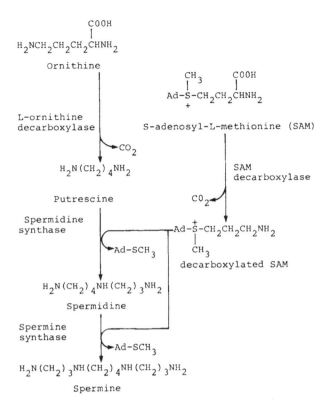

FIGURE 3. Biosynthetic pathway for polyamines.[3] (From Counsell, R. E. and Korn, N., *Principles of Radiopharmacology*, Vol. 1, Colombetti, L. G., Ed., CRC Press, Boca Raton, 1979, 215.)

IV. CONCLUSION

As mentioned, an ideal radiopharmaceutical for visualizing the pancreas or prostate has not been found, in spite of many sophisticated approaches to its development.

Part of the reason may be that research in this area has not been as concentrated as that of other areas in nuclear medicine. However, several approaches such as labeling antibodies show promise and significant improvements may be realized in the near future.

REFERENCES

1. **Risch, V. R.**, Radiopharmaceuticals in pancreatic imaging, in *The Chemistry of Radiopharmaceuticals*, Heindel, N. D., Burns, H. D., and Honda, T., Eds., Masson Publishing, New York, 1978, 53.
2. **Blau, M. and King, H. F.**, Radiopharmaceuticals for pancreas, prostate and adrenal imaging, in *Radiopharmaceuticals II*, Proc. 2nd Int. Symp. Radiopharmaceuticals, Seattle, Society of Nuclear Medicine, New York, 1979, 671.
3. **Counsell, R. E. and Korn, N.**, Biochemical and pharmacological rationales in radiotracer design, in *Principles of Radiopharmacology*, Vol. 1, Colombetti, L. G., Ed., CRC Press, Boca Raton, 1979, 215.
4. **Blau, M. and Manske, R. F.**, The pancreas specificity of 75 Se-Selenomethionine, *J. Nucl. Med.*, 2, 102, 1961.
5. **Hansson, E. and Blau, M.**, Incorporation of 75 Se-Selenomethionine into pancreatic juice proteins *in vivo*, *Biochem. Biophys. Res. Commun.*, 13, 71, 1963.

6. Cohen, Y. and Besnard, M., Fate of radiometabolites, in *Principles of Radiopharmacology*, Vol. 2, Colombetti, L. G., Ed., CRC Press, Boca Raton, 1979, 159.

7. Cowie, D. B. and Cohen, G. N., Biosynthesis by escherichia coli of active altered proteins containing selenium instead of sulfur, *Biochem. Biophys. Acta*, 26, 252, 1957.

8. Awwad, H. K., Potchen, E. J., Adelstein, S. J., and Dealy, J. B., 75 Se-Selenomethionine incorporation into human plasma proteins and erythrocytes, *Metabolism*, 15, 626, 1966.

9. Ben-Porath, M., Case, L., and Kaplan, E., The biological half life of 75 Se-selenomethione in men, *J. Nucl. Med.*, 9, 168, 1968.

10. Hoyte, R. M., Lin, S. S., Christman, D. R., Atkins, H. L., Hauser, W., and Wolf, A. P., Organic radiopharmaceuticals labelled with short-lived nuclides III ^{18}F-labelled phenylalanines, *J. Nucl. Med.*, 12, 280, 1971.

11. Melmed, R. N., Agnew, J. E., and Bouchier, I. A. D., Studies on the metabolism of 75 Se-L-selenocystine and 75 Se-Dl-mesoselenocystine with particular reference to their use in pancreatic scanning agents, *J. Nucl. Med.*, 10, 575, 1969.

12. Hatchette, J. B., Shuler, S. E., and Murison, P. J., Scintiphotos of the pancreas: analysis of 134 studies, *J. Nucl. Med.*, 13, 51, 1972.

13. Sodee, D. B., Radioisotope scanning of the pancreas with selenomethionine (75 Se), *Radiology*, 83, 910, 1964.

14. Agnew, J. E., Youngs, G. R., and Bouchier, I. A. D., Conventional and subtraction scanning of the pancreas: an assessment based on blind reporting, *Br. J. Radiol.*, 46, 83, 1973.

15. Landman, S., Polcyn, R. E., and Gottschalk, A., Pancreas imaging — is it worth it?, *Radiology*, 100, 631, 1971.

16. Atkins, H. L. and Som, P., Growth hormone and somatostatin effects on (^{75}Se) selenomethionine uptake by the pancreas, *J. Nucl. Med.*, 20, 543, 1979.

17. Cottrall, M. R. and Taylor, D. M., (^{75}Se) Selenomethionine uptake by the pancreas, *J. Nucl. Med.*, 20, 191, 1980.

18. Feller, P. A., Sodd, V. J., and Nishiyama, H., Absorbed dose comparison positron emitters ^{11}C, ^{13}N and ^{15}O versus gamma ray emitters, *Med. Phys.*, 6, 221, 1979.

19. Busch, H., Davis, J. R., Honig, G. R., Anderson, D. C., Nair, P. V., and Nyhan, W. L., The uptake of a variety of amino acids into nuclear proteins of tumor and other tissues, *Cancer Res.*, 19, 1030, 1959.

20. Washburn, L. C., Wieland, B. W., and Sun, T. T., 1 — ^{11}C DL-Valine, a potential pancreas imaging agent, *J. Nucl. Med.*, 19, 77, 1978.

21. Baalburg, W., Beerling-van der Molen, H. D., and Woldring, M. G., Evaluation of Carbon-11 labelled phenylglycine and phenylalanine pancreas scintigraphy, *Nucl. Med.*, 14, 60, 1975.

22. Hubner, K. F., Hayes, R. L., and Washburn, L. C., Scanning of the human pancreas with DL-valine-1-C11 and DL-tryptophan-1-C-11, *J. Nucl. Med.*, 19, 686, 1978.

23. Washburn, L. C., Sun, T. T., Byrd, B. L., Hayes, R. L., and Butler, T. A., DL-(Carboxyl-^{11}C) tryptophan, A potential agent for pancreatic imaging; production and preclinical investigations, *J. Nucl. Med.*, 20, 857, 1979.

24. Syrota, A., Comar, D., Cerf, M., Plummer, D., Maziere, M., and Kellershohn, C., (^{11}C Methionine pancreatic scanning with positron emission computed tomography, *J. Nucl. Med.*, 20, 778, 1979.

25. Spolter, L., Cohen, M. B., MacDonald, N., Chang, C. C., Takahashi, J., Neely, H., Huth, G., Meyers, S., and Bobinet, D., Synthesis and evaluation of N-13-L-alanine for pancreatic imaging, *J. Nucl. Med.*, 15, 535, 1974.

26. Lembares, N., Dinwoodie, R., Gloria, I., Harper, P., and Lathrop, K., A rapid enzymatic synthesis of 10 min ^{13}N-glutamate and its pancreatic localization, *J. Nucl. Med.*, 13, 786, 1972.

27. Stadalnik, R. C., Matolo, N. M., and Hein, L. J., L or D/L Tryptophan for pancreas scintigraphy?, *J. Nucl. Med.*, 19, 689, 1978.

28. Taylor, D. M. and Cottrall, M. F., Evaluation of amino acids labelled with ^{18}F for pancreas scanning, in *Radiopharmaceuticals and Labelled Compounds*, Vol. 1, International Atomic Energy Agency, Vienna, 1973, 433.

29. Atkins, H. L., Christman, D. R., Fowler, J. S., Hauser, W., Hoyte, R. M., Klopper, J. F., Lin, S. S., and Wolf, A. P., Organic radiopharmaceuticals labeled with isotopes of short half-life, V. ^{18}F-labeled 5 - and 6 - fluoro tryptophan, *J. Nucl. Med.*, 13, 713, 1972.

30. Kung, H. F., Gilani, S., and Blau, M., Pancreas specificity of iodo and bromo aliphatic amino acid analogs, *J. Nucl. Med.*, 19, 393, 1978.

31. Tisljar, U., Kloster, G., Ritzl, F., and Stocklin, G., Accumulation of radioiodinated L-Methyltyrosine, in pancreas of mice: concise communication, *J. Nucl. Med.*, 20, 973, 1979.

32. Boyd, C. M., Holcomb, G. M., Counsell, R. E., Beierwaltes, W. H., and Lieberman, L. M., Studies on the tissue distribution of ^{125}I-iodopropamide in the dog, *J. Nucl. Med.*, 12, 117, 1971.

33. Heindel, N. D., Risch, V. R., Burns, H. D., Honda, T., Brady, L. W., and Micalizzi, M., Syntheses and tissue distribution of 99mTc-sulfonylureas, *J. Pharm. Sci.*, 64, 687, 1975.

34. Heindel, N. D., Risch, V. R., Burns, H. D., Corley, E. G., Michener, E. A., and Honda, T., Synthesis and biological studies on an ^{125}I-analog of Diazoxide, in *Radiopharmaceuticals II*, Proc. 2nd Int. Symp. Radiopharmaceuticals, Society of Nuclear Medicine, New York, 1979, 697.

35. Scandellasci, C., Zaccaria, M., and Sicolo, N., Medical treatment of endogenous organic hyperinsulinism, *Horm. Metab. Res.*, Suppl., 6, 45, 1976.

36. Winstead, M. B., Parr, S., Lin, T. H., Khentigan, A., Lamb, J., and Winchell, H. S., Synthesis and *in vivo* distribution of ^{11}C-hydantoins, *J. Nucl. Med.*, 13, 479, 1972.

37. Balachandran, S., Beierwaltes, W. H., Ice, R. D., Fajans, S. S., Ryo, U. Y., Tedmond, M. J., Hetzel, K. R., Mosley, S. T., and Feldstein, B., Tissue Distribution of ^{14}C-, ^{125}I-, and ^{131}I-diphenylhydantoin in the toadfish, rat and human with insulinomas, *J. Nucl. Med.*, 16, 775, 1975.

38. Mosley, S. T., Kulkarni, P. V., and Parkey, R. M., Radiolabelled enzyme inhibitors as pancreatic scanning agents, in *Radiopharmaceuticals II*, Proc. 2nd Int. Symp. Radiopharmaceuticals, Society of Nuclear Medicine, New York, 1979, 709.

39. Risch, V. R., Honda, T., Heindel, N. D., Emrich, J. L., and Brady, L. W., Distribution of ^{99m}Tc-1-thioglucose in rats: effect of administration route on pancreatic specificity, *Radiology*, 124, 837, 1977.

40. Blau, M., Pancreas Scanning with Se-75 selenomethionine, in *Medical Radioisotope Scanning*, Vol. 2, International Atomic Energy Agency, Vienna, 1964, 275.

41. Mirand, E. A. and Bender, M., The selective uptake of zinc^{65} in normal and carcinomatous prostate glands of the rat, *Anat. Rec.*, 125, 618, 1965.

42. Johnston, G. S., Wade, J. C., Murphy, G. P., and Scott, W. W., ^{65}Zn and ^{69m}Zn studies in the dog, monkey and man: prostatic scanning, *J. Surg. Res.*, 8, 528, 1968.

43. Gold, F. M. and Lorber, S. A., Radioisotope ^{69m}Zn chloride prostate gland scan, *Invest. Urol.*, 8, 231, 1970.

44. Chisholm, G. D., Short, M. D., Chanadian, R., McRae, C. U., and Glass, H. I., Radiozinc uptake and scintiscanning in prostatic disease, *J. Nucl. Med.*, 15, 739, 1974.

45. Fruhling, J. and Coune, A., Radiozinc as a scintigraphic agent for the human prostate, *J. Nucl. Med.*, 16, 495, 1975.

46. Yano, Y. and Budinger, T. F., Cyclotron-Produced Zn-62: its possible use in prostate and pancreas scanning as a Zn-62 amino acid chelate, *J. Nucl. Med.*, 18, 815, 1977.

47. Rosenthal, S. M. and Tabor, C. W., The pharmacology of spermine and spermidine: distribution and excretion, *J. Pharmacol. Exp. Ther.*, 116, 131, 1956.

48. Bachrach, Y., *Function of Naturally Occurring Polycimines*, Academic Press, New York, 1973, 6.

49. Tabor, C. N. and Tabor, H., 1,4-Diaminobutane (putrescine) spermidine and spermine, *Annu. Rev. Biochem.*, 45, 285, 1976.

50. Pohjanpelto, P., Putrescine transport is greatly increased in human fibroblasts initiated to proliferate, *J. Cell. Biol.*, 68, 512, 1976.

51. Kano, K. and Oka, T., Polyamine transport and metabolism in mouse mammary gland, *J. Biol. Chem.*, 251, 2795, 1976.

52. Clark, R. B. and Fair, W. R., The selective *in vivo* incorporation and metabolism of radioactive putrescine in the adult male rat, *J. Nucl. Med.*, 16, 337, 1975.

53. Welch, M. J., Coleman, R. E., Straatmann, M. G., Asberry, B. E., Primeau, J. L., Fair, W. R., and Ter-Pogossian, M. M., Carbon-11-labelled methylated polyamine analogs: uptake in prostate and tumor in animal models, *J. Nucl. Med.*, 18, 74, 1977.

54. Ghanadian, R., Auf, G., Chaloner, P. J., and Chisholm, G. D., The use of methyltrienolone in the measurement of the free and bound cytoplasmic receptors for dihydrotestoterone in benign hypertrophied human prostate, *J. Steroid Biochem.*, 9, 325, 1978.

55. Shain, S. A. and Boesel, R. W., Androgen receptor content of the normal and hyperplastic canine prostate, *J. Clin. Invest.*, 61, 654, 1978.

56. deVoogt, H. J. and Dingjan, P., Steroid receptors in human prostatic cancer, a preliminary evaluation, *Urol. Res.*, 6, 151, 1978.

57. Sturman, M. F., Beierwaltes, W. H., Prakash, S., Ryo, U. Y., Ice, R. D., and Gitomer, W., Uptake of radiolabeled testosterone, 5-α-dihydrotestoterone, estradiol and pregnenolone by canine prostate, *J. Nucl. Med.*, 15, 94, 1974.

58. Szendroi, Z., Kocsar, L., Karika, Z., and Eckhardt, S., Isotope scanning of the prostate, *Lancet*, 7814, 1252, 1973.

59. Shida, K., Shimazaki, J., Kurihara, H., Ito, Y., Yamanaka, H., and Furuya, N., Uptake and scintiscanning of the prostate with ^{131}I Labelled estradiol phosphate, *Int. J. Nucl. Med. Biol.*, 3, 86, 1976.

60. Ghanadian, R., Waters, S. L., and Thakur, M. L., Studies with radioactive iodine labelled estrogens as prostatic imaging agents, *Int. J. Appl. Radiat. Isot.*, 26, 343, 1975.

61. Skinner, R. W. S., Pozderac, R. V., Counsell, R. E., Hsu, C., and Weinhold, P. A., Androgen receptor protein binding properties and tissue distribution of 2-selena-A-nor-5 α-androstan-17β-ol in the rat, *Steroids*, 30, 15, 1977.

Chapter 8

THE MECHANISMS OF LOCALIZATION, SPECIFICITY, AND METABOLISM OF ADRENAL GLAND IMAGING AGENTS

M.D. Gross, D.P. Swanson, D.M. Wieland, and W.H. Beierwaltes

TABLE OF CONTENTS

I. INTRODUCTION

The impetus for the development of radiopharmaceuticals that localize selectively within the adrenal glands stems from the difficulties encountered in the clinical evaluation of adrenal dysfunction. As a result of their retroperitoneal location and complicated biochemical functions, the adrenal glands do not lend themselves to simple anatomical or biochemical investigation. Methods (i.e., CAT, ultrasound) have been developed to provide anatomic information about the adrenal glands, but these modalities do not permit their functional evaluation. Adrenal scintigraphy, in a unique fashion, provides both functional and anatomic information concerning the status of the adrenal glands. This chapter will deal primarily with the development of adrenal imaging agents based on the cholesterol nucleus (Figure 1), including discussions regarding their specificity, metabolism, and mechanisms of localization at the cellular level.

II. RADIOCHOLESTEROL ADRENAL IMAGING AGENTS

A. Iodocholesterol Derivatives

Initial autoradiography and tissue distribution studies utilizing ^{14}C-cholesterol demonstrated a marked concentration of radioactivity in the adrenal glands of both mice and dogs at 1 to 2 days following intravenous administration.[1-3] The majority of this adrenal cholesterol was subsequently found to be present as ^{14}C-cholesterol esters,[3] an observation consistent with the knowledge that the adrenal cortex serves as a primary storage site for cholesterol of which 80 to 85% is in the esterified form.[4] As an important consideration for diagnostic imaging, it was also noted that the adrenal concentration of ^{14}C-cholesterol greatly exceeded that of the surrounding major organs, with adrenal:liver and adrenal:kidney radioactivity concentration ratios of 20:1 at 2 days postinjection.[3]

One of the earliest efforts to develop a radiolabeled cholesterol suitable for external imaging involved the direct iodination of cholesterol using sodium ^{131}I-iodide and chloramine-T.[5] Although the exact chemical nature of the final product was not determined, a diiodocholesterol derivative was most likely formed via addition of iodine across the 5 to 6 double bond. At 24 hr post injection, this ^{131}I-cholesterol resulted in an adrenal:liver radioactivity ratio of only 2.7:1, which was insufficient to permit distinct imaging of the adrenal glands.

In an attempt to produce a radiolabeled derivative which would not only retain the distribution characteristics of the parent cholesterol nucleus but also be stable in vitro and in vivo, Counsell et al.[6] synthesized [125]I-19-iodocholesterol (Figure 1) using previously developed methods for introducing iodine into the neopentyl, C-19 position of steroids. Subsequent [125]I-19-iodocholesterol (^{125}I-ICh) distribution studies in dogs demonstrated a concentration of radioactivity in the adrenal cortex which greatly exceeded that of surrounding organs. Adrenal:liver and adrenal:kidney radioactivity ratios at 2 days post administration were similar to those observed with ^{14}C-cholesterol.[3] In contrast to ^{14}C-cholesterol, the concentration of ^{125}I-ICh in the adrenal cortex did not continue to increase, but remained fairly constant over the 8-day interval studied. In addition, the extraadrenal concentrations of ^{125}I-ICh radioactivity cleared more rapidly (whole body T 1/2 biological = 6 days) than observed with ^{14}C-cholesterol (whole body T 1/2 biological = 62 days) resulting in adrenal:liver and adrenal:kidney ratios of over 100:1 and 200:1, respectively, at 8 days post injection. The thyroidal concentration of radioactivity from ^{125}I-ICh greatly exceeded that from ^{14}C-cholesterol, indicating that the radioiodinated derivative was undergoing in vivo metabolism and deiodination. However, when the adrenal cortices of dogs administered ^{125}I-ICh were

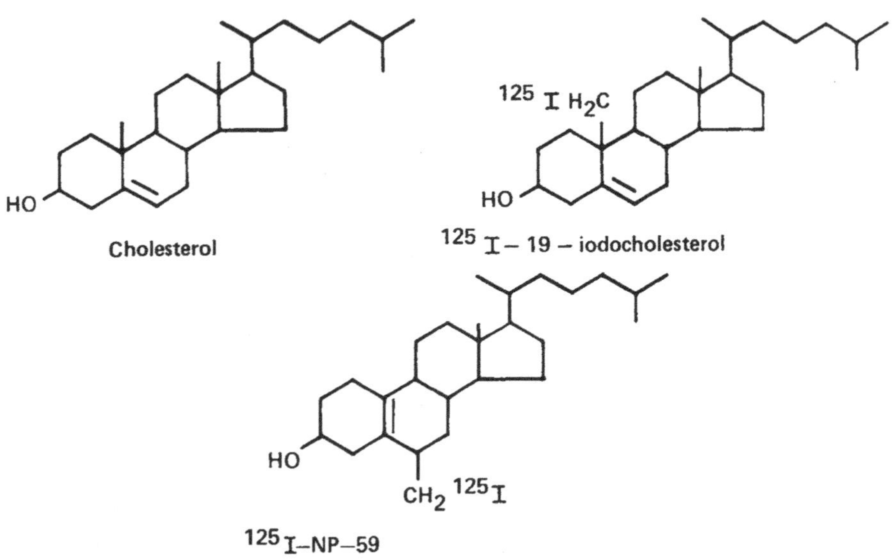

FIGURE 1.   The chemical structures of cholesterol, [125]I-19-iodocholest-5(6)-en-3β-ol ([125]I-19-io-docholesterol), and [125]I-6β-iodomethyl-19-norcholest-5(10)-en-3β-ol (^{125}I-NP-59).

extracted with chloroform, all of the radioactivity was recovered in the lipid phase, reflecting the absence of any significant levels of water-soluble free iodide in adrenal tissue.[3] Thin-layer chromatography of these lipid fractions suggested that at least a portion of the ^{125}I-ICh was in the ester form. This selective concentration of ICh in the adrenal cortex permitted the first external imaging of dog adrenals using a rectilinear scanner.[3]

During developmental research on the synthesis of ICh, a radiochemical impurity (other than free radioiodide) was consistently observed.[7,8] This impurity accounted for 10 to 20% of the radioiodinated ICh activity when acetone was used as the refluxing medium as per the original synthetic methodology.[6] If, however, ICh was refluxed in isopropanol or absolute ethanol, the "impurity" increased to 90%, leaving only 10% ICh. This "impurity" was subsequently isolated and identified as 6β-iodomethyl-19-norcholest-5(10)-en-3β-ol (NP-59, Figure 1).[7,8] The rearrangement of ICh to NP-59 apparently occurs as a result of the formation of two discrete homallylic isomeric cations of the steroid series; with ketonic solvents such as acetone or methyl ethyl ketone leading to a predominance of 19-iodosteroids, and alcoholic solvents such as ethanol or isopropanol giving a predominance of the 6β-iodomethyl-19-nor steroids.[8,10]

Tissue distribution studies in rats revealed that ^{125}I-NP-59 concentrated 5 to 10 times greater in the adrenal cortex than ^{125}I-ICh.[7,9] This elevated adrenal concentration remained relatively constant up to 15 days after administration.[9] The extraadrenal (with exception of thyroid) concentrations of ^{125}I-NP-59 peaked at one day and gradually decreased, but at a somewhat slower rate than that observed with ^{125}I-ICh. This latter fact may be a result of the greater in vivo stability of NP-59 as demonstrated by the thyroidal concentration of ^{125}I-iodine radioactivity, which was less than half the value observed with ^{125}I-ICh. In rats, ^{125}I-NP-59 adrenal:liver radioactivity ratios of 336:1 and adrenal:kidney ratios of 286:1 were reported at 5 days after administration.[9] Tissue distribution studies in dogs resulted in ^{125}I-NP-59 adrenal cortex:liver and adrenal cortex:kidney radioactivity ratios of approximately 100:1 at the same time interval.[9] As with ^{125}I-ICh, chloroform extraction of the dog adrenal cortices following ^{125}I-NP-59 administration showed no free radioiodide in the adrenal tissue with all radioactivity being recovered in the lipid phase.[9] Adrenal images of excellent quality were obtained on all dogs imaged with ^{131}I-NP-59.[9] Thus, in view of the higher adrenal uptake, equivalent target:background radioactivity ratios, superior images, and greater in vivo stability; ^{131}I-NP-59 became the agent of choice over ICh for adrenal imaging.

It should be noted that the initial distribution and imaging studies indicated above were performed with radioiodinated NP-59 which contained varying amounts of radioiodinated ICh as a radiochemical impurity. Likewise, initial studies performed using radioiodinated ICh contained varying amounts of NP-59 impurities. Recently, procedures have been developed for the synthesis of virtually pure ^{131}I-NP-59 and ^{131}I-ICh.[10-12] Tissue distribution data from rats injected with these high purity agents indicate that NP-59 concentrates in the adrenal gland 18 (Day 1), 51 (Day 3), and 27 (Day 7) times more avidly than ICh.[13] It is of further interest that these authors were unable to detect rat adrenal tissue by external imaging techniques up to 9 days post administration of pure ^{131}I-ICh. In addition, chemical and radiochemical stability studies on high purity NP-59 and ICh confirmed previous studies which indicated that radiolabeled NP-59 was more stable in vitro than radiolabeled ICh.[7,14] Either ^{125}I or ^{131}I-NP-59, in formulation and stored at 5°C, will remain greater than 97% radiochemically pure for at least 15 days; whereas radioiodinated ICh, under the same conditions, undergoes 20% decomposition after 3 weeks.

The demonstration that [131]I-6β-iodomethyl-19-norcholest-5(10)-en-3β-ol (NP-59) localizes to a greater extent in the adrenal gland than ICh challenged the basic theory that the exact cholesterol structure is a requirement for development of adrenal-specific radiopharmaceuticals. In an attempt to explain this phenomenon and to define the structural requirements essential for adrenal localization, Kojima et al. compared the rat tissue distribution of the parent compound of NP-59, [3]H-6β-methyl-19-norcholest-5(10)-en-3β-ol, with the distribution of [1α, 2α(n)-^{3}H] cholesterol.[15] The adrenal concentration of the [3]H-6β-methyl-19-norcholesterol was 4 to 8 times greater than ^{3}H-cholesterol at 3 and 7 days, respectively; with adrenal:liver radioactivity ratios of 251:1 for the [3]H-6β-isomer and only 4:1 for ^{3}H-cholesterol at 7 days. The fact that the adrenal activity of the 6β-methyl-19-nor derivative remained constant while the cholesterol activity decreased, provided additional evidence that the 6β-isomer was not metabolized by the same pathway. As a result of this study, it is apparent that the structural features of the 6β-methyl-19-nor-nucleus impart a greater adrenal affinity and decreased metabolic clearance than the native cholesterol nucleus.

In view of these studies, numerous derivatives of NP-59, ICh, and cholesterol have been subsequently synthesized in order to determine structural factors which may further affect adrenal affinity and accelerate the parameters of imaging. To determine the stereochemical requirements of the hydroxy group at the steroidal C-3 position, the radioiodinated 3-α-ol epimers of ICh and NP-59 were synthesized and their tissue distributions analyzed.[16] The 3α-epimer of NP-59 demonstrated 10 times less accumulation in the adrenal gland than the 3β-derivative, but 10 times greater adrenal localization than 3α-ICh, which underwent significant in vitro deiodination. It may, therefore, be concluded that a β-configuration of the C-3 steroidal hydroxyl group is an important requirement for adrenal gland specificity.

The substitution of radiobromine for radioiodine in the NP-59 structure (i.e., [82]-Br-6β-bromomethyl-19-norcholest-5(10)-en-3β-ol) resulted in an agent which showed selective adrenal gland concentration persisting for over 120 hr following administration.[17] This adrenal affinity was less than that of the original radioiodinated NP-59, but greater than radiolabeled ICh. The liver uptake of this brominated derivative was also observed to be slightly greater than either radioiodinated NP-59 or ICh.

Direct iodination of cholesterol at the C-6 position resulted in a radiopharmaceutical, ^{131}I-6-iodocholesterol (Figure 4), reported to be suitable for adrenal imaging in humans.[18] Although the adrenal:liver and adrenal:kidney ratios were lower than those

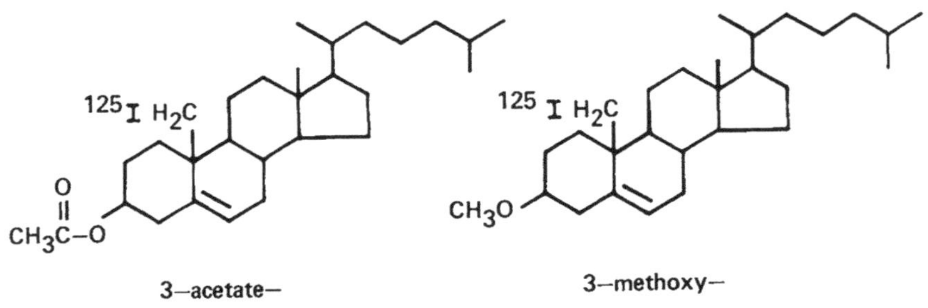

FIGURE 2.   3β-hydroxy substituted ester and ether derivatives of [125]I-19-iodocholesterol.

observed with NP-59, this agent may possess advantages over NP-59 in that it is reported to be stable to in vitro deiodination for up to 6 months at room temperature and demonstrates low in vivo thyroidal concentrations of liberated ^{131}I-iodine.

It has been shown that free and esterified forms of cholesterol exhibit distinct differences in regard to their uptake into the adrenal gland.[19] In view of this, various esters of radioiodinated ICh and NP-59 were synthesized, and their adrenal localization compared to the parent alcohols. The 3β-acetate of ICh (Figure 2) demonstrated the same adrenal localization as the parent compound, but rapidly deiodinated in vivo.[20,21] This may possibly be explained by the observation that this derivative underwent rapid hydrolysis in vivo to the parent compound.[20] In any event, the 3β-acetate derivative of NP-59 also showed nearly equal adrenal affinity as the parent radiopharmaceutical.[21] These studies are in contrast, however, to those performed by Counsell et al.,[22] wherein both the acetate and palmitate esters of ICh failed to show any appreciable adrenal uptake. In conjunction with this area of investigation, Szinai and Owoyale[20] substituted methoxy (Figure 2) and ethoxy groups for the 3β-hydroxy group of radioiodinated ICh. These derivatives also showed similar concentrations in the adrenal as ICh, and in addition appeared to be more stable. The propoxyl derivative formed an oil upon synthesis, and was not further evaluated. The results of these studies indicated that the 3β-hydroxy group of the radioiodinated steroids is not essential for adrenal localization, and that the 6β-methyl-19-nor derivatives continue to retain greater affinity for the adrenal than the 19-methyl analogs.

The often difficult multiple step synthetic procedures, the problems associated with separation and purification of final products, and the in vitro instability of radioiodinated steroids have prevented a thorough systematic structure-distribution study of this class of radiopharmaceuticals. A possibility of facilitating these studies was realized when Szinai and Owoyale[20] reported that the adrenal uptake of [125]I-3β-iodocholestene (Figure 3, Compound 1) was 2½ times greater than ^{131}I-ICh. Unfortunately, the adrenal:lung, adrenal:liver, and adrenal:spleen radioactivity ratios observed with [125]I-3β-iodocholestene in this study were lower than those observed with ^{131}I-ICh. Interestingly, a subsequent study using high specific activity [131]I-3β-iodocholestene and a formulation (1.8% Tween® 80, 8.2% absolute ethanol, dilute to volume with normal saline) that promoted in vivo solubility resulted in adrenal concentrations and target background ratios similar to the more adrenal specific NP-59.[23] These results were confirmed by another group who compared the adrenal specificity of radiolabeled 3β-fluoro(^{18}F), bromo ([77]Br), and iodo(^{131}I) derivatives of cholesterol with NP-59 and ^{131}I-ICh.[21] The 3β-iodo derivative of cholesterol accumulated in the adrenal more actively than the bromo or fluoro compounds; showing somewhat less localization than NP-59. The adrenal uptake increased steadily until 2 to 3 days for all of the agents studied with the exception of 3β-fluorocholesterol and ^{131}I-ICh, which reached peak adrenal concentrations at 2 hr post injection. To investigate the possibility that 3β-iodocholes-

STRUCTURE	COMPOUND NO.	R	RATIO $\left(\dfrac{\text{Adrenal Conc. NP–59}}{\text{Adrenal Conc. }3\beta\text{–iodo-cholestene}}\right)$
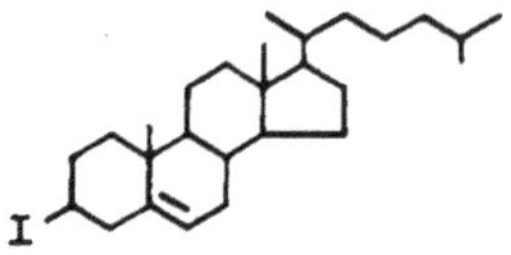	1	—	0.96
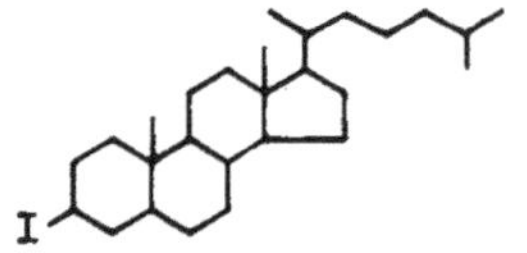	2	—	1.04
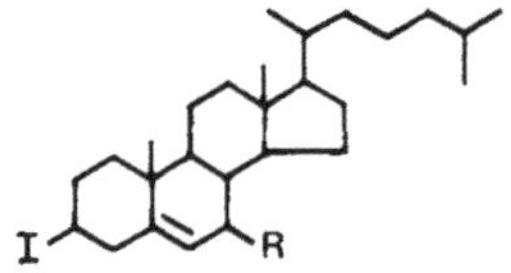	3	= O	unstable
	4	OH	unstable
	5	OCOCH₃	unstable
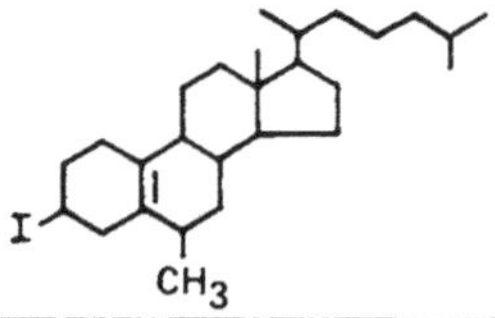	6	—	3.87
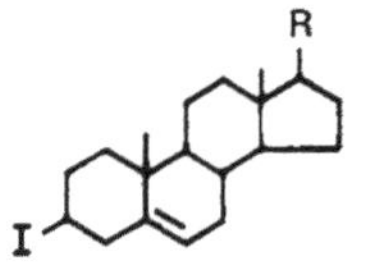	7	H	17.43 ⎫ maximum
	8	= O C₂H₅	10.17 ⎬ at 2 hrs. post injection
	9	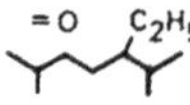	4.74 ⎭
	10	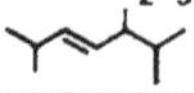 C₂H₅	10.84

FIGURE 3.   Comparison of adrenal affinity between NP-59 and various derivatives of 3β-iodocholestene at 5 days post injection.

terol undergoes adrenal metabolism to free iodine which subsequently binds to unsaturated fatty acids, these authors compared the retention of ^{125}I-3β-iodocholesterol and ^{14}C-3β-iodocholesterol in the adrenal. The clearance of ^{125}I-radioactivity from the adrenal was significantly faster than ^{14}C-activity, indicating negligible binding of free iodine.

Although adrenal images in dogs using ^{131}I-3β-iodocholestene were inferior to those obtained with NP-59,[24] a subsequent structure distribution study of 3β-iodo steroids was performed in consideration of the advantages associated with placing radioiodine at the C-3 position of cholesterol.[24] As discussed previously, the 3β-hydroxy group of cholesterol is not a requirement for adrenal localization. The radioiodinated 3β-iodocholestenes can be rapidly synthesized in a one step reaction from readily available precursors, and the alkyl iodide of 3β-iodocholestene demonstrates less than 2% in vitro deiodination after 12 days at room temperature.[24] The adrenal (dog) localization of several ^{125}I-3β-iodosteroids analyzed in this study are compared to NP-59 in Figure 3.[24] Reduction of the 5 to 6 double bond of cholesterol (Figure 3, Compound 2) had little effect on adrenal concentration, but a lower adrenal:liver ratio (i.e., increased liver uptake) was noted. Functionalization of the C-7 position with oxygen-containing moieties in an attempt to promote rapid hepatic clearance (similar to bile salt forma-

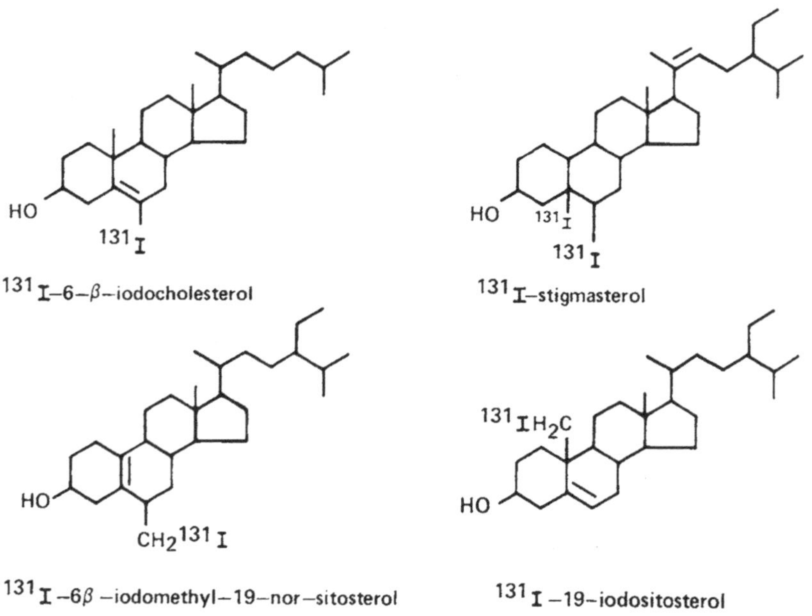

FIGURE 4. Other radioiodinated steroids proposed for adrenal imaging.

tion) resulted in derivatives (Figure 3, Compounds 3 to 5) which rapidly deiodinated in vitro. Interestingly, the adrenal uptake of the 6β-methyl-19-nor derivative of 3β-iodocholestene (Figure 3, Compound 6) was a factor of almost 4 less than 3β-iodocholestene. This observation in the 3β-iodocholestene series is in contradiction with the previously discussed studies wherein 6β-methyl-19-nor cholesterol shows greater adrenal affinity than cholesterol itself. The adrenal uptake of 3β-iodosteroid also demonstrated extreme sensitivity to modifications of the 17-β-side chain. Replacement of the lypophilic side chain of cholesterol with a hydrogen atom (Figure 3, Compound 7), or a keto group (Figure 3, Compound 8) completely abolished adrenal localization; an observation consistent with the low adrenal uptakes and target to background radioactivity ratios reported with the radioiodinated progesterone and pregnenolone derivatives,[25] and the ^{18}F-pregnenolone derivatives synthesized by Spitznagle and co-workers.[26] Even minor modifications of this lipophilic 17-β-side chain significantly reduced adrenal specificity as demonstrated by Compounds 9 and 10 (Figure 3).

In conjunction with this latter structural modification, the adrenal localization of ^{131}I-stigmasterol (Figure 4, synthesized by the chloramine-T method with probable diiodination across the 5 to 6 double bond) has been studied, with reported adrenal:liver and adrenal:kidney ratios of 7.5:1 and 44:1, respectively, at 2 days post administration.[5] Also, radioiodinated 6β-iodomethyl-19-norsitosterol (Figure 4) and 19-iodositosterol (Figure 4) have been synthesized and their tissue distributions analyzed.[27] The [131]I-6β-iodomethyl-19-norsitosterol showed a similar concentration (maximum at 3 days) in the adrenal as NP-59, but its liver uptake was approximately twice that of NP-59 over the 7-day period analyzed. In contrast, [131]I-19-iodositosterol underwent considerable deiodination in vivo. The peak adrenal concentration of radioactivity from the 19-iodo derivative was approximately 40 times less than the 6β-methyl-19-nor derivative. These results demonstrated a significantly lower adrenal specificity of radioiodinated 19-iodositosterol than studies on the same material by Counsell et al.[28] This discrepancy may again be due to the presence of the more adrenal specific [125]I-6β-iodomethyl-19-norsitosterol as a radiochemical impurity in Counsell's [125]I-19-iodositosterol.

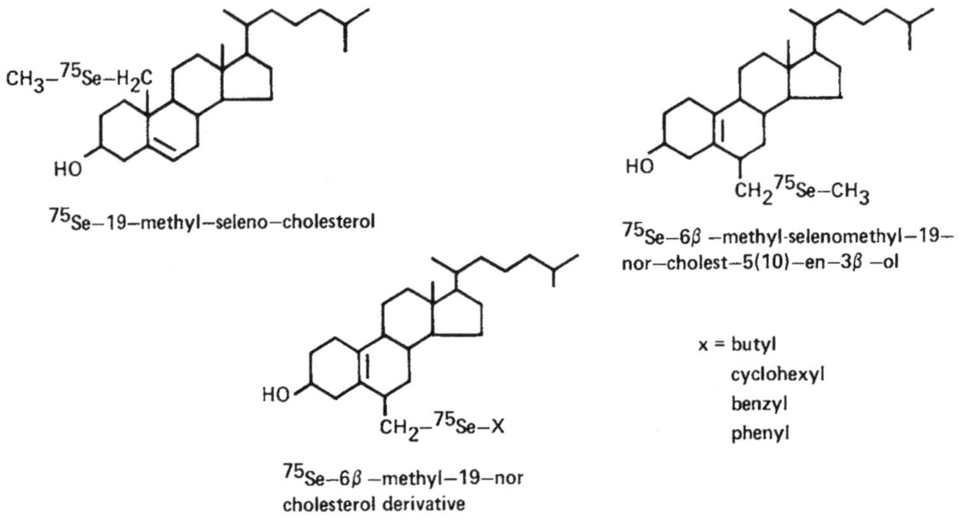

FIGURE 5.   Chemical structures of ^{75}Se-selenocholesterols proposed for adrenal imaging.

Although ^{131}I-NP-59 presently remains the radiolabeled steroid of choice for human adrenocortical imaging, there are some problems associated with its use. The estimated radiation dose (based on animal distribution studies) to the adrenals and ovaries following the administration of 1 mCi of NP-59 has been reported to be 150 rad and 8 rad, respectively.[29] However, using percent uptake data (0.16% administered dose/adrenal) obtained from clinical studies, the radiation dose to the adrenals is significantly lower, 27.5 rad/mCi.[30] A further decrease in adrenal dosimetry may be expected with the addition of dexamethasone suppression. The 8-day half-life of the I-131 radiolabel also presents limitations in regard to shelf-life and availability of the agent. In attempts to alleviate these problems, various other radionuclide-labeled derivatives of cholesterol, ICh, and NP-59 have been proposed for adrenal imaging.

A. Selenium-Cholesterol Derivatives

The use of a selenium-75 radiolabel may offer several advantages over iodine-131: a lower beta absorbed dose (7% that of I-131); a higher usable photon yield, which may permit a decreased administered dose; a longer physical half-life, which adversely affects the patient radiation dose, but also increases the shelf-life, and hence availability of the radiolabeled agent.[31] In view of these advantages, the synthesis and tissue distribution of ^{75}Se-19-methylseleno-cholest-5(6)-en-3β-ol (Figure 5), ^{75}Se-6β-methyl-selenomethyl-19-norcholest-5(10)-en-3β-ol (Figure 5), their esters, and other derivatives (Figure 5) have been reported.[31,32] A comparison of the adrenal concentrating ability of several of these ^{75}Se-selenosterols has been summarized by Riley.[32] Basically, the results of this study are in agreement with the previously discussed tissue distribution studies of the iodosterols. The ^{75}Se-6β-methyl-selenomethyl-19-nor derivative demonstrated approximately a 2.6 times greater adrenal affinity than ^{75}Se-19-methylseleno-cholesterol. The acetate esters of each of these agents only slightly altered their adrenal gland uptake. Addition of more lipid-soluble groups at the 6β-selenomethyl position (i.e., butyl-, cyclohexyl-, benzyl-, or phenyl-selenomethyl) reduced adrenal localization by a factor of 2 to 3 over ^{75}Se-6β-methyl-selenomethyl-19-nor-cholesterol. This latter agent has subsequently been developed and marketed in Europe as Scintidren®, and may have an additional advantage over radioiodinated steroids as a result of its greater in vivo stability, but this factor may also be responsible for the reported increase in background activity.[31]

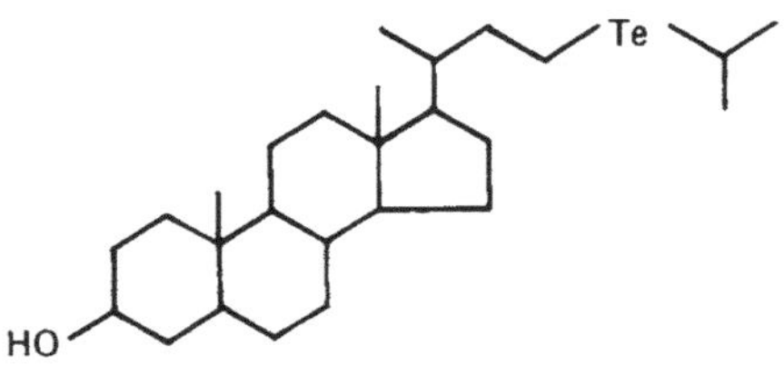

123m Te—cholesterol

FIGURE 6.   Chemical structure of [123]Te-24-
nor-23-(isopropyl telluro)-5α-cholane-3β-ol.

B. Tellurium-Cholesterol Derivatives

Tellurium-123m labeled steroids have also been developed as possible alternatives to NP-59 for adrenal imaging. Tellurium-123m decays by isomeric transition (T ½ = 120 days) with the emission of a 159 keV gamma radiation that is more suited for current imaging instrumentation than the radioactive emissions of either selenium-75 or iodine-131. Adrenal:liver radioactivity ratios of 25:1 at one day, increasing to 65:1 at 7 days post injection have been reported with [123m]Te-24-nor-23-(isopropyl tellura)-5α-cholan-3β-ol ([123m]Te-cholesterol, Figure 6) injected into rats.[33] Clear images of the rat adrenals were obtained at one day after injection. Chromatography of the lipid extracts of the adrenals revealed non-polar radioactive materials corresponding to [123]Te-steroid esters in addition to more polar radioactive components which may represent metabolites of the parent compound. A subsequent structure distribution study on [123]Te-steroids illustrated the following points: (1) as with the radioiodinated steroids, replacement of the 3β-hydroxy group with a 3α-hydroxy group abolished adrenal uptake; (2) shortening the 17-β-side chain of [123m]Te-cholesterol by two carbons also abolished adrenal localization; (3) the addition of an lipophilic 8 carbon group to the 17-β-side chain of [123m]Te-cholesterol resulted in very slow adrenal accumulation; (4) whereas increasing the length of the 17-β-side chain of [123m]Te-cholesterol by one carbon and replacing the 5 to 6 double bond slightly increased adrenal uptake; (5) however, the 3β-methoxy derivative of the latter agent decreased adrenal specificity by a factor of greater than five.[34]

III. MECHANISMS OF RADIOCHOLESTEROL ADRENAL GLAND LOCALIZATION AND METABOLISM

Although not the only agents demonstrating selective adrenal cortical specificity, the iodocholesterol derivatives have enjoyed the greatest clinical application since their introduction. The evolution of these iodocholesterol derivatives has been reviewed, and attention will now be focused upon their metabolism and known parameters of adrenal cortical uptake.

Within the mammalian system cholesterol is a ubiquitous substance. It serves primary functions in the structural maintenance of plasma membranes, and as a substrate nucleus upon which steroid hormones are synthesized.[1] In contrast to its universal structural function, the storage of cholesterol as cholesterol-ester is limited to only those tissues in which it is synthesized *de novo* or subsequently converted to steroid hormones.[4] These tissues are the adrenal cortex, the corpra lutea cells of the ovary, and the Leydig cells of the testes.[1,4] In each of these tissues, cholesterol is metabolized to produce the hormones characteristically released by these cells. Cholesterol is utilized by the luteal cells of the ovary to produce 17-β-estradiol and by the Leydig cells of the testes to synthesize testosterone.[4] The adrenal gland is a complex structure that is subdivided both histologically and biochemically. In the outer cortex (zona glomerulosa) the mineralocorticoid, aldosterone, is produced. The major glucocorticoid hormone, cortisol, is produced by a segment of the inner cortex, the zona fasiculata. The androgenic hormones, androstenedione and dihydroepiandrosterone, are produced by the innermost portion of the adrenal cortex, the zona reticularis.[4] The biosynthetic pathways of steroid production are depicted in Figure 7.

A. Transport of Cholesterol

The amount of cholesterol that ultimately reaches the tissues is an important factor for both the maintenance of cellular integrity and the performance of biosynthetic function, and in addition has considerable bearing upon the estimation of adrenal gland uptake of the iodocholesterol derivatives. The bulk of plasma cholesterol originates from dietary sources.[35] In the small intestine, cholesterol is assembled into lipid droplets of varying density.[35] As a result of the relative insolubility of cholesterol in plasma, carrier proteins or lipoproteins are found associated with cholesterol to facilitate its absorption and transport.[35,36] These complexes circulate to the liver where changes in the composition of the various lipoprotein components occur. They are then resecreted and circulated to other tissues to provide cholesterol for hormonal and structural functions (Figure 8). From previous investigations, information has emerged that the lipoprotein carriers responsible for the transfer of the majority of cholesterol to its tissues of metabolism (i.e., fibroblasts and adrenals) are species specific.[36] Moreover, in man, low density lipoproteins (LDL) appear to be the major carriers of cholesterol to adrenal cortical cells.[36-39]

In a recent study, Fukushi et al.[40] demonstrated that iodocholesterols, like native cholesterol, are bound to red blood cells and plasma lipoproteins. In this study using the rat as a model, it was shown that approximately 20% of either ^{131}I-NP-59 or ICh are bound to LDL. The other lipoproteins accounted for a small percentage of the total bound iodocholesterols (Table 1). The authors propose that β-lipoprotein, a unique lipoprotein characteristic of LDL, is an important factor for the adrenal gland uptake of these iodocholesterol agents. It is of interest, however, that a recent case report by Gordon et al.,[41] noted the absence of uptake of NP-59 in a patient with an expanded cholesterol pool and an elevated LDL concentration. After appropriate therapy for the hypocholesterolemia, adrenal uptake of radiotracer was observed. Thus, the contribution of β-lipoproteins to iodocholesterol uptake at this point remains unclear, and there appear to be other factors that regulate the uptake of iodocholesterol that may account for these recently described phenomena.

The importance of the endogenous cholesterol pool to adrenal gland iodocholesterol uptake has recently been evaluated in a group of patients with Cushing's syndrome. ^{131}I-NP-59 uptake (% administered dose/both adrenals) has been correlated with the endogenous circulating cholesterol pool as assessed by serum cholesterol measurements.[42] The method of calculating % administered dose/adrenals involves a semi-operator independent computer algorithm which allows an in vivo estimation of iodocholesterol uptake.[43] With this approach there appears to be a significant negative correlation between the level of serum cholesterol and the percent adrenal gland uptake of iodocholesterol in Cushing's syndrome. This observation is similar to the decrease in thyroidal ^{131}I uptake seen in patients with an expanded iodine pool.[44] Figure 9 illustrates this relationship.

As there exists an overlap of iodine uptake observed in patients with hyper- or hypothyroidism and normal, this problem also exists with reference to iodocholesterol adrenal gland uptake in patients with Cushing's syndrome. However, adrenal iodocholesterol uptake can be used to distinguish patients with ACTH-dependent (pituitary and ectopic ACTH syndrome) and ACTH-independent (adrenal adenoma or bilateral

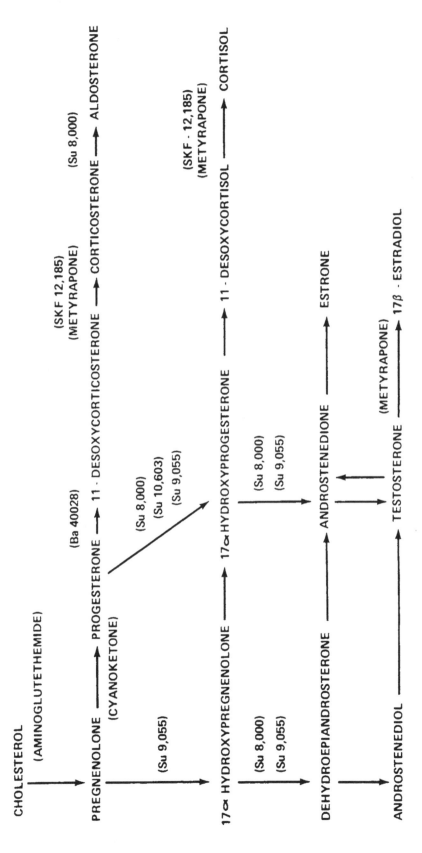

FIGURE 7. The pathways of steroid biosynthesis and their inhibitors.

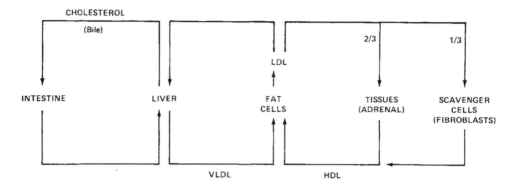

LDL   ·   LOW DENSITY LIPOPROTEINS
HDL   ·   HIGH DENSITY LIPOPROTEINS
VLDL ·   VERY LOW DENSITY LIPOPROTEINS

FIGURE 8.   The pathways of cholesterol metabolism.

Table 1
LIPOPROTEIN BINDING OF IODOCHOLESTEROL

		Plasma lipoprotein			
	RBC	HDL	LDL	IDL	VLDL
NM-145	61%	2.0	19.	9.8	5.9
NP-59	59%	2.5	19.	11.0	6.6

Note: At 2 hr post tracer administration.

From Fukushi, K., Irie, T., Nozaki, T., Ido, T., and Kasida, Y., *J. Labelled Compd. Radiopharm.*, 16, 49, 1979. With permission.

nodular hyperplasia) Cushing's syndrome (Figure 10).[45] Those patients with an ectopic or ACTH-independent etiology for their disease exhibit mean iodocholesterol uptakes that are significantly higher than the mean uptake observed in patients with pituitary ACTH-dependent Cushing's syndrome. The degree of the adrenal uptake overlap seen between these groups may be a result of the size of the endogenous cholesterol pool. A correction of iodocholesterol uptake based upon serum cholesterol levels may clarify this phenomenon and allow a clearer separation of patients with ACTH-dependent or pituitary Cushing's syndrome from ACTH-independent adrenal cortical hyperfunction, i.e., adrenal adenoma and bilateral nodular hyperplasia.

B. The LDL Receptor

The concept of the LDL receptor has been recently described by Brown and Goldstein.[37-39] The control of cholesterol uptake within the cultured fibroblast, and more recently within the isolated adrenal cortical cell, has shed much light upon mechanisms of cholesterol uptake and metabolism. Located on the plasma membranes of these cells are LDL receptors that bind LDL.[46] The affinity and apparently the numbers of these membrane-bound LDL receptors are under cellular and hormonal control. An increase in intracellular cholesterol concentration results in a decrease in LDL receptor affinity, as demonstrated in the fibroblast, and a decrease in the actual number of

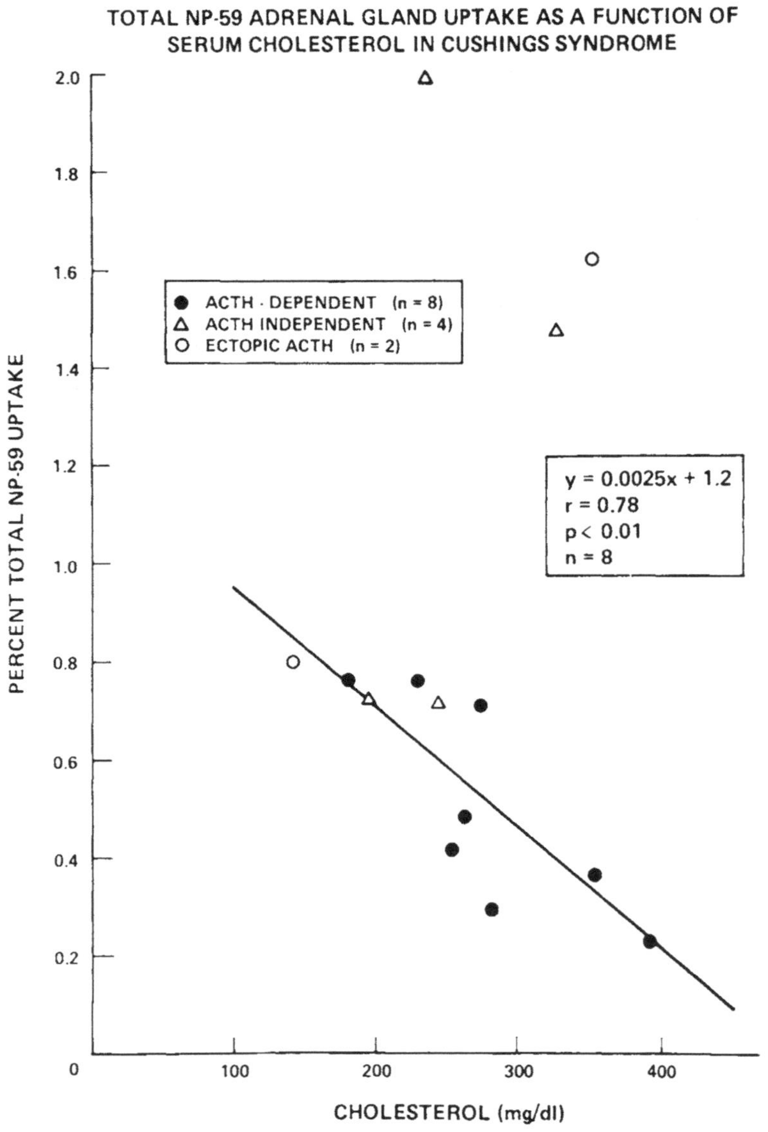

FIGURE 9. The inverse relationship of serum cholesterol and ^{131}I-6β-iodomethyl-norcholesterol adrenal gland uptake. (From Valk, T.W., Gross, M. D., Freitas, J. E., Swanson, D. P., Schteingart, D. E., and Beierwaltes, W. H., *J. Nucl. Med.*, 21, 1069, 1980. With permission.)

LDL receptors as shown in the isolated adrenal cortical cell.[38,39] These mechanisms allow for control of the rate at which cholesterol enters the cell. The specificity of the adrenal LDL receptor for LDL has been demonstrated by the observation of a lack of binding of high density lipoproteins.[38,39]

Not only does it appear that intracellular cholesterol is an important determinant of LDL-receptor activity, but numerous other mechanisms have recently been described that exert profound effects upon cellular LDL-cholesterol localization. Pituitary adrenocorticotrophic hormone (ACTH) has been shown to exert effects upon adrenal cholesterol uptake. In a rat model, Gwynne et al.,[46] have shown that lipoprotein intracellular transfer (in this case HDL being the major lipoprotein carrier of cholesterol

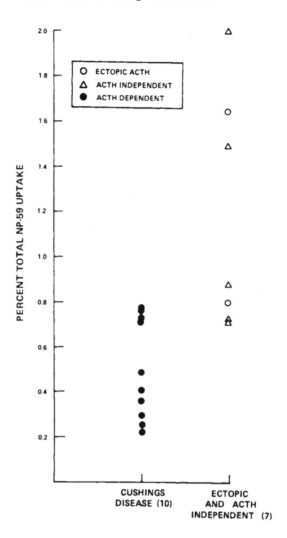

FIGURE 10.   Adrenal gland iodocholesterol uptake in Cushing's syndrome.

in the rat) was stimulated with the addition of ACTH. About 25% of cholesterol is found within the mitochondrial compartment of adrenal cortical cells. This cholesterol can then be converted to steroid hormones. These studies indicate that the adrenal cortex is responsive to hormonal influences, and the uptake of lipoprotein-bound cholesterol can be enhanced by ACTH. In a parallel study Faust et al.[38] found that cultured mouse adrenal cells exhibited ACTH stimulation of cellular uptake of lipoprotein-cholesterol. In this case LDL-cholesterol was found to be incorporated by the LDL-receptor mediated pathway. In the absence of LDL-cholesterol, the availability of cholesterol becomes rate-limiting in the control of steroid biosynthesis. With the addition of exogenous LDL-cholesterol, a large accumulation of intracellular cholesterol esters has been observed in conjunction with a fourfold stimulation of steroid biosynthesis. HDL-cholesterol was not observed to be effective as a carrier of cholesterol within this system.

The importance of lipoproteins has also been demonstrated with respect to the control of certain intracellular enzyme systems that are rate limiting in the control of *de novo* synthesis of intracellular cholesterol. In states of relative intraadrenal cholesterol depletion, *de novo* cholesterol synthesis can be demonstrated.[47] The enzymatic pathway of *de novo* cholesterol synthesis is rate limited by the enzyme 3-hydroxy-3-methylglutaryl coenzyme A reductase (3-HMG-CoA).[48] Balsubramaniam et al.[49] studied the effects of hypocholesterolemia upon the induction of 3-HMG-CoA activity, and showed that hypocholesterolemia activity increased 3-HMG-CoA by a factor of 150- to 200-fold. The restoration of plasma LDL-cholesterol resulted in a return of 3-HMG-CoA activity to basal levels, and a restoration of intracellular cholesterol concentrations.

The control of the conversion of cholesterol esters to cholesterol has also been observed to be under control of ACTH. Bechett and Boyd[51] have shown that the enzyme, cholesterol ester hydrolase, is activated by a cyclic AMP-dependent protein kinase that is ultimately under the control of pituitary ACTH. Davis and Garren[50] have confirmed this activation with ACTH.

In studies of the human adrenal cortical cell, Gwynne et al. [46] have demonstrated that the adrenocortical steroidogenic response is dependent upon an adequate supply of extracellular lipoprotein-derived cholesterol. The human adrenal cell preferentially uses cholesterol derived from LDL, and under ACTH stimulation the number of membrane-bound LDL binding sites increases.

Prostaglandin modulation of this system has been described by Honn and Chavin.[52] Prostaglandins of the E series (PGE_1, PGE_2) effect an adrenal cell increase in cyclic AMP (cAMP) levels, and subsequently increase adrenal cortisol output. Suppression of cAMP levels and cortisol secretion occurs with the prostaglandins of the F series (PGF_1, PGF_2). This system appears to be active earlier than that seen with ACTH alone and may function in both a facilitatory and inhibitory fashion upon ACTH-mediated steroidogenesis.

These complicated and interrelated mechanisms for the maintenance of adrenal cholesterol serve to maintain a fairly constant supply of cholesterol for not only structural, but the steroidogenic functions of the adrenal cortex. It does appear that iodocholesterol imaging agents are under the same control mechanisms of uptake as native cholesterol. Blair et al.[53] have shown that the adrenal uptake of ICh can be enhanced with the addition of ACTH. The uptake of iodocholesterol in cases of Cushing's syndrome is significantly higher than that seen in normal.[54,55] Dexamethasone suppression of pituitary ACTH results in a decrease of uptake to about 50% of that observed in the normal.[43] Successful adrenal tracer accumulation is dependent upon both the endogenous cholesterol pool and probably upon the carrier proteins available for cholesterol transport.[42,43] A proposed schema depicting these interrelationships affecting adrenal cortical cholesterol uptake is shown in Figure 11.

C. Iodocholesterol Metabolism

Once localized within the adrenal cortical cell, cholesterol is converted to its ester for storage and later conversion to pregnenolone for steroidogenesis.[55] The metabolic fate of iodocholesterol is not the same as that of native cholesterol. These agents apparently do not act as enzyme substrates since little if any iodinated steroids or iodinated steroid metabolites have been observed in the blood or urine of either animals or humans after their administration.[43] In studies examining the metabolic conversion of iodocholesterol, about 40% of administered ICh and 3.6% of NP-59 have been found to be esterified within the dog adrenal cortex.[56] Rizza et al.[57] have shown that the intraadrenal concentration of ICh can be increased after the administration of the metabolic inhibitor aminoglutethemide, but was not altered with ACTH. Thus, the iodocholesterols appear to exhibit impaired mobilization from the intra-adrenal cholesterol pool. These observations are consistent with the finding that ICh and NP-59 exhibit visualization of the adrenal glands in the normal for 10 to 14 days, and 18 to 24 days, respectively.[43]

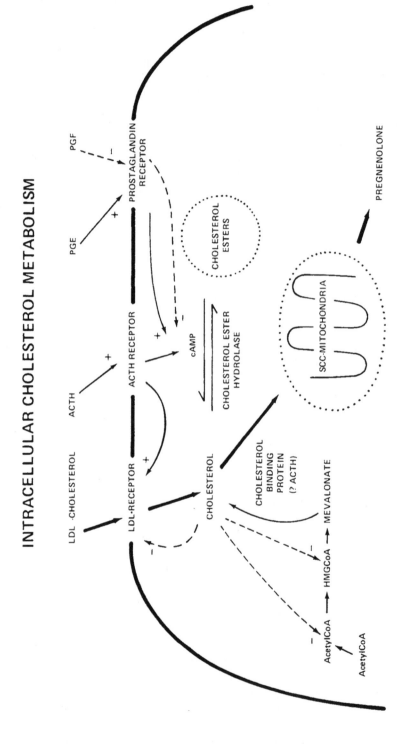

FIGURE 11.   A proposed schema of the control of adrenal cortical cholesterol uptake (NP-59).

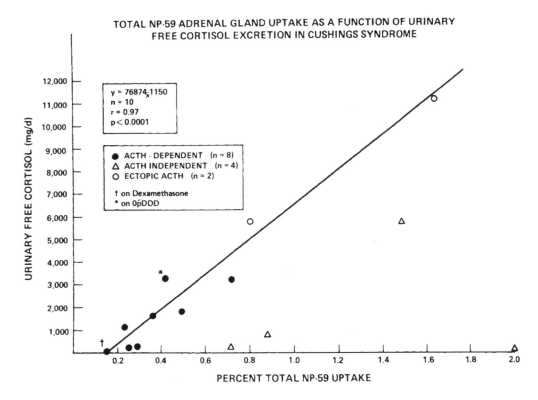

FIGURE 12. Iodocholesterol uptake as a function of urinary free cortisol excretion in Cushing's syndrome. (From Gross, M. D., Valk, T. W., Freitas, J. E., Swanson, D. P., and Beierwaltes, W. H., *J. Clin. Endocrinol. Metab.*, 52, 1062, 1981. With permission.)

IV. RELATIONSHIP OF IODOCHOLESTEROL ADRENAL GLAND UPTAKE TO PARAMETERS OF ADRENAL CORTICAL FUNCTION

A. Adrenal Hypercortisolism

The observation that the iodocholesterols are concentrated by the same mechanisms of uptake as native cholesterol supports the impression that the uptake of the adrenal imaging agents are a reflection of not only the anatomic status but the functional status of the adrenal glands. The degree to which the uptake of these agents reflects the biochemical processes involved has recently been assessed in a number of studies of adrenal hypercortisolism.

Of the glucocorticoids excreted in patients with Cushing's syndrome, urinary free cortisol (UFC) represents the most reliable parameter of cortisol production that is least affected by alterations in the plasma binding of cortisol and renal function. In a group of patients with Cushing's syndrome, the relation between calculated iodocholesterol uptake and UFC was assessed (Figure 12).[45] The relationship is significant with an r value of 0.91 ($p < .0001$). None of the other parameters examined; AM-PM ACTH, cortisol, dihydroepiandrosterone or urinary 17-hydroxysteroids have demonstrated a significant correlation. As mentioned above, iodocholesterol adrenal uptake can be used to separate patients with respect to their etiology of Cushing's syndrome. Although overlap of adrenal gland uptake does occur, it is intriguing that patients with ACTH independent Cushing's syndrome (adrenal cortical adenoma and bilateral nodular hyperplasia) have higher adrenal uptakes than those with ACTH dependent disease.

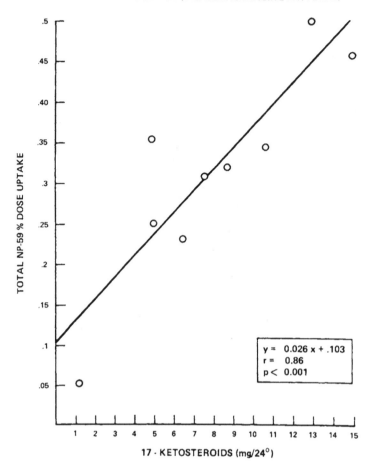

FIGURE 13.   Iodocholesterol (NP-59) uptake as a function of dexamethasone suppression (DS) 17-ketosteroid excretion in bilateral hyperplasia resulting in hyperandrogenism.

B. Adrenal Hyperandrogenism

In a parallel study, the uptake of iodocholesterol while under dexamethasone suppression was examined in patients with excessive production of adrenal-derived androgens. The dexamethasone suppressed 17-ketosteroid (17-KS) production was found to be significantly correlated with the calculated adrenal iodocholesterol uptake.[58] In those patients with bilateral adrenal disease (hyperplasia), the relationship between calculated percent uptake and 17-KS was significant ($r = 0.91$; $p < .001$) (Figure 13). In four cases of adrenal androgen producing adenomas a similar relationship was observed ($r = 0.87$; $p < .001$) (Figure 14). These studies indicate that iodocholesterol uptake is a reflection of adrenal androgen production in patients with hyperandrogenism while under dexamethasone suppression.

C. Relation of Serum Aldosterone to Iodocholesterol Adrenal Gland Uptake

The parameters of adrenal NP-59 uptake while under dexamethasone suppression have been examined in a dog model.[59] These studies were performed to examine the previously observed phenomenon of iodocholesterol adrenal gland visualization while

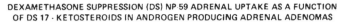

DEXAMETHASONE SUPPRESSION (DS) NP-59 ADRENAL UPTAKE AS A FUNCTION
OF DS 17 - KETOSTEROIDS IN ANDROGEN PRODUCING ADRENAL ADENOMAS

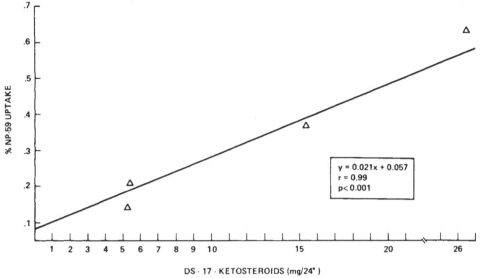

$$y = 0.021x + 0.057$$
$$r = 0.99$$
$$p < 0.001$$

DS - 17 - KETOSTEROIDS (mg/24°)

FIGURE 14.   Dexamethasone Suppression (DS) iodocholesterol (NP-59) uptake as a function of DS 17-ketosteroid excretion in hyperandrogenism from adrenal cortical adenomas.

Table 2
ALDOSTERONE AS A FUNCTION OF ^{131}I-6β-IODOMETHYL-NORCHOLESTEROL (NP-59) ADRENAL GLAND (TISSUE) UPTAKE IN CANINE MODEL

	Time	Control	Dexamethasone suppression (DS)		
			DS	DS (low Na)	DS (high Na)
Aldosterone	72	2.7 ± 0.26[a,b]	8.7 ± 2.6	49.0 ± 16.8[c]	4. ± 1.5[c,d]
(ng/dl)	120	4.73 ± 0.39	11.83 ± 3.8	53.3 ± 19.3[c]	3.5 ± 0.21[c,d]
^{131}I-NP-59	72	7.0 ± 0.59	3.77 ± 0.10	5.5 ± 0.4[c]	2.75 ± 0.38[c,d]
(%kg/dose/g)	120	9.15 ± 0.74	4.15 ± 0.18	4.94 ± 0.21[c]	3.56 ± 0.06[c,d]

Note: Cortisol: Control 1.5 mg/dl and All DS<0.5 mg/dl.

[a]   Mean ± SEM.
[b]   Three animals per interval.
[c] $p < 0.05$ vs. Control.
[d] $p < 0.05$ DS(High Va vs. Low Na).

From Gross, M. D., Grekin, R. J., Brown, L. R., Marsh, D. D., and Beierwaltes, W. H., *J. Clin. Endocrinol. Metab.*, 52, 612, 1981. With permission.

under dexamethasone suppression. Two groups of dogs were given NP-59, 250 μCi, intravenously. One of the groups received dexamethasone 30 vg/kg for 7 days prior to NP-59 administration and throughout a 5-day interval after tracer injection. Within the dexamethasone suppression group, one subgroup received a 150 mEq sodium chloride (NaCl) diet and 9α-fluorohydrocortisone, a second subgroup received a 10 mEq NaCl diet and furosemide 40 mg for three days to effect a salt depleted state, and a third subgroup received only dexamethasone. Serum aldosterone was monitored throughout the study (Table 2). Adrenal gland iodocholesterol uptake was measured

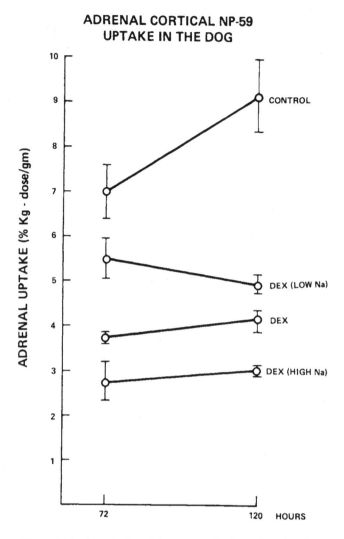

FIGURE 15.   NP-59 adrenal tissue uptake in the canine adrenal cortex (see text for details).

with adrenal cortical tissue samples obtained at either 3 or 5 days after tracer administration (Table 2). In a group of animals that received dexamethasone suppression and a normal salt diet, there was a significant fall of iodocholesterol uptake to about 50% of control values at 3 or 5 days, respectively. Dietary salt intake had a significant effect upon both aldosterone levels and adrenal gland NP-59 uptake (Figure 15). Both aldosterone levels and iodocholesterol uptake were stimulated with salt depletion and depressed with salt loading. These studies indicate that adrenal cortical activity is reflected in iodocholesterol adrenal uptake. The aldosterone-producing portion of the adrenal cortex can be functionally separated from the other zones (zona glomerulosa from zona fasiculata-reticularis) for plasma cortisol did not change while under dexamethasone suppression in either the low or high salt groups.

In a parallel study, Shapiro et al.[60] have demonstrated in an in vivo quantitative fashion that 6-methyl-Se-75-selenomethyl-9-norcholesterol (Scintadren®) shows uptake in both the normal adrenal cortex and the adrenal cortex of patients with adrenal cortical disease (Table 3). The observed adrenal gland uptake in this series is consistent with the studies using the iodinated derivatives.

Table 3

SE-75-SELENOCHOLESTEROL ADRENOCORTICAL
UPTAKE

	N	% Dose uptake
Normal	17	0.17% (0.07—0.3%)
Primary aldosteronism		
Solitary adenoma	8	0.47%
Bilateral hyperplasia	14	0.33%
Cushing's syndrome		
Solitary adenoma	5	0.90%
ACTH-dependent hyperplasia	3	0.58%
Ectopic-ACTH syndrome	7	0.69%
Congenital adrenal hyperplasia	1	3.61%
Adrenal carcinoma	3	0.008%

From Shapiro, B., Bitton, K. E., and Hawkins, I. A., *J. Nucl. Med.*,
21(Abstr.), 29, 1980. With permission.

Table 4

6β-IODOMETHYLNORCHOLESTEROL
UPTAKE IN METASTATIC ADRENAL
CARCINOMA

Tissue	Uptake (%dose/g)
Adrenal	1.08×10^{-2}
Liver metastases	6.5×10^{-3}
Normal liver	2.8×10^{-3}
Vertebral metastases	6.9×10^{-3}
Normal bone	0.9×10^{-3}

From Seabold, J. E., Hayne, T. P., Deasis, D. N., Samaan,
N. A., Glenn, H. J., and Jahns, M. F., *J. Clin. Endocrinol.
Metab.*, 45, 788, 1978. With permission.

V. IODOCHOLESTEROL LOCALIZATION IN ADRENAL CORTICAL TUMORS

The localization of adrenal cortical tumors with iodocholesterol represents the primary impetus for the development of these agents. However, iodocholesterol uptake in neoplastic adrenal tissues has shown variable results. Moses et al.[53] observed that the uptake of ICh was 0.002% dose/g in an adrenal carcinoma. This was too low to permit adequate scintigraphic visualization.[54] Seabold et al.[61] described the imaging of adrenal cortical neoplasms with ^{131}I-NP-59. The uptake of this agent was 0.01% dose/gm in those tumors with adequate uptake for visualization. Table 4 correlates the uptake of NP-59 with a metastatic adrenal tumor in bone and liver in that series. In a comprehensive study by Schteingart et al.,[62] ^{131}I-ICh uptake was measured in various adrenal neoplasms producing either cortisol, aldosterone or androgens, and in the surrounding "normal" adrenal cortex. ICh uptake, as expressed as % kg/dose/g, was less in the neoplasms than in the surrounding adrenal cortex in those cases of functioning adrenal cortical tumors that did not visualize (Table 5). Uptake was greater in the functioning neoplasms that did visualize than in the surrounding "normal cor-

Table 5

TISSUE UPTAKE OF ^{131}I-19-IODOCHOLESTEROL

(%dose/g tissue)

Cortisol secreting carcinoma	Tumor	Tumor (metastatic)	Normal adrenal cortex
1	3.6×10^{-4a}		NA
2	7.7×10^{-4}		7.1×10^{-3}
3	1.9×10^{-3}		NA
4	2.0×10^{-3}	1.7×10^{-3}	1.8×10^{-2}
5	1.4×10^{-4}	2.4×10^{-4}	1.2×10^{-3}
6	3.4×10^{-4}		2.2×10^{-3}
Androgen secreting carcinoma			
1	7.3×10^{-4}		1.5×10^{-2}
2	2.3×10^{-5}		NA
Aldosterone secreting adenomas (12)^b	$1—6 \times 10^{-2}$		$1—3 \times 10^{-2}$
Dexamethasone suppression aldosterone secreting adenomas (3)^b	$1—2 \times 10^{-2}$		$6—80 \times 10^{-4}$
Cortisol secreting adrenal adenomas (3)^b	$2.8—13 \times 10^{-2}$		$5—8 \times 10^{-3}$

^{a} Mean % dose/gm.
^{b} (N) Number of patients.

From Schteingart, D. E., Seabold, J. E., Gross, M. D., and Swanson, D. P., *J. Clin. Endocrinol. Metab.*, 52, 1156, 1981.

tex'' in that series (Table 5). These data also demonstrate that glucocorticoid suppression (either exogenous or endogenous) resulted in a 100-fold decrease in iodocholesterol (tissue) uptake in the "normal suppressed" adrenal tissue as compared to the tumorous adrenal tissue. This has not been confirmed with NP-59, and animal studies of adrenal cortical NP-59 uptake indicated a 50% fall in cortical uptake while on dexamethasone (glucocorticoid) suppression (vide supra).[59]

It is apparent that there are significant structural specificities for the iodocholesterol adrenal imaging agents. Small alterations in substituent groups and in the radionuclides incorporated within the cholesterol derivatives result in major changes in adrenal cortical uptake. Tissues with elevated secretory activity concentrate the iodocholesterols with greater affinity than the surrounding normal adrenal cortex. This accounts for the observation that clinical visualization of adrenal tumors is most successful in those cases of hormonally active neoplasms. In instances of adrenal gland hormone excess states, iodocholesterol (NP-59) uptake reflects the degree of functional activity of the adrenal cortex. Manipulations of adrenal cortical activity by dexamethasone and dietary salt alterations are reflected in iodocholesterol adrenal gland uptake.

Despite the extensive clinical exposure with the iodocholesterol derivatives (^{131}I-ICh and NP-59) the external imaging times necessary for diagnostic interpretation of studies and the slow adrenal clearance of these radiopharmaceuticals make them less than ideal for clinical adrenal cortical imaging. Little clinical information is available with either the selenium-labeled cholesterol,[60] and no clinical information is available with the tellurium-cholesterol derivatives.

VI. RADIOLABELED ENZYME INHIBITORS

The development of radiopharmaceuticals based on reversible and irreversible enzyme inhibitors has been extensively reviewed.[24,63] In regard to the use of this class of agents for adrenal imaging, the greatest success has been achieved with the reversible

Table 6

CONCENTRATION OF METYRAPONE DERIVATIVES IN THE ADRENAL
CORTEX OF DOGS

Compound	Peak Adrenal Cortex conc. (%kg−dose/g)	Adrenal cortex		
		Liver	Kidney	Blood
[³H(G)] - Metyrapone	2.75 (45 min)	6.3	9.8	16.2
[³H(G)] - Metyrapol	8.91 (1 hr)	17.1	46.9	89.1
¹²⁵I-5[B]-Iodometyrapol	0.94 (1 hr)	3.9	7.8	15.7

inhibitors of 11β-hydroxylase, metyrapone, and SKF-12185. See Figure 7 for the biochemical pathway involving the enzymatic conversion of cholesterol to cortisol, including known inhibitors of these respective enzymes.[24]

A. Metyrapone Derivatives

The pyridine derivative, metyrapone, has been shown to bind to the 11β-hydroxylase
enzyme system localized on the inner membrane of mitochondria of all zones of the
adrenal cortex.[64] The chemical structures, adrenal cortical concentration and target:
background radioactivity ratios of ³H-metyrapone and its alcohol derivative, ³H-metyrapol, are illustrated in Table 6. The low adrenal cortex:liver ratio achieved with ³H-
metyrapone is probably a result of this agent binding to hepatic cytochrome P-450,
the oxygen-activating protein portion of the 11β-hydroxylase enzyme system also responsible for adrenal cortex binding. The significantly enhanced adrenal cortex uptake
and adrenal:liver radioactivity ratios of ³H-metyrapol may be a reflection of the fact
that it has previously been shown in vitro to bind more avidly to adrenal 11β-hydrox-
ylase P-450 than metyrapone.[65,66]

Since ³H-metyrapol demonstrated the most favorable adrenal cortical imaging characteristics, attempts were made to synthesize a suitable gamma-emitting radioiodide
analog of this agent. Structure distribution studies involving modification of the ³H-
metyrapol side chain revealed that the pyridine rings were most likely the optimal position for incorporation of the radioiodine label without altering adrenal specificity.[24]
The ¹²⁵I-5(B)-iodometyrapol derivative (Table 7) has been synthesized,[67] however, subsequent tissue distribution studies revealed a fourfold lower concentration (Table 7) in
the adrenal cortex than the noniodinated parent compound. Unfortunately, incorpo-

Table 7

CONCENTRATION OF SKF-12185 DERIVATIVES IN THE ADRENAL CORTEX OF DOGS

Compound	Peak adrenal cortex conc. (%kg−dose/g)	Adrenal cortex		
		Liver	Kidney	Blood
[³H(G)] - SKF-12185	1.86 (30 min)	2.9	8.1	31.0
¹³¹I-3-iodo-(DL)-SKF-12185	2.90 (3 hr)	4.5	9.1	96.7
¹³¹I-3-iodo-(L)-SKF-12185	3.47 (3 hr)	8.7	17.4	46.3

ration of radioactive iodine into the 5-(A) pyridine ring presents a considerably greater synthetic problem which has not yet been overcome.

B. SKF-12185

In contrast to metyrapol, the radioiodinated derivative of SKF-12185 showed enhanced uptake (Table 8) into the adrenal cortex compared to the parent ³H-SKF-12185.[24] Although adrenal:liver radioactivity ratios with radioiodinated SKF-12185 were low (Table 8), successful images of the adrenals of dogs were obtained using ¹³¹I-SKF-12185. Subsequent attempts to image human adrenals with ¹²³I-SKF-12185 resulted in less than optimal results, probably due to the above mentioned low target:background ratios. Since these initial attempts at imaging human adrenals with ¹²³I-SKF-12185, the adrenal cortical localization of this agent has been further enhanced a factor of two and adrenal cortex:liver ratios increased to 8.7:1 (Table 8) by using the levorotatory form of radioiodinated SKF-12185 rather than the racemic mixture.[68] Unfortunately, we still have not succeeded in obtaining human adrenal images of comparable quality to NP-59.

VII. ADRENAL MEDULLA IMAGING AGENTS

A. Precursors of Epinephrine

The search for a clinically useful adrenomedullary imaging agent has been ongoing for nearly a decade.[69-78] These efforts have been primarily based on the observation that labeled catecholamines, especially dopamine, concentrate in the adrenal medulla. Morales and co-workers[69] studied the distribution and excretion of various ¹⁴C-catecholamines in dogs and demonstrated that the concentration of radioactivity in the adrenal medulla from intravenously administered ¹⁴C-dopamine exceeded that of all other epinephrine precursors. A selective concentration of radioactivity from ¹⁴C-dopamine has also been observed in the normal human adrenal medulla,[79] neuroblastoma,[79] and pheochromocytoma.[72]

Fowler and co-workers[73,74] synthesized ¹¹C-dopamine and observed localization in

Table 8

COMPARISON OF K, AND ADRENAL MEDULLA UPTAKE OF PNMT INHIBITORS

PNMT Inhibitors	Specific activity	$K_I(M)$	Adrenal medulla up-take (5 min)[a]
$^3H(G)$- (structure)	1.0 mCi/mg	3×10^{-9}	1.85
	0.1 mCi/mg	8×10^{-9}	0.72
	Carrier free	20×10^{-9}	1.58
	Carrier free	700×10^{-9}	0.99

[a]   % kg Dose/g; 2 dogs sacrificed for each compound.

dog adrenal medulla at short time intervals (< 2 hr), but attempts to image the adrenals were not reported. With carrier-free ^{11}C-dopamine, it was observed that the concentration in the adrenal medulla increased approximately fourfold. Dopamine labeled in the 6-position with radioiodine (^{123}I and ^{131}I) surprisingly showed adrenal medullary concentrations nearly identical to that obtained with ^{11}C-dopamine.[81] Also significant was that only minor in vivo deiodination occurred; other workers have observed extensive in vivo deiodination of radioiodinated tyramines and phenylethylamine.[69,82] Fowler and associates concluded that the uptake of radioiodinated 6-iododopamine in the adrenal medulla was probably not high enough to permit imaging, though they suggested that carrier-free radioiodine labeled dopamine, if it could be synthesized, would show enhanced uptake and might be useful as a scanning agent.

A possible alternative to labeling dopamine directly on the catechol ring was prompted by the finding that the NHSO$_2$R moiety can be used as a bioisosteric substitute for the 3-OH group of catecholamine.[82] Ice, Wieland, and co-workers[84] synthesized the seven radiolabeled sulfonanilide analogues of dopamine shown in Figure 16. Compound I (R = ^{35}SO$_2$CH$_3$) localized in the adrenal of both the rat and dog, with uptake and target-to-nontarget values similar to dopamine itself. However, the remaining analogues (II-VII) containing iodo-, aryl-, and alkylaryl groups failed to localize in the adrenals, thus attesting to the low bulk tolerance in this region of the molecule. Other bioisoteric substitutes for the 3-OH group are known,[85] some of which may be more amenable to γ-label incorporation than are the sulfonanilide analogs. These catecholamine analogues do not act as substrates for catechol-O-methyltransferase (COMT), and their enhanced metabolic stability suggests that peak adrenal uptake may occur at longer time intervals than observed with labeled catecholamines.[86]

B. Enzyme Inhibitors

In view of the success with radiolabeled inhibitors of adrenocortical enzymes, a sim-

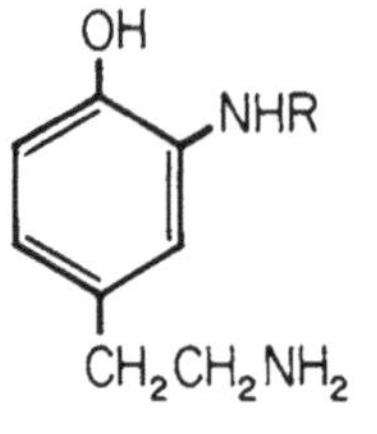

COMPOUND	R
I	$^{35}SO_2CH_3$
II	$^{35}SO_2Ph\text{-}I\text{-}\underline{p}$
III	$SO_2Ph\text{-}^3H\text{-}\underline{p}$
IV	$SO_2CH_2Ph\text{-}^3H\text{-}\underline{p}$
V	$SO_2(CH_2)_2Ph\text{-}^3H\text{-}\underline{p}$
VI	$SO_2(CH_2)_3Ph\text{-}^3H\text{-}\underline{p}$
VII	$SO_2CH_2{}^{125}I$

FIGURE 16.   The seven radiolabeled sulfonanilide analogues of dopamine.

ilar approach has been applied to the development of a scanning agent for the adrenal medulla. The enzymes responsible for the sequential conversion of tyrosine to epinephrine have been characterized and are shown in Figure 17.[87,89] Of the four enzymes involved, only phenylethanolamine-*N*-methyltransferase (PNMT) is unique to the adrenal medulla (although, a small amount is found in the brain). However, levels of the other three enzymes, especially tyrosine hydroxylase and dopamine-β-hydroxylase, can be expected to be higher in the adrenal medulla than in surrounding adrenergically innervated organs.

In our laboratory, the radiolabeled tyrosine hydroxylase inhibitor, I-125-3-iodo-α-methyltyrosine, failed to show selective uptake in the dog adrenal medulla, possibly a result of rapid in vivo deiodination.[90] Kloss and co-workers[91] recently evaluated 26 tyrosine derivatives labeled with I-131. Although a number of the tyrosine derivatives showed selective uptake in the adrenal glands of mice, most of the compounds failed to show similar localization in dogs. The species variability encountered may in part be linked to differences in metabolism including the action of dehalogenases.

Although many in vitro inhibitors of PNMT have been developed over the years, only recently has an inhibitor (SKF-64139) been found effective in vivo.[92] The presence of the lipophilic chlorine atoms in the 7 and 8 positions of SKF-64139 suggested an approximate bioisosteric substitution of a single iodine atom for the two chlorine atoms.[93] As shown in Table 8, the inhibitory constants (K_i) of three of the four compounds show a qualitative correlation with their respective in vivo concentrations in the dog adrenal medulla; i.e., the more potent the inhibitor, the higher the concentration. The exception is 8-iodotetrahydroisoquinoline, which, based on its 8 nm K_i value, should give the highest adrenomedullary concentration of the three iodo isomers evaluated. The reason for the anomaly may be the low specific activity (0.1 mCi/mg) of the compound.

Although the I-125-labeled PNMT inhibitors were specific for the dog adrenal medulla, their low absolute uptake and rapid washout from the adrenals discouraged further investigation.

C. Neuronal Blocking Agents

A new approach to adrenomedullary imaging was suggested by the work of Korn et al.,[78] who in 1977 reported that radioiodinated analogs of the neuronal blocking agent, bretylium, concentrate in the canine adrenal medulla. No attempts at adrenal imaging were reported as the absolute uptake and target-to-nontarget concentration ratios were not sufficiently high to warrant it. The tissue distribution data revealed that for maximum adrenal medullary concentration, two conditions are necessary: (1) the iodine

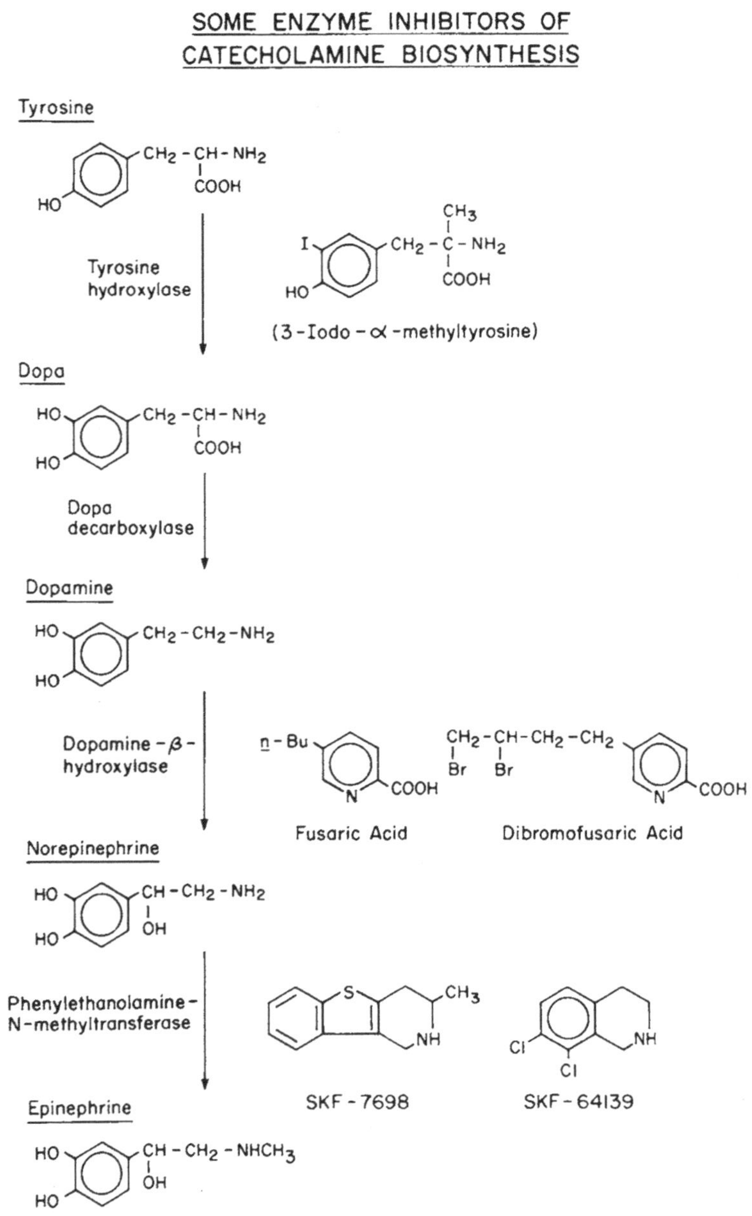

SOME ENZYME INHIBITORS OF
CATECHOLAMINE BIOSYNTHESIS

FIGURE 17. Enzymes responsible for conversion of tyrosine to epinephrine including known enzyme inhibitors.

must be in the para position, and (2) the quaternary head should have three methyl groups.

These results were surprising in light of the structure of the bretylium molecule itself, and the fact that Boura and co-workers,[94] the developers of bretylium, showed as early as 1959 that either para substitution or a trimethyl quaternary head strikingly lowered pharmacological activity. It is thought that the uptake of bretylium into adrenergic nerves is a prerequisite for its effective action. Thus, if one considers the adrenal medulla a large, specialized adrenergic nerve, then the iodo analog most similar to bretylium, ortho-iodobenzyldimethylethyl ammonium (Figure 18), might be expected to show the highest adrenal medullary localization. An explanation for the low adrenal

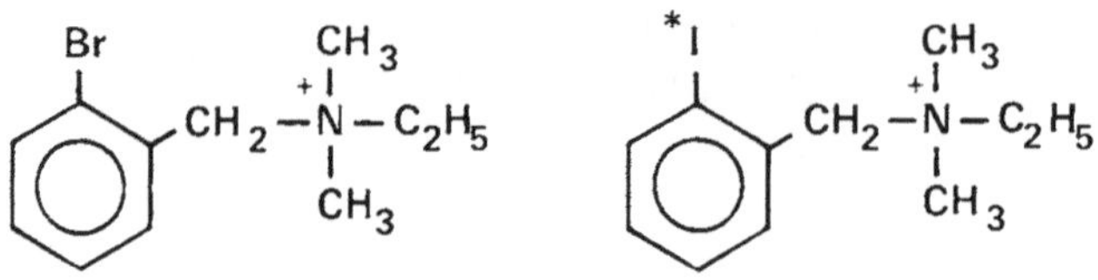

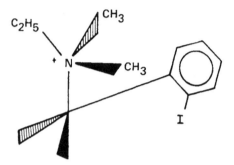

FIGURE 18.   Structures of bretylium and ortho-iodobenzyldimethylethyl ammonium.

FIGURE 19.   Staggered conformation of ortho-iodobenzyldimethyl ammonium.

medulla uptake of ortho-iodobenzyldimethylethyl ammonium, suggested by Counsell et al.,[77] was that steric repulsion between the iodine atom and the bulky quaternary head weakened the aromatic carbon-iodine bond leading to rapid in vivo deiodination. They pointed to the high thyroid radioactivity values as evidence. No comparative in vitro stability data were reported.

Since the favored ground state conformation of the ortho-iodobretylium analogues most likely has the benzylic hydrogens anti to the ortho iodine, as shown in Figure 19, the steric bulk of the quaternary head should have little or no effect on the strength of the carbon-iodine bond. In our estimation, more plausible explanations for the high thyroid values might be: (1) a remarkably selective dehalogenase action, or (2) the presence of considerable free radioiodide as an in vitro impurity. The latter possibility is tenable in view of the lack of radio-TLC data for the compound.[77]

To obtain definitive answers to these questions, we repeated and expanded the study of the radioiodinated bretylium analogs.[94] As shown in Table 9, our results are in direct contrast to the findings of Counsell.[78] From this preliminary structure-distribution-relationship (SDR) study, the following conclusions can be drawn in regard to optimizing the adrenal medulla uptake of iodobenzyl quaternary compounds: (1) Aromatic pattern: ortho $\gg$ para, and (2) R: $CH_2CH_2OH \sim> CH_2CH_3 \gg CH_3 > CH_2CH_2CH_3$. These patterns of medullary uptake exactly parallel the reported neuron blocking potencies of the analogous bromobenzyl quaternaries reported by Boura.[93]

The previously unreported ethylhydroxy derivative shown in Table 9, although very similar to the ethyl analog in peak adrenal medulla concentration, generally had higher target-to-nontarget concentration ratios; background activities in the liver and kidney were initially lower and declined more rapidly with time. This compound, labeled with I-131, was chosen for scintigraphy studies. The first discernible images of the dog adrenal medullae were obtained 3 days after i.v. injection of 1.5 mCi of the compound. The adrenal medullae showed clear delineation from background beginning at 4 days post injection. The major deterrent to visualization of the adrenals in less than 3 days

Table 9
SDR STUDY OF I-125-BRETYLIUM
ANALOGS IN DOGS

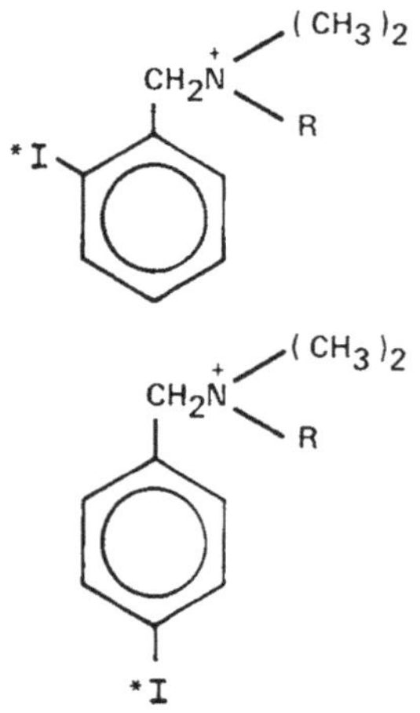

Position of Radioiodine	R	Maximum Adrenal Medulla conc.[a]
Ortho	CH$_3$	0.79
Ortho	CH$_2$CH$_3$	3.74
Ortho	CH$_2$CH$_2$CH$_3$	0.37
Ortho	CH$_2$CH$_2$OH	3.55
Para	CH$_3$	0.42
Para	CH$_2$CH$_3$	0.18

[a]   Concentrations are in % kg dose/g.

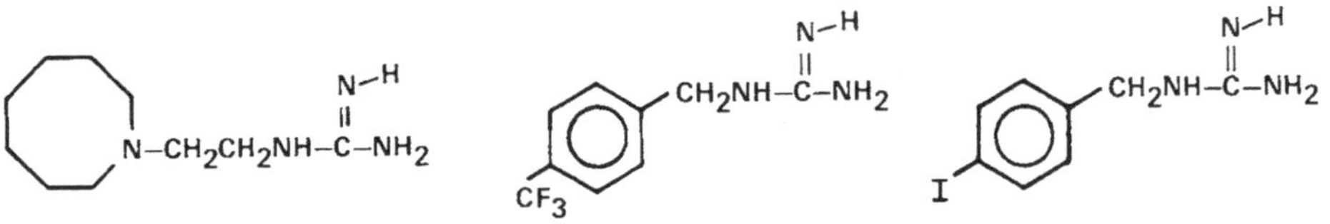

FIGURE 20.   Structures of guanethidine, para-trifluoromethylbenzylguanidine, and para-iodobenzyl-guanidine.

was the activity in the gall bladder and bowel. Liver activity presented minimal problems, even at 24 hr post injection.

A short clinical trial with this I-131-labeled agent in three patients suffering from suspected adrenal medullary disease failed to produce images of the human adrenal medulla. Nonetheless, the successful imaging of the canine adrenal medulla prompted further work to develop a more adrenophilic agent.

Efforts were then directed towards the adrenergic neuron blocking agent, guanethidine (Figure 20). Despite early controversy over its mode of action, guanethidine is generally thought to block adrenergic transmission by the same mechanism as bretylium.[95] Only inferential evidence exists for the blocking mechanism, but it is known to involve binding to some cytoplasmic component in the adrenergic neuron. However, guanethidine has an additional characteristic not shared by bretylium, and that is its ability to displace norepinephrine from the neuronal storage vesicles. Thus, a greater

Table 10

TISSUE DISTRIBUTION OF I-125-
IODOBENZYLGUANIDINES IN THE DOG AT 48 hr

$$\text{}^{125}\text{I} - \text{C}_6\text{H}_4 - \text{CH}_2\text{NH} - \overset{\overset{\displaystyle N^H}{\|}}{C} - \text{NH}_2$$

Position of Iodine	Adrenal Medulla	Adrenal Cortex	Liver	Blood	Spleen	Kidney	Thyroid
Para	14.3	0.17	0.04	0.02	0.27	0.04	13.8
Ortho	1.3	0.01	0.01	0.00	0.00	0.01	9.4
Meta	13.6	0.10	0.02	0.01	0.19	0.03	1.3

Note: Concentrations are in % kg-dose/g.

Table 11

RESERPINE BLOCKING STUDY
(%kg dose/g)

	No. of Dogs	Time (hr)	[Adr Med]
Control	2	2	4.86 ± 0.46
Reserpine	2	2	0.85 ± 0.19

Note: Reserpine (1 mg/kg Serpasil) or placebo were given i.m. 4 hr later 100 μCi of I-125-Meta-Io-dobenzylguanidine were given i.v. and dogs sacrificed 2 hr later.

accumulation of guanethidine than bretylium might be anticipated in adrenergic nerves as well as in the adrenal medulla.

Guanethidine itself cannot be easily labeled with radioiodine. But aralkylguanidines, which are aromatic analogs of guanethidine, can be conveniently radioiodinated in high yield. Many of the aralkylguanidines are potent neuron blocking agents. The para-trifluoromethylbenzylguanidine (Figure 20) was the most potent neuron blocker of a series of benzylguanidines evaluated by Short and Darby.[96] The para-iodobenzyl-guanidine (Figure 20) was of lesser potency but still active. The ortho-iodobenzylguan-idine was not active and the meta isomer was not reported. Thus, we decided to eval-uate the radioiodo-benzylguanidines in hopes that they would show higher adreno-medullary uptake than was observed with the iodo-bretylium analogs.[97]

Table 10 summarizes the tissue concentrations of three isomeric I-125-iodobenzyl-guanidines in seven selected tissues of the dog. Both the meta and para isomers show strikingly high concentrations in the adrenal medulla — four times greater than the maximum adrenomedullary uptake of the best bretylium analogs and nearly an order of magnitude higher than that obtained with labeled dopamine.

Blocking studies (Table 11) with reserpine, which is known to selectively block vesi-cular uptake, show that most of the adrenomedullary activity is localized in the storage vesicles. This result has been further substantiated by subcellular fractionation studies.

Three dogs injected with I-131-para-iodobenzylguanidine all gave clear and nearly identical images of the adrenal medulla 3 days post injection. Further improved images were obtained using I-131-meta-iodobenzylguanidine: clear adrenomedullary images appeared at 1 or 2 days post injection. The imaging superiority of the meta isomer is likely due to the lower background activity in non-target organs such as spleen, liver and GI tract (see Table 10). The greater in vivo stability of the carbon-iodine bond of the meta isomer, as evidenced by its lower thyroid radioactivity levels, may partly account for the improved images; i.e., the intestinal mucosa is known to sequester radioiodide.

Current efforts are directed at expanding the structure distribution relationship study of the I-125-iodo-aralkylguanidines. Preliminary scintigraphy studies with rhesus monkeys using I-131-meta-iodobenzylguanidine have produced the first adrenomedullary images in a primate.[97]

REFERENCES

1. Applegren, L. E., Sites of steroid hormone formation: Autoradiographic studies using labeled precursors, *Acta Physiol. Scand. Suppl.*, 301, 1, 1967.
2. Beierwaltes, W. H., Varma, U. M., and Lieberman, L. M., Percent uptake of labeled cholesterol in the adrenal cortex, *J. Nucl. Med.*, 10, 387, 1969.
3. Blair, R. J., Beierwaltes, W. H., Lieberman, L. M., Boyd, C. M., Counsell, R. E., Weinhold, P. A., and Varma, V. M., Radiolabeled cholesterol as an adrenal scanning agent, *J. Nucl. Med.*, 12, 176, 1971.
4. Samuels, L. T. and Uchikaua, T., Biosynthesis of adrenal steroids, in *The Adrenal Cortex*, Eisenstein, A. B., Ed., Little, Brown, Boston, 1967, 61.
5. Nagai, T., Solis, B. A., and Koh, C. S., An approach to developing adrenal gland scanning agents, *J. Nucl. Med.*, 9, 576, 1969.
6. Counsell, R. E., Ranade, V. V., Blair, R. J., Beierwaltes, W. H., and Weinhold, P. A., Tumor localizing agents. IX. Radioiodinated cholesterol, *Steroids*, 16, 317, 1970.
7. Kojima, M., Maeda, M., Ogawa, H., Nitta, K., and Ito, T., New adrenal-scanning agent, *J. Nucl. Med.*, 16, 666, 1975.
8. Basmadjian, G. P., Hetzel, K. R., Ice, R. D., and Beierwaltes, W. H., Synthesis of a new adrenal cortex imaging agent 6β-^{131}I-iodomethyl-19-norcholest-5(10)-en-3β-ol (NP-59). *J. Labelled Compd.*, 11, 427, 1975.
9. Sarkar, S. D., Beierwaltes, W. H., Ice, R. D., Basmadjian, G. P., Hetzel, K. R., Kennedy, W. P., and Mason, M. M., A new and superior adrenal scanning agent, NP-59, *J. Nucl. Med.*, 16, 1038, 1975.
10. Maeda, M. and Kojima, M., Homoallylic rearrangement of 19-iodocholest-5-en-3β-ol: new adrenal scanning agent, *Steroids*, 26, 241, 1975.
11. Scott, K. N., Couch, M. W., Mareci, T. H., and Williams, G. M., Synthesis and purification of radioactive 6β-iodomethyl-19-norcholest-5(10-en-3β-ol, *Steroids*, 28, 295, 1976.
12. Couch, M. W., Scott, K. N., and Williams, C. M., New method for the synthesis of radioactive 19-iodocholest-5-en-3β-ol, *Steroids*, 27, 451, 1976.
13. Couch, M. W. and Williams, C. M., Comparison of 19-iodocholesterol and 6-iodomethyl-norcholesterol as adrenal scanning agents, *J. Nucl. Med.*, 18, 724, 1977.
14. Scott, K. N., Mareci, T. H., Couch, M. W., and Williams, C. M., Chemical and radiochemical stability of the adrenal-scanning agents, 6β-iodomethyl-19-norcholest-5(10)-en-3β-ol and 19 Iodocholest-5-en-3β-ol, *Steroids*, 30, 511, 1977.
15. Kojima, M., Maeda, M., Ogawa, H., Nitta, K., Ito, T., and Omeda, F., Comparison of uptake of 6β-[^{3}H]methyl-19-norcholest-5(10)-en-3β-ol and [1α, 2α(n)−^{3}H] cholesterol by rat adrenal, *Radioisotopes*, 25, 222, 1976.
16. Kojima, M., Komatsu, H., Shimoirisa, H., Morita, H., Sone, H., and Maeda, M., Structural relationship between radioactive 6β-iodomethyl-19-norcholest-5(10)-en-3β-ol and its analogue for adrenal accumulation, *J. Labelled Compd.*, 16, 169, 1979.
17. Kojima, M., Maeda, M., Komatsu, H., Shimoirisa, H., Ogawa, H., Nitto, K., and Ito, T., Radiobromine labeled norcholesterol analogs. Synthesis and tissue distribution study in rats of bromine-82 labeled 6β-bromomethyl-19-norcholest-5(10)-en-3β-ol, *Steroids*, 29, 443, 1977.

18. Jixiao, M. and Ruisen, Z., ^{131}I-6-iodocholesterol, an agent for imaging the adrenal gland, *Chinese Med. J.*, 92, 237, 1979.
19. Dexter, R. N., Fishman, L. M., and Ney, R. L., Stimulation of adrenal cholesterol uptake from plasma by adrenocorticotrophin, *Endocrinology*, 87, 836, 1970.
20. Szinai, S. S. and Owoyale, J. A., Adrenal scanning agents, paper presented before the Medicinal Chemistry Division of the American Chemical Society, 169th National Meeting, Philadelphia, April 1975.
21. Fukushi, K., Irie, T., Nozako, T., Ido, T., and Kasida, Y., Adrenal affinity and plasma lipoprotein binding of radiohalogeno derivatives of cholesterol, *J. Labelled Compd.*, 16, 49, 1979.
22. Counsell, R. E., Ranade, V. V., Kulkarni, P. G., and Afiatpour, P., Potential organ or tumor-imaging agents. Esters of 19-radioiodinated cholesterol, *J. Nucl. Med.*, 14, 777, 1973.
23. Yu, T., Wieland, D. M., Ice, R. D., and Beierwaltes, W. H., Synthesis of labeled 3-iodocholesterols for adrenal imaging, *J. Labelled Compd.*, 13, 274, 1977.
24. Beierwaltes, W. H., Wieland, D. M., Yu, T., Swanson, D. P., and Mosley, S. T., Adrenal imaging agents: rationale, synthesis, formulation and metabolism, *Semin. Nucl. Med.*, 8, 5, 1978.
25. Counsell, R. E., Hong, B. H., Willette, R. E., and Ranade, V. V., Tumor localizing agents. 5. Radioiodinated pregnanes, *Steroids*, 11, 817, 1968.
26. Spitznagle, L. A., Eng, R. R., and Marino, C. A., Design of radiopharmaceuticals for adrenal imaging: Effect of structure and metabolism on the distribution of H-3 and F-18 labeled pregnenolones, in *Radiopharmaceuticals II*, Proc. 2nd Int. Symp. Radiopharmaceuticals, Society of Nuclear Medicine, New York, 1979, 685.
27. Ito, T., Yamauchi, S., Maeda, M., Komatsu, H., and Kojima, M., Tissue distribution studies in rats with two ^{131}I-labeled adrenal scanning agents: Comparison between ^{131}I-1-6-iodomethyl-19-norsitost-5(10)-en-3β-ol (NST-6-^{131}I) and 19-iodositost-5-en-3β-ol (ST-19-^{131}I), *Int. J. Nucl. Med. Biol.*, 6, 163, 1979.
28. Counsell, R. E., Kulkarni, P. G., Afiatpour, P., and Ranade, V. V., Potential organ or tumor-imaging agents: 19-radioiodinated sterols, *J. Med. Chem.*, 16, 945, 1973.
29. Ice, R. D., Kircos, I. T., Coffey, J. L., Watson, E., Sarkar, S. D., and Beierwaltes, W. H., Radiation dosimetry of ^{131}I-6β-iodomethylnorcholesterol, *J. Nucl. Med.*, 17, 540, 1976.
30. Carey, J. E., Thrall, J. H., Freitas, J. E., and Beierwaltes, W. H., Absorbed dose to the human adrenals from iodomethylnorcholesterol (^{131}I) "NP-59": Concise communication, *J. Nucl. Med.*, 20, 60, 1979.
31. Sarkar, S. D., Ice, R. D., Beierwaltes, W. H., Gill, S. P., Balachandran, S., and Basmadjian, G. P., Selenium-75-19-selenocholesterol - A new adrenal scanning agent with high concentration in the adrenal medulla, *J. Nucl. Med.*, 17, 212, 1976.
32. Riley, A. L. M., The development of selenium-75 cholesterol analogues, *J. Labelled Compd.*, 16, 28, 1979.
33. Knapp, F. F. and Ambrose, K. R., Tellurium-123m labeled 24-nor-23-(isopropyl tellura)-5α-cholan-3β-ol: A potential adrenal imaging agent, *J. Nucl. Med.*, 18, 600, 1977.
34. Knapp, F. F., Ambrose, K. R., and Callahan, A. P., The effects of nuclear and sidechain modifications of the adrenal concentration of steroids labeled in the sidechain with Te-123m, *J. Labelled Compds. Radiopharm.*, 16, 35, 1979.
35. Jackson, R. L., Morrisett, J. D., and Gotto, A. M., Lipoprotein structure and metabolism, *Physiol. Rev.*, 56, 259, 1976.
36. Gwynne, J., Hess, B., Roundtree, R., and Wolf, W., Enhancement of steroidogenesis by serum lipoproteins in human, rat and mouse adrenal cells, Abstr. 294, The Endocrine Society, Bethesda, Md., June 1978.
37. Brown, M. S. and Goldstein, J. L., Receptor-mediated control of cholesterol metabolism, *Science*, 191, 150, 1976.
38. Faust, J. R., Goldstein, J. L., and Brown, M. S., Receptor-mediated uptake of low density lipoprotein and utilization of its cholesterol for steroid synthesis in cultured moust adrenal cells, *J. Biol. Chem.*, 252, 4861, 1977.
39. Brown, M. S. and Goldstein, J. L., Regulation of the activity of the low density lipoprotein receptor in human fibroblasts, *Cell*, 6, 307, 1975.
40. Fukushi, K., Adrenal Affinity and Plasma Lipoprotein Binding of Radiohalenoderivatives of Cholesterol, Abstr. 23, in 2nd Int. Symp. Radiopharmaceutical Chem., Oxford, U.K., 1978.
41. Gordon, L., Mayfield, R. K., Levine, J. H., Lopes-Virella, M. F., Sagel, J., and Buse, M. G., Erroneous negative adrenal imaging with 131-I-iodomethyl-19-norcholesterol (131-I-chol), *J. Nucl. Med.*, 20 (Abstr.), 677, 1979.
42. Valk, T. W., Gross, M. D., Freitas, J. E., Swanson, D. P., Schteingart, D. E., and Beierwaltes, W. H., The relationship of serum lipids to adrenal gland uptake of 6β-{^{131}I}-iodomethyl-19-norcholesterol in Cushing's Syndrome, *J. Nucl. Med.*, 21, 1069, 1980.

43. Thrall, J. H., Freitas, J. E., and Beierwaltes, W. H., Adrenal scintigraphy, *Semin. Nucl. Med.*, 8, 23, 1978.

44. Grayson, R. R., Factors which influence the radioactive iodine thyroidal uptake test, *Am. J. Med.*, 28, 397, 1960.

45. Gross, M. D., Valk, T. W., Freitas, J. E., Swanson, D. P., Schteingart, D. E., and Beierwaltes, W. H., The relationship of adrenal iodomethylnorcholesterol uptake and cortisol secretion in Cushing's Syndrome, *J. Clin. Endocrinol. Metab.*, 52, 1062, 1981.

46. Gwynne, J. T., Mahattee, D., Brewer, A. B., and Ney, R. L., Adrenal cholesterol uptake from plasma lipoproteins: regulation by corticotrophin, *Proc. Natl. Acad. Sci.*, 73, 4329, 1976.

47. Anderson, J. M. and Dietschy, J. M., Cholesterologenesis: depression in extrahepatic tissues with 4-aminopyrazolo-3,4-ol pyridimine, *Science*, 193, 903, 1976.

48. Clinkenbeard, K. D., Sugiyama, T., Reed, W. D., and Lake, M. D., Intracellular localization of 3-hydroxy - 3 methylglutaryl coenzyme A enzymes in liver. Separate cytoplasmic and mitochrondrine 3-hydroxy-3-methylglutaryl coenzyme - A generating system for cholesterolgenesis and steroidogenesis, *J. Biol. Chem.*, 250, 3108, 1975.

49. Balsubramaniam, S., Goldstein, J. L., Faust, J. R., Brunschede, G. Y., and Brown, M. S., Lipoprotein-mediated regulation of 3-hydroxy-3-methylglutaryl coenzyme. A reductace activity and cholesteryl ester metabolism in the adrenal gland of the rat, *J. Biol. Chem.*, 252, 1771, 1977.

50. Davis, W. W. and Garren, L. D., Evidence for the stimulation of adrenocorticotrophic hormone on the conversion of cholesterol esters to cholesterol in the adrenal, *in vivo*, *Biochem. Biophys. Res. Commun.*, 24, 805, 1966.

51. Beckett, J. G. and Boyd, G. S., Purification and control of bovine adrenal cortical cholesterol ester hydralase evidence for actuation of the enzyme by a phosphorylation, *J. Biochem.*, 72, 223, 1977.

52. Honn, K. V. and Chavin, W., Prostaglandin modulation of the mechanism of ACTH action on the human adrenal, *Biochem. Biophys. Res. Commun.*, 73, 164, 1976.

53. Moses, D. C., Schteingart, D. E., Sturman, M. F., Beierwaltes, W. H., and Ice, R. D., Efficacy of radiocholesterol imaging of the adrenal glands in Cushing's syndrome, *Surg. Gynecol. Obstet.*, 139, 1, 1974.

54. Barbarino, A., DeMarinis, L., Leberale, I., and Menini, E., Evaluation of steroid laboratory tests and adrenal gland imaging with radiocholesterol in the aetiological diagnosis of Cushing's syndrome, *Clin. Endocrinol.*, 10, 107, 1979.

55. Gordon, L., Mayfield, R. K., and Levine, J. H., Failure to visualize adrenal glands in a patient with bilateral adrenal hyperplasia, *J. Nucl. Med.*, 21, 49, 1980.

56. Fukushi, S., Nakajima, K., Miura, T., Haruyama, K., Kida, T., Saito, M., and Nakamura, M., Comparative study of adrenal scanning agents, 6β-iodomethyl-19-norcholest-5(10)-en-3β-ol^{131}I (NCh-6-^{131}I) and 131-I-19-iodocholesterol (CL-19-^{131}I), *Jpn. J. Nucl. Med.*, 13, 775, 1976.

57. Rizza, R. A., Wahner, H. W., Spelsberg, T. C., Northcutt, R. L., and Moses, H. L., Visualization of nonfunctioning adrenal adenomas with iodocholesterol: possible relationship to subcellular distribution of tracer, *J. Nucl. Med.*, 19, 458, 1978.

58. Gross, M. D., Valk, T. W., Freitas, J. E., Swanson, D. P., Woodbury, M. G., Schteingart, D. E., and Beierwaltes, W. H., Dexamethasone suppression (DS) ^{131}I-iodomethylnorcholesterol (NP-59) adrenal uptake is a measure of 17-ketosteroid excretion in hyperandrogenism, *J. Nucl. Med.*, 21, (Abstr.), 28, 1980.

59. Gross, M. D., Grekin, R. J., Brown, L. R., Marsh, D. D., and Beierwaltes, W. H., The relationship of adrenal iodocholesterol uptake to adrenal zona glomerulosa function, *J. Clin. Endocrinol. Metab.*, 52, 612, 1981.

60. Shapiro, B., Bitton, K. E., and Hawkins, L. A., Results of quantitative adrenal imaging in 62 cases using selenium-75-selenocholesterol, *J. Nucl. Med.*, 21 (Abstr.), 29, 1980.

61. Seabold, J. E., Hayne, T. P., Deasis, D. N., Samaan, N. A., Glenn, H. J., and Jahns, M. F., Detection of metastatic adrenal carcinoma using ^{131}I-6β-iodomethyl-19-norcholesterol total body scans, *J. Clin. Endocrinol. Metab.*, 45, 788, 1978.

62. Schteingart, D. E., Seabold, J. E., Gross, M. D., and Swanson, D. P., I^{131}-19-iodocholesterol adrenal tissue uptake and imaging in adrenal carcinoma, *J. Clin. Endocrinol. Metab.*, 52, 1156, 1981.

63. Beierwaltes, W. H., Wieland, D. M., and Swanson, D. P., Radiolabeled enzyme inhibitors for imaging, in *Principles of Radiopharmacology*, Vol. 2, Colombetti, L. G., Ed., CRC Press, Boca Raton, Fla., 1979, 41.

64. Gower, D. B., Modifiers of steroid-hormone metabolism: a review of their chemistry, biochemistry, and clinical application, *J. Steroid Biochem.*, 5, 501, 1974.

65. Williamson, D. G. and O'Donnell, V. J., The interaction of metopirone with adrenal mitochondria cytochrome P-450. A mechanism for the inhibition of adrenal steroid 11β-hydroxylation, *Biochemistry*, 8, 1306, 1969.

66. Satre, M. and Vignais, P. V., Steroid 11β-hydroxylation in beef adrenal cortex mitochondria. Binding affinity and capacity for specific [^{14}C] steroids and for [^{3}H] metyrapol, an inhibitor of the 11β-hydroxylation reaction, *Biochemistry*, 13, 2201, 1974.

67. Wieland, D. M., Kennedy, W. P., Ice, R. D., and Beierwaltes, W. H., Radioiodinated pyridines - Potential adrenocortical imaging agents, *J. Labelled Compd.*, 13, 229, 1977.

68. Wu, J. L., Wieland, D. M., Beierwaltes, W. H., Swanson, D. P., and Brown, L. E., Radiolabeled enzyme inhibitors - enhanced localization following enantiomeric purification, *J. Nucl. Med.*, 19, 677, 1978.

69. Morales, J. O., Beierwaltes, W. H., Counsell, R. E., and Meier, D. H., The concentration of radioactivity from labeled epinephrine and its precursors in the dog adrenal medulla, *J. Nucl. Med.*, 8, 800, 1967.

70. Counsell, R. E., Smith, T. D., DiGiulio, W., and Beierwaltes, W. H., Tumor localizing agents. VIII. Radioiodinated phenylalanine analogs, *J. Pharm. Sci.*, 57, 1958, 1968.

71. Lieberman, L. M., Beierwaltes, W. H., Varma, V. M., Weinhold, P., and Ling, R., Labeled dopamine concentration in human adrenal medulla and in neuroblastoma, *J. Nucl. Med.*, 10, 93, 1969.

72. Anderson, B. G., Beierwaltes, W. H., Harrison, T. S., Ansari, A. N., Buswink, A. A., and Ice, R. D., Labeled dopamine concentration in pheochromocytomas, *J. Nucl. Med.*, 14, 781, 1973.

73. Fowler, J. S., Ansari, A. N., Atkins, H. L., Bradley-Moore, P. R., MacGregor, R. R., and Wolf, A. P., Synthesis and preliminary evaluation in animals of carrier-free ^{11}C-1-dopamine hydrocholoride. X, *J. Nucl. Med.*, 14, 867, 1973.

74. Fowler, J. S., Wolf, A. P., Christman, R. D., MacGregor, R. R., Ansari, A., and Atkins, H., Carrier-free ^{11}C-labeled catecholamines, in *Radiopharmaceuticals*, Subramanian, G., Rhodes, B. A., and Cooper, J. T., Eds., Society of Nuclear Medicine, New York, 1975, 196.

75. Ice, R. D., Wieland, D. M., and Beierwaltes, W. H., Concentration of dopamine analogs in the adrenal medulla, *J. Nucl. Med.*, 16, 1147, 1975.

76. Fowler, J. S., MacGregor, R. R., and Wolf, A. P., Radiopharmaceuticals. 16. Halogenated dopamine analogs. Synthesis and radiolabeling of 6-iododopamine and tissue distribution studies in animals, *J. Med. Chem.*, 19, 356, 1976.

77. Counsell, R. E., Yu, T., Ranade, V. V., and Buswink, A., Potential organ- or tumor-imaging agents. 14. Myocardial scanning agents, *J. Med. Chem.*, 16, 1038, 1973.

78. Korn, N., Buswink, A., Yu, T., Carr, E. A., Carroll, M., and Counsell, R. E., A radioiodinated bretylium analog as a potential agent for scanning the adrenal medulla, *J. Nucl. Med.*, 18, 87, 1977.

79. Lieberman, L. M., Beierwaltes, W. H., Varma, V. M., Weinhold, P., and Ling, R., Labeled dopamine concentration in human adrenal medulla and in neuroblastoma, *J. Nucl. Med.*, 10, 93, 1969.

80. Fowler, J. S., MacGregor, R. R., and Wolf, A. P., Radiopharmaceuticals. Halogenated dopamine analogs: synthesis and radiolabeling of 6-iododopamine and tissue distribution studies in animals, *J. Med. Chem.*, 19, 356, 1976.

81. Counsell, R. E., Smith, T. D., Ranade, V. V., Noronha, O. P., and Desai, P., Potential organ or tumor imaging agents. XI. Radioiodinated tyramines, *J. Med. Chem.*, 16, 684, 1973.

82. Larsen, A. A., Gould, W. A., Roth, H. R., Comer, W. T., Uloth, R. H., Dungan, K. W., and Lish, P. M., Sulfonanilides. II. Analogs of catecholamines, *J. Med. Chem.*, 10, 462, 1967.

83. Ice, R. D., Wieland, D. M., Beierwaltes, W. H., Lawton, R. G., and Redmond, M. J., Concentration of dopamine analogs in the adrenal medulla, *J. Nucl. Med.*, 16, 1147, 1975.

84. Kaiser, C., Schwartz, M. S., Colla, D. F., and Wardell, J. R., Adrenergic agents. III. Synthesis and adrenergic activity of some catecholamine analogs bearing a substituted sulfonyl or sulfonylalkyl group in the meta position, *J. Med. Chem.*, 18, 674, 1975.

85. Uloth, R. H., Kirk, J. R., Gould, W. A., and Larsen, A. A., Sulfonanilides. I. Monoalkyl- and Aryl-sulfonamidophenethanolamines, *J. Med. Chem.*, 9, 88, 1966.

86. Melmon, K. L., Catecholamines and the adrenal medulla, in *Textbook of Endocrinology*, 5th ed., William, R. H., Ed., W. B. Saunders, Philadelphia, 1974, 283.

87. Sandler, M. and Rushsen, C. R. J., The biosynthesis and metabolism of catecholamines, in *Progress in Medicinal Chemistry*, Vol. 6, Ellis, C. P. and West, C. S., Eds., Butterworths, London, 1964, 200.

88. Axelrod, J., Purification of phenylethanolamine-N-methyl-transferase, *J. Biol. Chem.*, 237, 1657, 1962.

89. Wieland, D. M., Brown, L. E., and Beierwaltes, W. H., unpublished results.

90. Kloss, G. and Leven, M., Accumulation of radioiodinated tyrosine derivatives in the adrenal medulla and in melanomas, *Eur. J. Nucl. Med.*, 4, 179, 1979.

91. Pendleton, R. G., Kaiser, C., and Gessner, G., Studies of adrenal phenylethanolamine-N-methyltransferase (PNMT) with SKF-64139, a selective inhibitor, *J. Pharmacol. Exp. Ther.*, 197, 623, 1976.

92. Yu, T., Wieland, D. M., Brown, L. E., and Beierwaltes, W. H., Synthesis of radiolabeled inhibitors of phenylethanolamine-N-methyltransferase, *J. Labelled Compd.*, 16, 173, 1979.

93. Boura, A., Copp, F., and Green, A., Animal physiology: New antiadrenergic compounds, *Nature (London)*, 184, 70, 1959.
94. Wieland, D. M., Swanson, D. P., Brown, L. E., and Beierwaltes, W. H., Imaging the adrenal medulla with an I-131-labeled antiadrenergic agent, *J. Nucl. Med.*, 20, 155, 1979.
95. Nickerson, M. and Collier, B., Drugs inhibiting adrenergic nerves and structures innervated by them, in *The Pharmacological Basis of Therapeutics*, 5th ed., Goodman, L. S. and Gilman, A., Eds., Macmillan, New York, 1975, 553.
96. Short, J. H. and Darby, T. D., Sympathetic nervous system blocking agents. III. Derivatives of benzylguanidine, *J. Med. Chem.*, 10, 833, 1967.
97. Wieland, D. M., Wu, J. L., Brown, L. E., Manger, T. J., Swanson, D. P., and Beierwaltes, W. H., Radiolabeled adrenergic neuron blocking agents: adrenal medulla imaging with [^{131}I]-iodobenzylguanidine, *J. Nucl. Med.*, 20, 349, 1980.

Chapter 9

TRAPPING AND METABOLISM OF RADIOIONS BY THE THYROID

Y. Yano

TABLE OF CONTENTS

I. INTRODUCTION

The development of radiotracer methodology and the subsequent growth of in vivo nuclear medicine imaging techniques were immeasurably influenced by the unique avidity of the thyroid gland for iodine. It is significant that the potential advantages of radioisotopes (radioions) as tracers in the study of transport and metabolism were recognized within a few years after the initial production of radioiodine by neutron activation of stable iodine in 1935.

In 1938, Hertz et al.,[1] using ^{128}I (25 min half-life) found a ninefold greater uptake of radioiodine in the thyroid relative to the liver. A few years later, 8 d ^{131}I was produced in the cyclotron developed by Ernest Lawrence. The new radioisotope of iodine was reported by Livingood and Seaborg[2] and by Tape and Cork.[3]

Hamilton and Soley at Berkeley were the first to administer radioiodine for studies of the human thyroid.[4] Since that time, radioisotopes of iodine have played a predominant role in understanding the function of the thyroid gland, and in diagnosis and therapy of thyroid diseases.[5]

A number of excellent publications reviewing the structure, function, and physiology of the thyroid gland, and their relationships to the trapping and metabolism of radioions by the thyroid, have been used in the preparation of this material.[5-22] Emphasis is on the mechanism of localization and cellular function, especially as these are related to trapping and metabolism of radioiodine. Other radioions such as ^{99m}TcO$_4^-$, ^{131}Cs$^+$, ^{201}Tl$^+$, ^{67}Ga^{+3}, and ^{75}Se selenomethionine are also discussed.

II. THYROID GLAND

A. Anatomy

The thyroid gland is the largest of the endocrine glands (average wt. 25 to 35 g). It is H-shaped with two lateral lobes and an isthmus, and it is located at the third and fourth tracheal cartilage. The thyroid is grey or reddish brown in color and is highly vascular with the highest rate of blood flow per gm of tissue of any organ in the body. An amount of blood equal to the total blood volume passes through it every hour. Fibrous septa form structural lobes or macroscopic disks which are composed of clusters of 20 to 40 follicles inside of a connective tissue sheath. These form the basic unit of the thyroid lobule.[12]

The thyroid is made up of multiple acni or follicles. Each spherical follicle is surrounded by a single layer of cells filled with pink-staining proteinaceous material called "colloid". In the inactive thyroid the colloid is abundant, the follicles are large, and the lining cells are flat. In the active thyroid the follicles are small, the cells cuboidal or columnar, and the edge of the colloid is scalloped forming many small resorption lacunae.[7] The thyroid cells rest on basal lamina that separate them from adjacent capillaries.

B. Physiology and Biochemistry

The biochemistry of iodine in the thyroid was elucidated most clearly by the use of ^{131}I.[14] The first step is the concentration of iodide in the thyroid many times greater than the concentration in the blood. It was found that thiocyanates given before the administration of radioiodine would suppress the uptake of ^{131}I in the thyroid; if the thiocyanate is administered after the radioiodine, the ^{131}I that had accumulated in the thyroid is flushed out. The second step is the assumption that iodide is oxidized to iodine because proteins will combine only with the oxidized form. Third, the iodine combines with thyroglobulin and the synthesis of thyroxine begins.

The thyroid cells have a dual function. These are to (1) collect and transport iodine

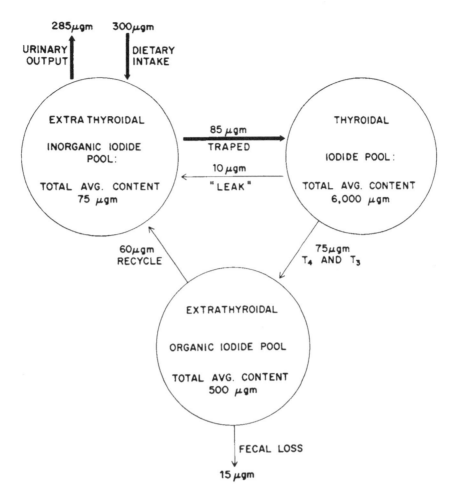

FIGURE 1. Daily turnover of iodine. (From Rosenfield, R. L., Refetoff, S., Hoffer, P. B., Gottschalk, A., and DeGroot, L. G., in *Pediatric Nuclear Medicine,* James, A. E., Wagner, H. N., Jr., and Cooke, R. E., Eds., W. B. Saunders, Philadelphia, 1974, chap 10. With permission.)

for the synthesis of thyroglobulin, and (2) to secrete thyroglobulin into the colloid for the release of the thyroid hormones tri-iodothyronine (T_3) and the tetraiodothyronine (T_4) or thyroxine into the circulation.[7]

Physiologically and biochemically the thyroid gland is dedicated to the production of the protein thyroglobulin.[8] To accomplish this, the thyroid avidly extracts iodine from the circulating vascular pool. Broadly defined there are three major iodine compartments: (1) the extra-thyroidal inorganic iodine pool, (2) the thyroidal iodine pool, and (3) the extrathyroidal organic iodine pool.[18,22]

The normal daily turnover of iodine in intercompartmental exchange is seen in Figure 1.[18] Ingested iodide is rapidly absorbed and enters the extrathyroidal iodide pool which receives an additional 60 µg/day of iodine from breakdown of thyroid hormones and 10 µg "leak" of nonhormonal iodine from the thyroid gland. This pool normally contains 75 µg of iodide which turns over five times per day.

Iodine from the extrathyroidal pool is extracted primarily by the kidneys and thyroid. In normal individuals about 85 µg of iodide are taken up by the thyroid each day and the remainder is excreted by the kidneys. About 10 to 35% of the extrathyroid iodide pool is normally taken up in the thyroid, and the iodide is rapidly organified

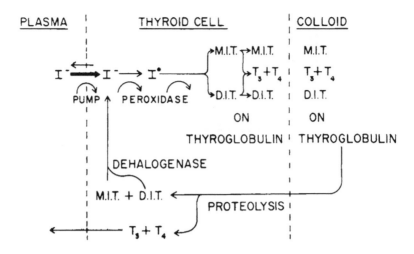

FIGURE 2.   Intrathyroidal iodine metabolism. (From Rosenfield, R. L., Refetoff, S., Hoffer, P. B., Gottschalk, A., and DeGroot, L. G., in *Pediatric Nuclear Medicine*, James, A. E., Wagner, H. N., Jr., and Cooke, R. E., Eds., W. B. Saunders, Philadelphia, 1974, chap. 10. With permission.)

and bound to tyrosine to form iodotyrosines. About 75 μg of iodine per day are then secreted in the form of T_3 and T_4.[18] These thyroid hormones are nearly completely protein-bound. This extrathyroidal organic iodine pool contains more T_3 than T_4. After degradation of these hormones, about 60 μg per day are returned to the extrathyroidal inorganic iodine pool, and 15 μg are lost by fecal excretion.[18]

1. Intrathyroidal Metabolism of Iodine[8,9,18]

Figure 2 depicts intrathyroidal iodine metabolism.[18] Active transport mechanism concentrates I⁻ in thyroid to 20 times iodide level in plasma. Iodide is oxidized by peroxidase and H_2O_2 to I° which is bound to tyrosyl in preformed thyroglobulin. Iodination gives mono-iodotyrosine and di-iodotyrosine, MIT and DIT, respectively. A small part of the iodinated tyrosyl molecule in thyroglobulin is enzymatically coupled to give iodothyronines. Either T_3, tri-iodothyronine, and T_4, tetraiodothyronine (thyroxine), is formed.

Thyroglobulin has a molecular weight of 650,000, and contains 120 tyrosyl groups and 25 iodine atoms. In normal thyroglobulin the ratio of T_4 to T_3 is 4 to 1. Thyroglobulin containing large amounts of MIT and DIT is stored in colloid form within the follicle for 50 to 100 days.

Eventually, follicular colloid is pinocytosed at apical thyroid cell border. Colloid droplets formed within the cell fuse with lysosomes to form phagosomes through which thyroglobulin is completely digested to its amino acid components by proteolytic enzymes and peptidases. T_3 and T_4 are secreted into the blood, and the iodinated tyrosines (MIT and DIT) are deiodinated by deiodase. The liberated iodide is returned to the intrathyroid iodine pool. This iodine constitutes a large part of the iodine used in the production of thyroid hormones.

2. Thyroid Hormones

Thyroid hormones are metabolically active, and are secreted into the circulation. The synthesis of thyroid hormones is regulated by thyroid stimulating hormone (TSH) from the pituitary gland. The release of TSH from the pituitary is stimulated by thyrotropin releasing factor (TRF) which is produced in the hypothalamus. The feedback mechanisms of the thyroid-pituitary-hypothalamus interaction are shown in Figure 3.[18]

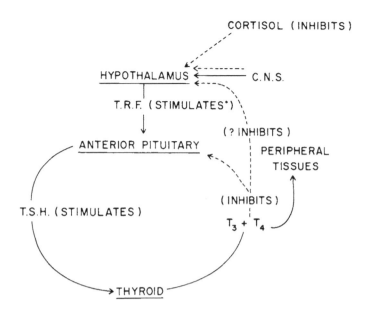

CORTISOL (INHIBITS)

HYPOTHALAMUS ← ── C.N.S.

T.R.F. (STIMULATES*)

(? INHIBITS)

ANTERIOR PITUITARY

PERIPHERAL
TISSUES

(INHIBITS)

T.S.H. (STIMULATES)

$T_3 + T_4$

THYROID

*EFFECT BLOCKED BY EXCESS $T_3 + T_4$
SOLID LINES INDICATE STIMULATORY
PATHWAY
BROKEN LINES INDICATE INHIBITORY
PATHWAY

FIGURE 3. Thyroid-pituitary-hypothalamus axis. (From Rosenfield, R. L., Refetoff, S., Hoffer, P. B., Gottschalk, A., and DeGroot, L. G., in *Pediatric Nuclear Medicine*, James, A. E., Wagner, H. N., Jr., and Cooke, R. E., Eds., W. B. Saunders, Philadelphia, 1974, chap. 10. With permission.)

3. Serum Transport of Thyroxine and Tri-iodothyronine[18,19,22]

Thyroxine (T_4) is carried on three major serum proteins; thyroxine-binding globulin (TBG), thyroxine-binding prealbumin (TBPA), and albumin. Most of the circulating T_4 is bound to serum proteins with only a small fraction circulating as free hormone in dynamic equilibrium with the protein bound fraction. The free T_4, although insignificant quantitatively, is of prime biochemical importance. It is presumed that only the free T_4 can penetrate the cell and activate metabolic processes. The "free" T_4 is believed to be responsible for both physiologic effects of thyroid hormone and the feedback regulation of hormone production and release.

The biological half-life is 6.5 days for T_4 and 0.8 day for T_3. The thyroidal secretion ratio of T_4 to T_3 is about 4:1; however, because of the rapid degradation rate for T_3, the ratio in blood is 40:1.

About 40 to 50% of extrathyroidal thyroxine and 85 to 90% of extrathyroidal T_3 is intracellular. The nature of the cell binding sites is poorly understood.

4. Action of Thyroid Hormone[18]

Thyroid hormone affects activity of virtually every organ system within the body. Effects can be demonstrated on growth and development, temperature control, muscle activity, and protein, fat, and carbohydrate metabolism. The exact mechanism of its mode of action is unknown. The stimulation of macromolecular synthesis may be the primary intracellular mechanism of thyroid hormone effects.

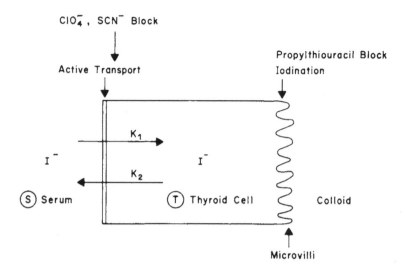

FIGURE 4.   Trapping ability of thyroid cell, K_1 is iodide trapped and K_2 is efflux of iodide. (From Rapoport, B. and DeGroot, L. J., *Semin. Nucl. Med.*, 1, 265, 1971. With permission.)

III. THYROID TRAPPING OF IODIDE[22,25]

The thyroid can trap and concentrate iodide to many times that of the plasma concentration. This mechanism is distinct from the binding or organification mechanism by which thyroidal iodide is incorporated into covalent linkages for thyroidal iodoamino acids. Normally, this organification of trapped iodide is low. Thus, to measure the trapping mechanism, the binding mechanism is blocked with a goitrogen of the thiouracil type. However, there is an iodide pool in the thyroid that can be displaced by perchlorate. There appear to be two "pools" of thyroidal iodide which apparently are different kinetically: one represents the iodide pool trapped from the blood, and the other the iodide pool formed by deiodination of iodotyrosine.[8,9] The latter pool is the larger of the two.

There is active transport in concentrating iodine in the thyroid colloid from the circulation. The trapping mechanism could be similar to the Na^+/K^+ pump. The thyroid is 50 mV negative to the interstitial and colloid area. Thus, I^- is pumped into the cell at its base against an electrical gradient and then diffuses down the electrical gradient into the colloid.[7,25,32]

Iodide accumulation or extraction by the thyroid is the result of two processes: the first is the entry process also variously called the one-way I^- clearance, "active" I^- transport, the I^- pump or the I^- trap; the second is the exit process. The overall process can be called the I^- concentrating mechanism.[22,25] Figure 4 shows the iodide influx-efflux. Iodide entering the thyroid cell from the plasma is rapidly incorporated into the tyrosine residues of preformed thyroglobulin, and is stored outside the cell in the follicle. Thus, the level of inorganic iodide within the cell is low.

Hence, the entry of iodide into the thyroid cell is normally the rate-limiting step in thyroid hormone formation. By blocking organification of iodide with propylthiouracil, the maximal ability of the thyroid cell to concentrate iodide from the plasma can be determined. The thyroid to serum inorganic iodide ratio (T/S) represents the trapping ability of the thyroid cell. Iodide trapping is stimulated by TSH which stimulates K_1, the influx, and not by diminishing K_2, the efflux.[22]

Other anions such as TcO_4^-, SCN^-, and ClO_4^- are known inhibitors of iodide trapping. Both TcO_4^- and ClO_4^- are competitive inhibitors while SCN^- apparently increases K_2, the efflux, of iodide from the cell.[23]

Iodide trapping is an active process requiring energy. Agents which inhibit or uncouple oxidative phosphorylation, such as 2,4-dinitrophenol, cyanide, and anoxia also abolish the active transport of iodide.[24] Iodide accumulation is associated with Na^+, K^+, and ATPase-dependent transport. Both systems are inhibited by cardiac glycosides.[25] However, there are differences between the two systems.

Other tissues which concentrate I^- (extrathyroidal iodide space) against a concentration gradient are salivary glands, stomach, gastric mucosa, mammary gland, small intestines, and choroid plexus.[28-30] In addition, kidneys share properties of iodide-concentrating tissues although the isolated organ does not concentrate I^-. The iodide concentrating tissues are similar to thyroid tissue in concentrating iodide, i.e., inhibition by anions such as SCN^-, ClO_4^-, BF_4^-, and NO_3^- which are also concentrated in thyroid; inhibition by poisons such as 2,4-dinitrophenol and cardiac glycosides; requirement for K^+, and half-saturation with iodide near 3.1×10^{-5} M I^-. Some dissimilar features are: failure of thyroid to concentrate SCN^- to the same extent as saliva and gastric juices, and failure to respond to thyrotropin.[31] Iodide transport is oriented around each follicle so that the concentration gradient from the capillary across the basal membrane of the parenchymal cells is built against an electrochemical gradient.[32]

Under ordinary environmental conditions the capacity for iodine accumulation is apparently not the limiting factor in quantitative regulation of hormone synthesis.[33-35] However, it is known that the fundamental property of the thyroid to bind iodine is more easily saturated with increasing concentration of iodide in the extracellular milieu than is the transport capacity.[36] A low concentration of iodide in the plasma indicates a small quantity of intrathyroidal iodine as iodide.[37] In contrast, a high level of plasma iodide is indicative of an increase in the fraction of intrathyroidal iodine present as inorganic iodide.[38]

The effective or apparent clearance of iodide from the thyroid is a balance between transport of iodide from the capillaries into the cells, and the bidirectional movement of iodide from the cell into the folicular lumen or back into the capillaries.[25] With physiological concentrations of iodide in the blood, the value for "effective" iodide clearance and that for unidirectional flow from capillaries to cells are nearly the same.[39]

Under conditions of high plasma iodide levels, a large part of the transported iodide reenters the circulation from the gland.[40] The organic binding of transported iodine does not occur immediately; therefore, the capacity of transport mechanism is not the limiting factor for hormone synthesis.[41] In this connection it is significant that the stimulation of thyroxine synthesis by TSH does not depend upon increased iodide transport into the gland.[42,43]

In relation to the adaptation of the thyroid gland to iodine deficiency, it is important to note that an insufficient supply of iodine can be a rate limiting factor for hormone syntheses in spite of greatly increased capacity of iodide transport system. This is the case:

1. If the iodide available to be transported sinks below a certain threshhold
2. If the transport mechanism is not functioning
3. If the transport mechanism is inhibited by drugs such as thiocyanate, perchlorate or ouabain[44,45]

Iodide transport is not an autonomous phenomenon, but is dependent upon other metabolic processes occurring in the cell. Intact aerobic glycolysis is essential for normal iodide transport. Inhibition of K-Na pump inactivates the iodide transport mechanism. The Na-K pump is linked to adenosine triphosphatase in the cell membrane. An inhibition of ATPase activity is associated with a parallel decrease of iodide transport.

The thyroid can concentrate I^- 10 to 100 times the concentration in the surrounding medium. The higher values are obtained when organification is blocked by thiouracil although significant gradients are obtained in glands free to form organic iodine. Concentration of I^- does not require organization into follicles but does require cellular integrity.

There appear to be two entirely different control systems to regulate the iodide-concentrating activity of the thyroid.[26] One responds to the levels of circulating hormones produced by the thyroid, the other responds to the availability of the precursor, iodide, for the production of the hormones. When the levels of the circulating thyroid hormone declines, TSH secretion increases to stimulate the thyroidal iodide pump. Increased iodide intake, however, leads to a depression of the iodide concentrating mechanism. The latter control mechanism is autoregulatory since the iodide controls its own rate of transport which can be independent of the TSH effect.

Iodide concentrating activity is the result of the levels of TSH (the dominant factor) which increases the V_{max} and the iodide supply which increases the K_m of the iodide pump.[26]

A unique peculiarity of the thyroidal iodide transport system is that it is spatially linked with enzyme systems that continuously synthesize amino acids, assembles them as highly specialized proteins which bind iodine into amino acids and eventually produces hormonal iodothyronines. Amino acid synthesis and iodination proceed independently of each other.

Organic iodine formation is prevented by antithyroid agents or excess iodide. In certain abnormal thyroid tissue there is an inability to oxidize iodide to iodine which results in an increase of iodide concentration. Radioautographic studies in which the organic iodine formation was blocked with thiouracil suggests a very rapid transport of ^{131}I into the follicular lumen.

In isolated thyroid cells prepared by trypsin treatment there was an accumulation of iodide against a concentration gradient. When organic iodine formation was blocked the cell/medium ratios of iodide ion were 6-9. The accumulated $^{131}I^-$ was released by ClO_4^- indicating the presence of the anion $^{131}I^-$.[35] These results indicate that the iodide concentrating ability lies in the thyroid cell and not in follicular structure.[22]

IV. THYROIDAL HORMONE SYNTHESIS

Synthesis of thyroid hormones requires the synthesis of thyroglobulin and the oxidation of iodide in the thyroid cell for addition of iodine to a tyrosine residue of thyroglobulin to form MIT and DIT. Both MIT and DIT while still coupled to thyroglobulin combine to form the iodothyronines T_3 and T_4. The iodinated thyroglobulin is then stored in the follicular lumen.

Normally, the synthesis of thyroglobulin is linked to the availability of reactive iodine. Synthesis occurs in the ribosomes of the endoplasmic reticulum. Thyroglobulin iodination occurs in three steps, see Figure 5. In the first step H_2O_2 is generated within the cell; next, the peroxidase in the cell combines with H_2O_2 and oxidizes the iodide; finally, the oxidized iodide attaches covalently to the 3 and 5 position of the tyrosine residue.[8]

Two main theories of the coupling mechanism exist for the formation of T_3 or T_4. Generally, it is believed that two iodotyrosine molecules combine while part of the thyroglobulin amino acid sequence. An alternate view to the coupling reaction states that one molecule of free DIT is deaminated by transaminase to *p*-hydroxy-3,5-diio-

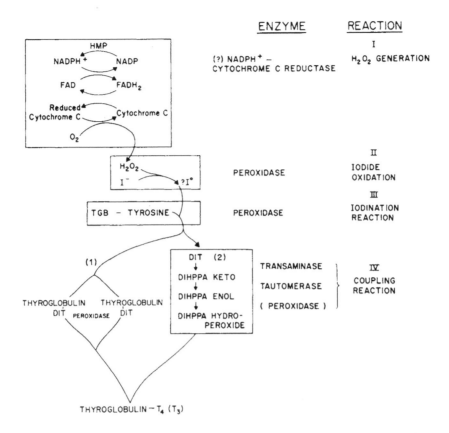

FIGURE 5.   Present concept of thyroglobulin iodination. (From Rapoport, B., and DeGroot, L. J., *Semin. Nucl. Med.*, 1, 265, 1971. With permission.)

dophenylpyruvic acid (DIHPPA), which undergoes tautomerization and subsequent peroxidation to free DIHPPA-hydroperoxide.[46] The latter compound couples with DIT in thyroglobulin to form T_4. However, it is currently believed that peroxidase catalyzes all three steps in thyroid biosynthesis, iodide oxidation, binding to tyrosine, and lastly to iodotyrosine coupling.

Earlier it was believed that iodination occurred in the follicular lumen because the labeled protein bound ^{131}I appeared within that space only a few seconds after giving the tracer. Contrary evidence shows that the biosynthetic enzymes are particulate bound and have not been found in the follicular colloid.

Current opinion is that iodination occurs at the microvillous interface between colloid and thyroid cell.[46] It appears likely that thyroglobulin synthesized on ribosomal RNA template, passes via the endoplasmic reticulum to the Golgi apparatus where it is wrapped into membrane-bound vesicles. These are extruded into the follicular lumen with iodination occurring at the apical cell border as the thyroglobulin leaves the cell. It is believed that there is a continual recirculation of thyroglobulin vesicles between the follicular lumen and the cell, with progressive iodination of thyroglobulin near the cell border with each cycle.[48]

V. THYROID HORMONE BINDING AND TRANSPORT[22]

Thyroid hormone secreted by the thyroid enters a pool of hormone in the bloodstream which is in equilibrium with an intracellular pool of thyroid hormone. Thyroid binding proteins serve as a large reservoir of thyroid hormone circulating with the bloodstream, which are not lost by glomerular filtration, but are available for inflow into the cell for metabolic action.

Three recent major advances in understanding thyroid hormone distribution and turnover are

1.   Characterization of thyroid hormone binding proteins in the plasma.
2.   The concept that the metabolic effect of thyroid hormones is a function of the minute quantity of unbound "free" hormone which is in equilibrium with a much larger pool of bound hormone.
3.   That a large portion of extrathyroidal hormone (30 to 40%) is loosely bound intracellularly in a freely exchangeable pool with plasma thyroid hormone.

Three thyroid hormone binding proteins in plasma — thyroid binding globulin (TBG), thyroid binding prealbumin (TBPA), and thyroid binding albumin — together bind 99.95% of the thyroxine (T_4) in the plasma.[22,49] T_3 is bound to a lesser degree and none is bound to TBPA. Electrophoresis of T_4 shows 70% bound to TBG, 10% bound to TBPA and 20% bound to albumin.[49]

Thyroid binding globulin has a high affinity for T_4, but is present in small amounts so that its low capacity is easily saturated.[22,49] This glycoprotein has a molecular weight of 59,000 and a plasma concentration of 1.5 mg/100 mℓ. It is able to bind 16 to 24 μg of thyroxine per 100 mℓ of plasma and it has a low affinity for T_3.[22] Thyroid binding prealbumin is present in plasma at a concentration of 23 to 25 mg/100 mℓ and has a molecular weight of 70,000. Despite its 20 times greater concentration, TBPA binds less than half the amount of T_4 compared to TBG. Because of its greater concentration, its capacity at saturation, about 300 μg T_4/100 mℓ, is greater than that of TBG. TBPA does not bind T_3. The affinity of albumin for T_4 is even lower but because of its high concentration its bonding capacity is very large.[22]

VI. METABOLISM OF THYROID HORMONES[7,20,22]

Thyroxine and tri-iodothyronine are deaminated and deiodinated in many tissues. Deamination of T_3 and T_4 produces pyruvic acid analogs while decarboxylation produces acetic acid analogs. In the liver the thyroid hormones are conjugated to form sulfates and glucuronides. About 5% per day of iodide loss results from breakdown of the conjugates in the bile and intestines. Deiodination is the principal metabolic pathway with over 80% of radioactive iodine found in the urine after ingestion of I-131 labeled thyroxine.

VII. REGULATION OF THYROID FUNCTION[7,22]

Homeostasis of thyroid hormone synthesis and secretion is maintained both by intrathyroidal and extrathyroidal control mechanisms. The extrathyroidal control mechanism consists of a negative feedback of thyroid hormone on hypothalamic and pituitary regulation of thyroid function. There is direct feedback of thyroid hormone on the pituitary, but such a feedback has not been shown for the hypothalamus. It is currently believed that the hypothalamus modulates the inhibitory effect of thyroid hormone on the pituitary. The efferent component of the feedback system consists of two hormones operating in series. Thyrotropin releasing factor (TRF) is secreted by the hypothalamus and is transported to the adenohypophysis. There it stimulates secretion of TSH into the systemic circulation, and possibly the synthesis of TSH which is a prime regulator of thyroid function. It stimulates all aspects of thyroid metabolism, such as increased glucose oxidation via the hexosemonophosphate shunt and increased phospholipid turnover. Thyroid-stimulating hormone also stimulates thyroidal trapping of iodide, but only after a latent period of a few hours. It also stimulates iodination of tyrosine.

TSH appears to enhance all phases of protein synthesis, and cyclic nucleotides may be intermediates in the expression of all actions of TSH. The nucleotides effect the loading of thyroid follicle with stored thyroid hormone and triggering its release by another mechanism.

Other agents such as prostaglandin E_1 (PGE$_1$) produce certain actions of TSH. The two adenylcyclase activators TSH and PGE$_1$ and the prostaglandin antagonist 7-oxa-13-prostynoic acid (PY$_1$) were used in iodide trapping experiments with isolated bovine thyroid cells. The results showed: (1) no additive effects of TSH and PGE$_1$, (2) PY$_1$ abolished PGE$_1$ effects and markedly inhibited TSH effects and (3) PY$_1$ inhibited activation of adenyl cyclase. These results suggest that TSH effects on iodide trapping are mediated via adenyl cyclase activation, and that there may be a single TSH related prostaglandin receptor in thyroid which must be activated to obtain TSH stimulation.

VIII. SUMMARY OF THYROID METABOLISM OF IODINE

Thyroid metabolism of iodine can be divided into four phases:

1. Iodide ion trapping
2. Organic binding of iodine
3. Storage of thyroid hormone
4. Release of thyroid hormone

Nearly one-fifth of the iodide in the blood that perfuses the thyroid is extracted (trapped) in a single passage through the gland.[17] The trapping of iodide follows saturation kinetics. After trapping of iodide, rapid synthesis of thyroid hormone occurs. Organic radioactivity is demonstrated within 5 sec after intravenous administration of I-125 in experimental animals.[20,21] Secretion of radioiodine-labeled thyroxine begins within 2 hr after intravenous injection in humans.[20] Because of continuous trapping, organification, storage and release functions, the saturation kinetics of the iodide continue to change with an overall increase in radioiodine in the gland until the rate of hormone release exceeds the other functions, at which time the peak thyroid uptake is seen.

The advantage of radioiodine for thyroid imaging is an ability to visualize the pathophysiological sequences as opposed to Tc-99m whose role is confined to the trapping phase.[17,51]

IX. OTHER RADIOIONS AND THE THYROID GLAND

A. Tc-99m Pertechnetate (TcO$_4^-$)

1. Trapping

The uptake of both I$^-$ and TcO$_4^-$ is by trapping but TcO$_4^-$ is not organified; thus, the images from the two radionuclides might not be the same.[51]

Thyroid uptake curves for ^{131}I and ^{99m}TcO$_4$ are markedly different. The activity of ^{131}I in the gland rose sharply from 1 to 24 hr in contrast to ^{99m}TcO$_4$ which decreased with time. The activity of ^{99m}TcO$_4$ in the plasma decreased more rapidly than ^{131}I. Thus, the behavior of the two isotopes cannot be compared on the basis of thyroidal concentration alone. A more accurate picture can be obtained by using the ratio of thyroid activity to blood activity (T/B). The T/B is higher for ^{131}I than for ^{99}TcO$_4^-$ because of the rapid organic binding of iodine. Pertechnetate trapping in the thyroid is blocked by KClO$_4$ and stimulated by TSH. There appears to be some slight further metabolism of TcO$_4$ in the thyroid.[52] Thyroid uptake and turnover of pertechnetate were determined in normal and hyperthyroid patients.[53,54] Thyroid clearance is the rate of change of thyroid uptake divided by plasma activity. With radioiodide there is a reasonably constant clearance value with time. However, with pertechnetate the clearance falls rapidly with time until negative values are reached. It is clear that pertechnetate returns from the thyroid to plasma probably by passive diffusion. The unidirectional clearance and the intrathyroidal space were markedly different in euthyroid and hyperthyroid subjects. However, the intrathyroidal turnover rate was similar in the two groups of subjects.[53,54]

Washout curves of ^{99m}TcO$_4$ in hyper and euthyroid studies show a single exponential washout curve for euthyroid and a double exponential washout curve for hyperthyroid.

Dual tracer experiments were done with ^{131}I$^-$ and ^{99m}TcO$_4^-$ to gain insight into movement of anions between gland and circulation which might otherwise be obscured by organic binding of radioiodide.[55] Thyroid does accumulate ^{99m}TcO$_4^-$ from the circulation and releases it upon administration of ClO$_4^-$. At normal levels of circulating iodide nearly all the iodide entering the thyroid is rapidly organified; thus, there was no large change in the efflux rate after injection of TSH. On the other hand, at high levels of circulating iodide the thyroid is able to trap more I$^-$ than can be rapidly organified and results in an increase of trapped iodide. When TSH was administered there was a rapid decline in iodide clearance as well as in pertechnetate clearance. Within 20 min, however, the glandular iodide clearance increased above pre-TSH levels while the release of pertechnetate continued. The reason for these changes lies in increased organification for iodide but not for pertechnetate.[55]

In studies done on the metabolism of ^{99m}TcO$_4^-$ by the thyroid it was concluded that measurement of ^{99m}TcO$_4^-$ transport may be used as a qualititative, but not quantitative, index of alterations in the activity of thyroidal iodide transport mechanism.[56] This is because of the formation of origin material from TcO$_4$ by the thyroid as determined by electrophoresis which leads to the conclusion that TcO$_4$ undergoes some chemical changes in the thyroid.

2. Kinetics of the Human Thyroid Trap[57,60]

Because ^{99m}TcO$_4^-$ does not undergo significant binding to thyroglobulin and because ^{99m}TcO$_4^-$ uptake is competitively inhibited by excess iodide, pertechnetate is a good model for studying the iodide trap. The three compartmental thyroid trap model, Figure 6, proposed by Hays et al.,[57,59] shows the sites of iodide effects, TSH effect, and propylthiouracil (PTU) effect.

In normal functioning thyroid the following results were obtained from the ^{99m}TcO$_4$ derived model: (1) delayed effects of iodide, (2) early and intermediate effects of TSH, and (3) early and delayed effects of PTU. In Graves's Disease V_2 increased 7-fold, λ_{21} flow from plasma to follicular cells (TcO$_4$ clearance from plasma) increased 22-fold, and V_3 (colloid volume) increased 20-fold. Compartments 2 and 3 are relatively independent and 2 is follicular cells. At least two of the rate constants λ_{21} (plasma clearance) and λ_{23} (the exit from colloid) are independent of each other. After radioiodine treatment V_3 and λ_{21} decreased.

There is complete inhibition of λ_{21} by iodide or perchlorate. A good fit of the model is obtained by setting λ_{21} to zero. Parameters for thyroidal trap fit more nearly a logarithmic than a linear distribution.

B. Thyroidal Uptake of Other Radioions

Other radionuclides used in thyroid studies include thallium-201, ^{75}Se-selenomethionine, gallium-67, and cesium-131. Most of these studies were done in an effort to differentiate benign from malignant thyroid nodules which were "cold" to radioiod-

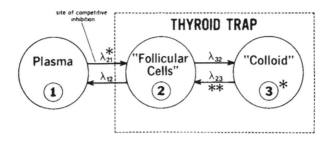

Three-compartment model of thyroidal trap, illustrating sites of iodide effect. * Reduced (p < 0.05); ** Increased (p < 0.01).

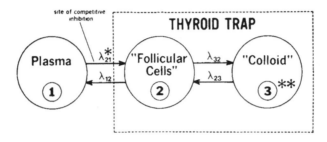

Model illustrating sites of TSH effect. * Increased (p < 0.05); ** Increased (p = 0.02).

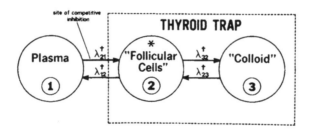

Model illustrating sites of PTU effect. † Reduced * Reduced (p < 0.05).

FIGURE 6.   Three compartment thyroid trap model. (From Hays, M. T., *J. Nucl. Med.*, 20, 944, 1979. With permission.)

ine.[20,51] Most malignancy is found in "cold" nodules (12.4% of total population) compared to only 1 to 2% of total population for "hot or warm" nodules.[20]

Thallium-201 concentrated in thyroid carcinoma when ^{131}I and ^{99m}Tc failed to concentrate.[60] All patients with goiters also showed marked uptake with no uptake in normals.[61] The mechanism for uptake is similar to potassium but the exact mechanism is still obscure. Thallium-201 is similar to Cs-131 which is a good agent to predict a benign thyroid lesion when no accumulation occurs in a "cold" nodule.[20,62] In this case the probability of malignancy is down to 5%. However, with positive uptake of Tl-201 or Cs-131 there is still a 78% probability of benignancy.[20]

Selenium-75-selenomethionine is not a suitable agent for slow growing tumors such as most thyroid tumors because of the slow rate of protein synthesis. The exception to this is anaplastic thyroid carcinoma.[20]

The role of ^{67}Ga is similar to Se-75-selenomethionine in predicting malignant thyroid lesions, a questionable one.[20] Tissue distribution studies showed the highest concentration in nodules of Hashimoto's thyroiditis and anaplastic carcinoma.[63]

ACKNOWLEDGMENTS

The author thanks Ms. Linda Lutgens for assistance in preparation of this manuscript.

REFERENCES

1. Hertz, S., Roberts, A., and Evans, R. D., Radioactive iodine as an indicator in the study of thyroid physiology, *Proc. Soc. Exp. Biol. Med.*, 38, 510, 1938.
2. Livingood, J. J. and Seaborg, G. T., Radioactive isotopes of iodine, *Physiol. Rev.*, 54, 775, 1938.
3. Tape, G. F. and Cork, J. M., Induced radioactivity in tellerium, *Physiol. Rev.*, 53, 676, 1938.
4. Hamilton, J. G., Soley, M. G., and Eichorn, M. B., Deposition of radioactive iodine in human thyroid tissue, *Univ. Calif. Berkeley Publ. Pharmacol.*, 1, 339, 1940.
5. Beierwaltes, W. H., The history of the use of radioactive iodine, *Semin. Nucl. Med.*, 9, 151, 1979.
6. Studer, H. and Greer, M. A., The regulation of thyroid function, in *Iodine Deficiency*, Hans Huber, Bern, 1968.
7. Ganong, W. F., *Review of Medical Physiology*, Lange, Los Altos, Calif., 1977.
8. DeGroot, L. J., Current views on formation of thyroid hormones, *N. Engl. J. Med.*, 272, 243, 1965.
9. DeGroot, L. J., Kinetic analysis of iodine metabolism, *J. Clin. Endocrinol.*, 26, 149, 1966.
10. Halmi, N. S. and Pitt-Rivers, R., Iodide pools of rat thyroid, *Endocrinology*, 70, 660, 1962.
11. Nagataki, S. and Ingbar, S. H., Demonstration of second thyroidal iodide pool in rat thyroid glands by double isotope labeling, *Endocrinology*, 74, 479, 1963.
12. Derman, H., Anatomy and histology of the thyroid, in *Evaluation of Thyroid and Parathyroid Functions*, Sunderman, F. W. and Sunderman, F. W., Jr., Eds., J. B. Lippincott, Philadelphia, 1963, chap 2.
13. Lewallen, C. G., Selected features of iodine and thyroxine metabolism, in *Evaluation of Thyroid and Parathyroid Functions*, Sunderman, F. W. and Sunderman, F. W., Jr., Eds., J. B. Lippincott, Philadelphia, 1963, chap. 3.
14. Owens, A. O., Jr. and Flock, E. V., Synthesis, transport, and degradation of thyroid hormones, in *The Thyroid*, Hazard, J. B. and Smith, D. E., Eds., William & Wilkins, Baltimore, 1964, chap. 3.
15. Studer, H. and Greer, M. A., The regulation of thyroid function, in *Iodine Deficiency*, Hans Huber, Bern, 1968, chap. 1-4.
16. Pitt-Rivers, R. and Tata, J. R., in *The Chemistry of Thyroid Diseases*, Charles C Thomas, Springfield, 1960, chap. 1 and 2.
17. Beierwaltes, W. H., Physiology of the thyroid gland, in *Principles of Nuclear Medicine*, Wagner, H. N., Ed., W. B. Saunders, Philadelphia, 1968, 308.
18. Rosenfield, R. L., Refetoff, S., Hoffer, P. B., Gottschalk, A., and DeGroot, L. G., Diagnosis of thyroid disease in pediatrics, in *Pediatric Nuclear Medicine*, James, A. E., Wagner, H. N., Jr., and Cooke, R. E., Eds., W. B. Saunders, Philadelphia, 1974, chap. 10.
19. Oppenheimer, J. H., Role of plasma proteins in the binding distribution and metabolism of the thyroid hormones, *N. Engl. J. Med.*, 278, 1153, 1968.
20. Nishiyama, H., Radiopharmaceuticals for thyroid imaging: a review, in *Proc. 2nd Int. Symp. on Radiopharmaceuticals*, Sorenson, J. A., Ed., The Society of Nuclear Medicine, New York, 1979, 655.
21. Pitt-Rivers, R., Some biological reactions of iodine, in *Further Advances in Thyroid Research*, Vol. 1, Fellinger, K. and Hoffer, R., Eds., Verlag der Wiener Mediziniochem Akademie, Wein, 1971, 15.
22. Rapoport, B. and DeGroot, L. J., Current concepts of thyroid physiology, *Semin. Nucl. Med.*, 1, 265, 1971.
23. Scranton, J. R., Nissen, W. M., and Halmi, N. S., The kinetics of the inhibition of thyroidal iodide accumulation by thiocyanate, a re-examination, *Endocrinology*, 85, 603, 1969.

24. Freinkel, N. and Ingbar, S., The relationship between metabolic activity and iodide-concentrating capacity of surviving thyroid slices, *J. Clin. Endocrinol.*, 15, 442, 595, 1955.
25. Wolff, J., Transport of iodide and other anions in the thyroid gland, *Physiol. Rev.*, 44, 45, 1964.
26. Sherwin, J. R. and Tong, W., The actions of iodide and TSH on thyroid cells showing a dual control system for the iodide pump, *Endocrinology*, 94, 1465, 1974.
27. Braverman, L. E. and Vagenshis, A. G., The thyroid, in *Nuclear Medicine*, Wagner, H. N., Jr., Ed., H. P. Publishing, New York, 1975, chap. 13.
28. Brown-Grant, K., Extrathyroidal iodide concentrating methods, *Physiol. Rev.*, 41, 189, 1961.
29. Wolff, J. and Maurey, J. R., Thyroidal iodide transport. II. Comparison with non-thyroid-iodide-concentrating tissues, *Biochim. Biophys. Acta*, 47, 467, 1961.
30. Myant, N. B., Corbett, B. D., Honour, A. J., and Pochin, E. E., Distribution of radioiodide in man, *Clin. Sci.*, 9, 405, 1950.
31. Halmi, N. S., Thyroidal iodide transport, *Vitam. Horm. (N.Y.)*, 19, 133, 1961.
32. Andros, G. and Wollman, S., Autoradiographic localization of iodide-125 in the thyroid epithelial cell, *Proc. Soc. Exp. Biol. Med.*, 115, 775, 1964.
33. Berson, S. A. and Yalow, R. S., The iodide trapping and binding functions of the thyroid, *J. Clin. Invest.*, 34, 186, 1955.
34. Ingbar, S. H., Simultaneous measurement of the iodide-concentrating and protein binding capacities of the normal and hyper functioning human thyroid gland, *J. Clin. Endocrinol.*, 15, 238, 1955.
35. Ingbar, S. H. and Freinhel, N., Concentration gradients for radioiodide in unblocked thyroid glands of rats: effect of perchlorate, *Endocrinology*, 58, 95, 1956.
36. Rosenberg, I. N., Altmans, J. C., and Davis, J. R., Effect of methimazole upon thyroid clearance of circulating iodide, *Endocrinology*, 75, 89, 1964.
37. Rosenberg, I. N., Altmans, C. J., and Bekar, A., Thyrotropin-induced release of iodide from the thyroid, *Endocrinology*, 69, 438, 1961.
38. Braverman, L. E. and Angbar, S. H., Changes in thyroidal function during adaptation to large doses of iodide, *J. Clin. Invest.*, 42, 1216, 1963.
39. Wollman, S. H., Kinetics of accumulation of radioiodine by thyroid gland: longer time intervals, *Am. J. Physiol.*, 202, 189, 1962.
40. DeGroot, L. J., Thyroid iodide kinetics and evidence for an iodide leak, *J. Clin. Invest.*, 44, 1039, 1965.
41. DeGroot, L. J. and Davis, A. M., The early stage of thyroid hormone function: studies on rat thyroids in vivo, *Endocrinology*, 69, 695, 1961.
42. Shimoda, S. and Greer, M. A., Stimulation of in vitro iodothyronine synthesis in thyroid lobes from rats given a low iodide diet, propylthiouracil or TSH *in vivo*, *Endocrinology*, 78, 715, 1966.
43. Shimoda, S., Kendall, J. W., and Greer, M. A., Acute effects of thyrotropin on thyroid hormone biosynthesis in the rat, *Endocrinology*, 79, 921, 1966.
44. Turkington, R. W., Effect of ouabain on thyrotropin-stimulated respiration of thyroid slices, *J. Biol. Chem.*, 238, 3463, 1963.
45. Wolff, J. and Halmi, N. S., Thryoidal iodide transport. V. The role of $Na^+- K^+$ activated, Ouabain sensitive adenosinetriphosphatase activity, *J. Biol. Chem.*, 238, 847, 1963.
46. Blasi, F., Fragomele, F., and Covelli, I., Enzymatic pathway for thyroxine synthesis through p-hydroxy-3,5,diiodophenylpyruvic acid, *Endocrinology*, 85, 542, 1969.
47. Paston, I., Certain functions of isolated thyroid cells, *Endocrinology*, 68, 924, 1961.
48. Rapoport, B., Niepomniszize, H., Bigazzi, M., Hati, R., and DeGroot, L. J., Studies on the pathogenesis of poorly iodinated thyroglobulin in nontoxic multonodular goiter, *J. Clin. Endocrinol.*, 34, 822, 1972.
49. Lutz, J. H. and Gregerman, R. I., pH dependence of the binding of thyroxine to prealbumin in human serum, *J. Clin. Endocrinol.*, 29, 487, 1969.
50. Burke, G., Kowalski, K., and Barbiarz, B., Effects of thyrotropin, prostaglandin E_1 and a prostaglandin antagonist on iodide trapping in isolated thyroid cells. II, *Life Sci.*, 10, 513, 1971.
51. Keys, J. W., Jr., Thrall, F. H., and Carey, J. H., Technical considerations in *in vivo* thyroid studies, *Semin. Nucl. Med.*, 8, 43, 1978.
52. Papadopoulas, S., MacFarlane, S., and Harden, R. McG., A comparison between the handling of iodine and technetium by the thyroid gland of the rat, *J. Endocrinol.*, 38, 381, 1967.
53. Shimmins, J., Hilditch, T., Harden, R. McG., and Alexander, W. D., Thyroidal uptake and turnover of the pertechnetate ion in normal and hyperthyroid subjects, *J . Clin. Endocrinol.*, 28, 575, 1968.
54. Burke, G., Halko, A., Silverstein, G. E., and Hilligoss, M., Comparative thyroid uptake studies with ^{131}I and $^{99m}TcO_4^-$, *J. Clin. Endocrinol.*, 34, 630, 1972.
55. Issacs, G. H. and Rosenberg, I. N., Effect of thyrotropin on thyroid clearance of iodide and pertechnetate: comparative observation at normal and high plasma iodide concentrations, *Endocrinology*, 81, 981, 1967.

56. Socolow, E. L. and Ingbar, S. H., Metabolism of pertechnetate by the thyroid gland of the rat, *Endocrinology*, 80, 337, 1967.

57. Hays, M. T., Kinetics of the human thyroid trap: experience in normal subjects and in thyroid disease, *J. Nucl. Med.*, 20, 219, 1979.

58. Hays, M. T., Kinetics of the human thyroid trap: a compartment model, *J. Nucl. Med.*, 19, 789, 1978.

59. Hays, M. T., Kinetics of the human thyroid trap: effects of iodide, thyrotropin, and propylthiouracil, *J. Nucl. Med.*, 20, 944, 1979.

60. Fukuchi, M., Kido, A., Hondo, K., Tachibana, K., Onoue, K., Morita, T., and Nagai, K., Uptake of thallium-201 in enlarged thyroid glands: concise communication, *J. Nucl. Med.*, 20, 827, 1979.

61. Fukichi, M., Hyodo, J., Tachibana, K., Nishikawa, A., Kido, A., and Nagai, K., Marked thyroid uptake of thallium-201 in patients with goiter: case report, *J. Nucl. Med.*, 18, 1199, 1977.

62. Charkes, N. D., Sklaroff, D. M., and Cantor, R. E., Tumor scanning with radioactive Cs-131, in *Recent Advances in Nuclear Medicine*, Croll, M. N. and Brady, L. W., Eds., Appleton-Century-Crofts, New York, 1966, 235.

63. Soins, J. S., Quantitative estimation of ^{67}Ga-citrate in hypofunctioning thyroid nodules, *Radiology*, 122, 243, 1977.

INDEX

D